FAIRNESS 2016

AA004765

Journal of Physics: Conference Series Volume 742

Garmisch-Partenkirchen, Germany
14 - 19 February 2016

ISBN: 978-1-5108-3104-9
ISSN: 1742-6588

Printed from e-media with permission by:

Curran Associates, Inc.
57 Morehouse Lane
Red Hook, NY 12571

Some format issues inherent in the e-media version may also appear in this print version.

Copyright© (2016) by the Institute of Physics
All rights reserved. The material featured in this book is subject to
IOP copyright protection, unless otherwise indicated.

Printed by Curran Associates, Inc. (2016)

For permission requests, please contact the Institute of Physics
at the address below.

Institute of Physics
Dirac House, Temple Back
Bristol BS1 6BE UK

Phone: 44 1 17 929 7481
Fax: 44 1 17 920 0979

techtracking@iop.org

Additional copies of this publication are available from:

Curran Associates, Inc.
57 Morehouse Lane
Red Hook, NY 12571 USA
Phone: 845-758-0400
Fax: 845-758-2633
Email: curran@proceedings.com
Web: www.proceedings.com

Table of contents

Volume 742

FAIRNESS 2016
14–19 February 2016, Garmisch-Partenkirchen, Germany

Accepted papers received: 9 August 2016
Published online: 5 October 2016

Preface

Papers

FAIRNESS 2016 was the fourth edition in a series of workshops designed to bring together excellent international young scientists with research interests focused on physics at FAIR (Facility for Antiproton and Ion Research) and was held on February 14-19 2016 in Garmisch-Partenkirchen, Germany. The topics of the workshops cover a wide range of aspects in both theoretical developments and current experimental status, concentrated around the four scientific pillars of FAIR. FAIR is a new accelerator complex with brand new experimental facilities, that is currently being built next to the existing GSI facility close to Darmstadt, Germany.

The spirit of the conference is to bring together young scientists, e.g. young non-tenured scientists, postdocs and advanced PhD students to present their work, to foster active informal discussions and build up networks. Every participant in the meeting with the exception of the organizers gives an oral presentation, and all sessions are followed by an hour long discussion period. During the talks, questions are anonymously collected in a box to stimulate discussions. The broad physics program at FAIR is reflected in the wide range of topics covered by the workshop:

- Atomic and plasma physics, biophysics, material sciences and applications
- Nuclear structure, astrophysics and reactions
- Physics of hot and dense nuclear matter, QCD phase transitions and critical point
- Hadron Spectroscopy, Hadrons in matter and Hypernuclei
- Experimental programs APPA, CBM, HADES, PANDA, NUSTAR, as well as BES, NICA and the RHIC beam energy scan

For these different areas one invited speaker was selected to give a longer introductory presentation. The write-ups of the talks presented at FAIRNESS 2016 are the content of this issue of Journal of Physics: Conference Series and have been refereed according to the IOP standard for peer review. This issue constitutes therefore a collection of the forefront of research that is dedicated to the physics at FAIR.

June 2016,

Organizers of FAIRNESS 2016: Marco Destefanis, Tetyana Galatyuk, Fernando Montes, Diana Nicmorus, Claudia Ratti, and Laura Tolos.

Support for holding the conference was provided by:

HELMHOLTZ ASSOCIATION

EMMI

FAIR

Content from this work may be used under the terms of the Creative Commons Attribution 3.0 licence. Any further distribution of this work must maintain attribution to the author(s) and the title of the work, journal citation and DOI.
Published under licence by IOP Publishing Ltd

Peer review statement

All papers published in this volume of *Journal of Physics: Conference Series* have been peer reviewed through processes administered by the proceedings Editors. Reviews were conducted by expert referees to the professional and scientific standards expected of a proceedings journal published by IOP Publishing.

Content from this work may be used under the terms of the Creative Commons Attribution 3.0 licence. Any further distribution of this work must maintain attribution to the author(s) and the title of the work, journal citation and DOI.
Published under licence by IOP Publishing Ltd

Journal of Physics: Conference Series **742** (2016) 012001 doi:10.1088/1742-6596/742/1/012001

Binding in light nuclei: Statistical NN uncertainties vs Computational accuracy[1]

R. Navarro Pérez[1], **A. Nogga,**[2] **J. E. Amaro**[3] **and E. Ruiz Arriola**[3]

[1] Nuclear and Chemical Science Division, Lawrence Livermore National Laboratory Livermore, California 94551, USA

[2] Forschungszentrum Jülich, Institut für Kernphysik (Theorie), Institute for Advanced Simulation, Jülich Center for Hadron Physics and JARA - High Performance Computing, D-52425 Julich, Germany.

[3] Departamento de Física Atómica, Molecular y Nuclear and Instituto Carlos I de Física Teórica y Computacional, Universidad de Granada E-18071 Granada, Spain

E-mail: navarroperez1@llnl.gov,a.nogga@fz-juelich.de, amaro@ugr.es,earriola@ugr.es

Abstract. We analyse the impact of the statistical uncertainties of the the nucleon-nucleon interaction, based on the Granada-2013 np-pp database, on the binding energies of the triton and the alpha particle using a bootstrap method, by solving the Faddeev equations for ^{3}H and the Yakubovsky equations for ^{4}He respectively. We check that in practice about 30 samples prove enough for a reliable error estimate. An extrapolation of the well fulfilled Tjon-line correlation predicts the experimental binding of the alpha particle within uncertainties.

Nuclear structure *ab initio* calculations are notoriously difficult and computationally demanding and have thus so far been limited to light nuclei, although recently, these calculations have been extended to more complex systems [1, 2, 3]. Besides giving important input for applications such as astrophysically relevant nuclear reactions, these calculations are important tests of current nuclear interactions. To this aim, not only the result itself is important but also the uncertainty (see e.g. the special issue [4].) From a theoretical point of view and the inferred predictive power uncertainties can be grouped into three main categories

- The input information: the basic nucleon-nucleon (NN) interaction should describe a relevant piece of the NN scattering data and the simplest two-body bound state: the deuteron. We will call this the statistical uncertainty for reasons to be justified below.
- The solution method: the way the multinucleon problem is solved once the NN interaction is represented. This requires some sufficiently high precision which makes computations costly. We will call these the numerical uncertainty.
- The representation problem: the way the input NN data are represented theoretically. Normally potentials are used, but the form of the potential in the short range region, below $2-3$fm, is generally not universal, and they are often tailored to make the solution of the many body problem as simple as possible. We will call these the systematic uncertainty [2].

[1] Presented by RNP at Workshop for young scientists with research interests focused on physics at FAIR 14-19 February 2016 Garmisch-Partenkirchen (Germany)

[2] This includes in particular any theoretically based expansion of the interaction rooted or inspired by QCD such as chiral perturbation theory or large N_c expansions were some renormalization scheme dependence is unavoidable.

Content from this work may be used under the terms of the Creative Commons Attribution 3.0 licence. Any further distribution of this work must maintain attribution to the author(s) and the title of the work, journal citation and DOI.
Published under licence by IOP Publishing Ltd

Journal of Physics: Conference Series **742** (2016) 012001 doi:10.1088/1742-6596/742/1/012001

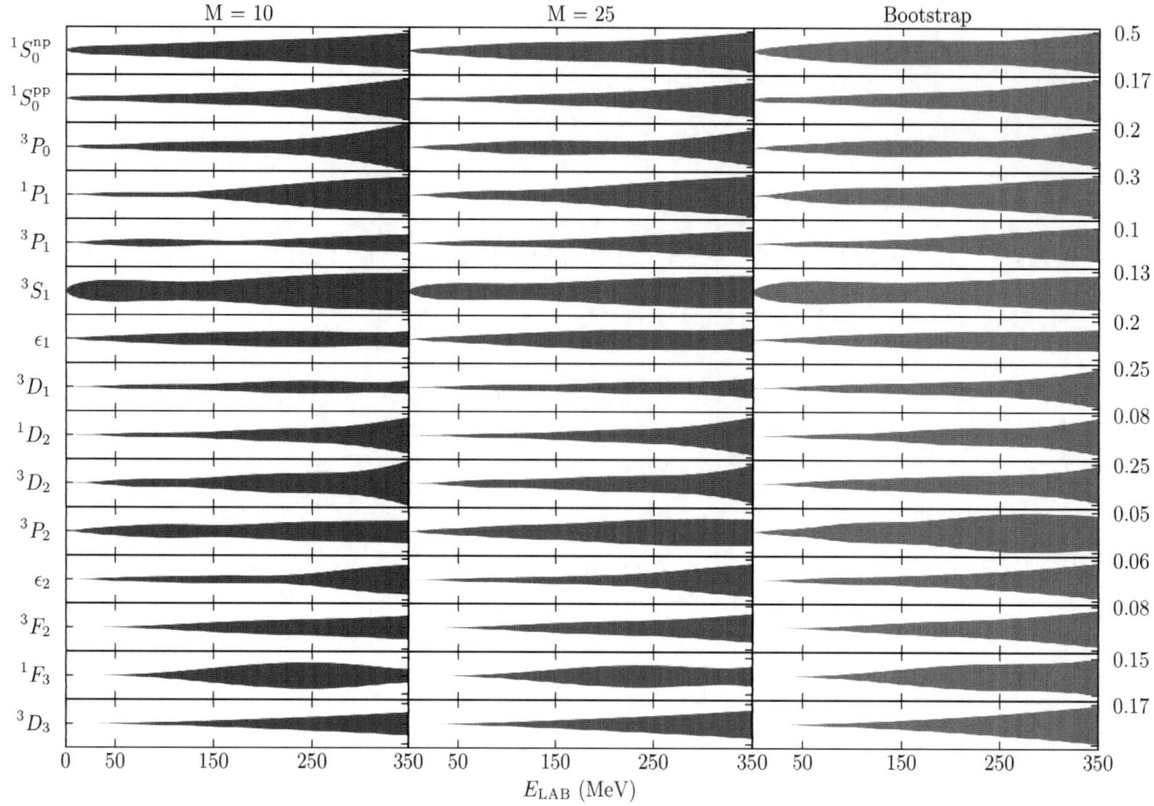

Figure 1. (Color online) Phaseshift statistical error bands (in degrees) for the δ-shell potential [5]. The error bands on the first two columns where obtained using a MonteCarlo family of δ-shell potentials where potential parameters are random numbers following the multivariate normal distribution determined by the original fit covariance matrix. The columns use a sample size of $M = 10$ (left column) and $M = 25$ (middle column). The right column is the error bar obtained from the Bootstrap to experimental data presented in [6] with $M = 1000$. All phase shifts are np unless otherwise indicated.

Assuming that these sources of error are independent of each other, we expect the total uncertainty to be given, as usual, by

$$\Delta E^2 = \Delta E^2_{\rm stat} + \Delta E^2_{\rm num} + \Delta E^2_{\rm syst} \tag{1}$$

Clearly, the total error is dominated by the largest one. So, it makes sense either to reduce the largest source of uncertainty or to tune all uncertainties to a similar level. This sets the limit of predictive power in *ab initio* calculations. While numerical accuracy has been a goal in itself in few-body calculations, the *physical* accuracy is given by *all* possibles sources of uncertainties.

In this talk, we discuss the relation between the statistical uncertainties stemming from the finite experimental accuracy of NN scattering data [7, 8, 9, 10] and the currently available numerical accuracy with which the few body problem can be solved. A pioneering work was carried out in [11] where the so-called statistical regularization was used to evaluate the impact of errors on the binding energies of the $A = 3, 4$ systems. The analysis was based on the Paris potential which has $\chi^2/\text{d.o.f.} \sim 2$.

The recent Granada-2013 3σ-self consistent database comprises 6713 np and pp scattering data below $E_{\rm LAB} = 350\text{MeV}$ and has a $\chi^2/\text{d.o.f} = 1.04$ [12, 5]. The procedure to propagate uncertainties is based in spirit on the bootstrap analysis proposed in [6] where the 6713 np and pp scattering data are randomized and multiple ($M = 1020$) χ^2-fits yield a multivariate distribution of fitting parameters. This provides a sample enabling a random evaluation of any observable. We monitor the size M of the needed sample by looking for statistical stability of the output. The result for the errors in the corresponding phase shifts is compared in Fig. 1 for different Monte Carlo generated sample sizes following a gaussian multivariate distribution dictated by the parameter's covariance matrix. As we see $M = 25$ already gives a result rather close to the full bootstrap method.

We have built a simple and smooth gaussian potential which can be used in most few- and many-body calculational schemes and which provides an acceptable $\chi^2/\text{d.o.f.} = 1.06$ [13], so it can be considered to be statistically equivalent to the original delta-shells potential [8, 5]. As we will put forward here, and in agreement with previous findings using either the hyperspherical harmonics (HSH) method for $A = 3$ [14] and no-core full configuration shell model calculations [15], these estimates already suggests that the numerical accuracy is close to optimal given the statistical uncertainty. We will use here the Faddeev equations for the $A = 3$ case and the Yakubovsky equations for the $A = 4$ situation. As a first step we will consider only NN forces explicitly and leave out 3N and 4N forces for future developments. The multiple evaluations for the triton are shown in Fig. 2. As in [14] we bin the distribution according to the numerical accuracy, $\Delta E_t^{\text{num}} \sim 1\text{keV}$ [3].

In a Monte Carlo approach many variations of the parameters produce irrelevant changes. A principal component analysis looks for eigenvalues and eigenvectors of the computed observable and provides valuable information on the most relevant changes of the input parameters but has seldomly been investigated in nuclear physics (see however Ref. [16] and references therein). In Fig. 3, we show the results of such an analysis applied to the coefficients of the gaussian potential of Ref. [8] implemented in a Monte Carlo fashion. We found that the number of principal components to obtain most of the uncertainty in E_t is around 10. This indicates that regarding ΔE_t^{stat} a fit to the NN scattering data base could be done with less parameters, if the fit were to be designed in terms of relevant parameters only.

The tiny error band suggests that the discrepancy between our number $E_t^{\text{th}} = -7.6669 \pm 0.0124\text{MeV}$ and the $E_t^{\text{exp}} = -8.4820 \pm 0.0001 MeV$ has to be sought in missing three-nucleon forces (3NFs). It is well known that 3NFs give an important contribution to nuclear bindings [17, 18]. This raises the question of how much of this statistical uncertainty will be absorbed into variations in the parameters of the 3NFs. In order to implement some 3N information, we invoke the empirical linear correlation displayed by the Tjon line [17] [4].

In the Monte Carlo method, any choice of parameters $\mathbf{p}$ determines a value of the triton binding energy. Given the variations of the triton binding energy, we expect, when determining the α particle binding energy, a Tjon-like linear correlation of the form $E_\alpha(\mathbf{p}) = aE_t(\mathbf{p}) + b$. The values found are $a = 4.7(1)$ and $b = 11.4$. Thus we expect a Tjon-like correlation would give $\Delta E_\alpha^{\text{stat}} = 4.7(1) \times \Delta E_t^{\text{stat}} = 50(5)\text{keV}$ which is mainly determined by the channels involving relative S-waves. In Fig. 4, we show our results for the ^{3}H and ^{4}He binding energy. The yellow band shows the fit including the uncertainty. The error bars show the numerical uncertainty. Whereas the variation of the binding energies is rather large, the linear correlation indicates that most of this variation will be eventually absorbed into a properly adjusted 3NFs. We take the band width as an indication for the remaining error induced by the uncertainty of NN data. As one can see, the band width and the numerical errors are comparable. Therefore, we deduced

[3] One of the advantages of the present method over the HSH expansion employed in [14] is that the determination of avoided crossings are difficult to identify rigorously; an effect which has to be assisted by visual inspection and, if mistaken, has a repulsive effect. This explains the longer tails of the triton binding energies distributions. We thank Eduardo Garrido for noting this.

[4] See [19, 20] for a Similarity Renormalization Group analysis yielding the simple formula $E_\alpha = 4E_t - 3E_d$.

Journal of Physics: Conference Series **742** (2016) 012001 doi:10.1088/1742-6596/742/1/012001

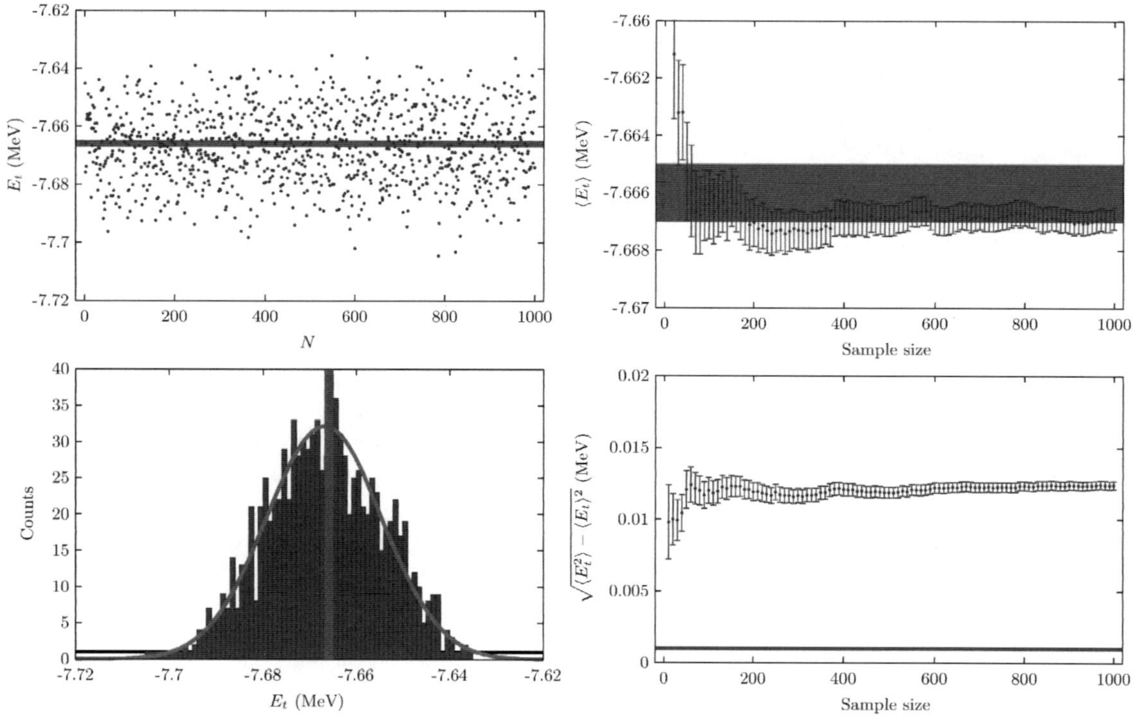

Figure 2. (Color online) Distribution (top left) and histogram (bottom left) representing the triton binding energy (in MeV) for a sample of 1000 Monte Carlo generated gaussian potential parameters. Finite sample estimates for the population mean (top right) and population standard deviation (bottom right) of the triton binding energy as a function of the sample size M of the gaussian potential parameters are also shown. In all panels the red band represents the value obtained with the most likely parameters $E_t(\mathbf{p}_0) = 7.666 \pm 0.001$MeV and its width is the numerical error. We take the bin size equal to the numerical precision.

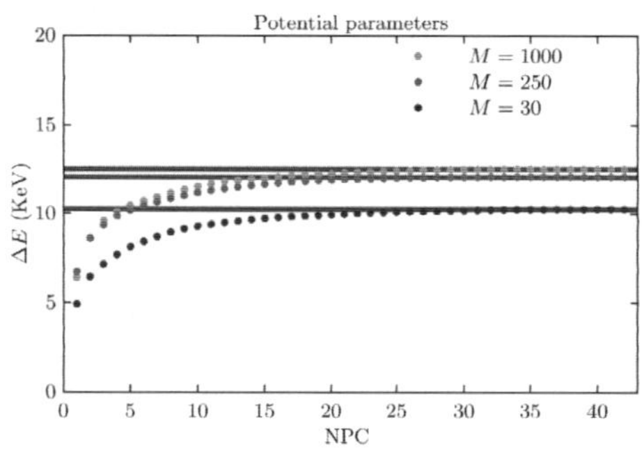

Figure 3. (Color online) Number of principal components contributions of gaussian parameters for samples of size $M = 30, 250, 1000$ (from bottom to top) to the triton binding energy uncertainty.

that this uncertainty is comparable to the statistical one. Strong efforts to increase the numerical accuracy are therefore not desired. For this analysis, we used a moderate sample of only M=30, the smallness of which is justified from the analysis of Fig. 2, as far as uncertainty estimates are concerned. In order to have a tighter predicted extrapolated band one would need to reduce the numerical error in E_α in harmony with the Tjon slope $\Delta E_\alpha^{\rm num} \sim 4.7 \Delta E_t^{\rm num} \sim 5$keV. Further details will be presented in a forthcoming publication.

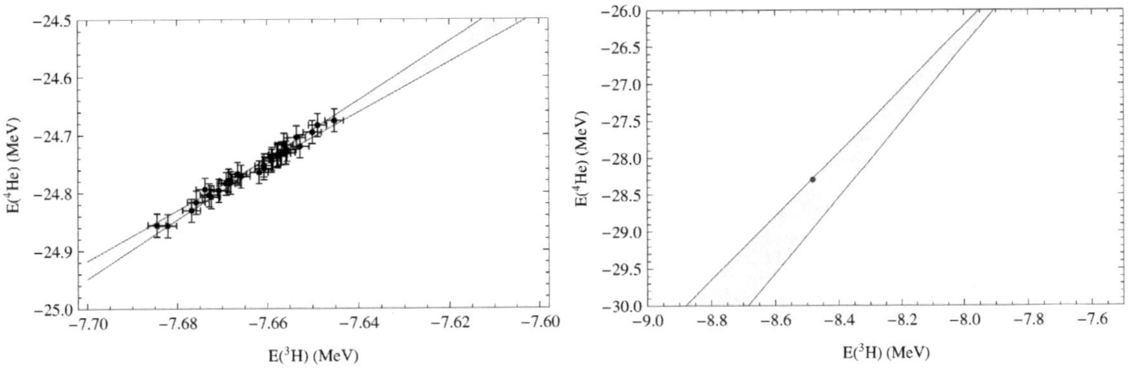

Figure 4. (Color online) Tjon type analysis of the ^{4}He binding energy vs the ^{3}H binding energy. We show the fit to the sample of $N = 30$ Monte Carlo generated binding energies both in a small scale (left panel) and extrapolated in a larger scale (right panel) compared with the experimental point (blue dot). We take $\Delta E_t^{\rm num} = 1\,{\rm keV}$ and $\Delta E_\alpha^{\rm num} = 20\,{\rm keV}$.

Acknowledgements
This work is supported by Spanish DGI with Feder funds (grant FIS2014-59386-P) and Junta de Andalucía (grant FQM225), the U.S. Department of Energy by Lawrence Livermore National Laboratory under Contract No. DE-AC52-07NA27344, the U.S. Department of Energy, Office of Science, Office of Nuclear Physics under Award No. DE-SC0008511 (NUCLEI SciDAC Collaboration). The numerical calculations have partly been performed on JUQUEEN, JUROPA and JURECA of the JSC, Jülich, Germany.

References
[1] Quaglioni S, Hupin G, Calci A, Navratil P and Roth R 2015 (*Preprint* 1509.09009)
[2] Meissner U G 2014 *Nucl. Phys. News* **24** 11–15 (*Preprint* 1505.06997)
[3] Dytrych T, Maris P, Launey K D, Draayer J P, Vary J P, Langr D, Saule E, Caprio M A, Catalyurek U and Sosonkina M 2016 (*Preprint* 1602.02965)
[4] Ireland D and Nazarewicz W 2015 *(Editors) Journal of Physics. G, Nuclear and Particle Physics* **42**
[5] Navarro Pérez R, Amaro J E and Ruiz Arriola E 2013 *Phys. Rev.* **C88** 064002
[6] Navarro Pérez R, Amaro J E and Ruiz Arriola E 2014 *Phys. Lett.* **B738** 155–159
[7] Navarro Pérez R, Amaro J E and Ruiz Arriola E 2012 (*Preprint* 1202.6624)
[8] Navarro Pérez R, Amaro J E and Ruiz Arriola E 2013 *Phys. Lett.* **B724** 138–143
[9] Navarro Pérez R, Amaro J E and Ruiz Arriola E 2012 *PoS* **QNP2012** 145
[10] Navarro Pérez R, Amaro J E and Ruiz Arriola E 2016 (*Preprint* 1601.08220)
[11] Adam R, Fiedeldey H, Sofianos S and Leeb H 1993 *Nuclear Physics A* **559** 157–172
[12] Navarro Perez R, Amaro J and Ruiz Arriola E 2013 Granada Database http://www.ugr.es/~amaro/nndatabase/ accessed: 2016-04-04
[13] Navarro Pérez R, Amaro J E and Ruiz Arriola E 2014 *Phys. Rev.* **C89** 064006
[14] Navarro Pérez R, Garrido E, Amaro J E and Ruiz Arriola E 2014 *Phys. Rev.* **C90** 047001
[15] Navarro Pérez R, Amaro J E, Ruiz Arriola E, Maris P and Vary J P 2015 *Phys. Rev.* **C92** 064003
[16] Al-Sayed A 2014 *Nucl. Phys.* **A933** 154–164
[17] Hammer H W, Nogga A and Schwenk A 2013 *Rev. Mod. Phys.* **85** 197
[18] Kalantar-Nayestanaki N, Epelbaum E, Messchendorp J G and Nogga A 2012 *Rept. Prog. Phys.* **75** 016301
[19] Ruiz Arriola E, Szpigel S and Timoteo V S 2014 *Few Body Syst.* **55** 971–975
[20] Ruiz Arriola E, Szpigel S and Timoteo V S 2016 (*Preprint* 1601.02360)

XYZ: the case of $Z_c(3885)/Z_c(3900)$

M. Albaladejo[1], **F. K. Guo**[2,3], **C. Hidalgo-Duque**[1], **and J. Nieves**[1]

[1] Instituto de Física Corpuscular (IFIC), Centro Mixto CSIC-Universidad de Valencia, Institutos de Investigación de Paterna, Aptd. 22085, E-46071 Valencia, Spain

[2] State Key Laboratory of Theoretical Physics, Institute of Theoretical Physics, Chinese Academy of Sciences, Beijing 100190, China

[3] Helmholtz-Institut für Strahlen- und Kernphysik and Bethe Center for Theoretical Physics, Universität Bonn, D-53115 Bonn, Germany

Abstract. The resonant structure $Z_c(3885)/Z_c(3900)$ is studied in a unitary, $J/\psi\pi$–$D^*\bar{D}$ coupled channel T-matrix formalism. The two experimental spectra ($D^*\bar{D}$, $J/\psi\pi$) in which this state is seen are reproduced in two different scenarios. In both scenarios the two observations, $Z_c^\pm(3900)$ and $Z_c^\pm(3885)$, are shown to have the same common origin. In the first one, the Z_c originates from a resonance with a mass close to and above the $D\bar{D}^*$ threshold. In the second one, the Z_c peak stems from a virtual state. Precise measurements of the line shapes around the $D\bar{D}^*$ threshold are called for in order to understand the nature of this state.

1. Introduction

In this talk we report on our recent work [1] about the nature of the $Z_c(3900)/Z_c(3885)$ state(s). The resonant-like structure $Z_c(3900)^\pm$ was first seen simultaneously by the BESIII and Belle collaborations [2, 3] in the $J/\psi\pi$ spectrum produced in the $e^+e^- \to Y(4260) \to J/\psi\pi^+\pi^-$ reaction. It was confirmed in an analysis [4] based on CLEO-c data. A similar structure, $Z_c(3885)^\pm$, with quantum numbers favored to be $J^P = 1^+$, has also been reported by the BESIII collaboration [5,6] in the $\bar{D}^*D$ spectrum of $e^+e^- \to \bar{D}^*D\pi$. It is generally accepted that the two structures have indeed the same common origin. We generically denote it here as Z_c.

The state $Z_c(3900)$ is an exotic one, since it couples strongly to charmonium and yet it is charged, thus its minimal constituent quark content should be four quarks, $c\bar{c}u\bar{d}$ (for Z_c^+). Many different interpretations of its nature have been given, most of which can be traced back from references in Ref. [1]. It has also been studied in Lattice QCD (see Ref. [7] for references). A simultaneous study of the two reactions in which the Z_c structure has been seen was performed in Ref. [1].[1] From this analysis, information about this seemingly resonant intriguing structure is extracted. In Sec. 2 a $\bar{D}^*D$, $J/\psi\pi$ coupled channel formalism is settled. The resulting T-matrix is used in the calculation of the amplitudes for the reactions $Y(4260) \to J/\psi\pi\pi, \bar{D}^*D\pi$, assuming that the $Y(4260)$ state is dominantly a $D_1(2420)\bar{D}$ + c.c. bound state [9, 10]. The results and conclusions are presented in Sec. 3.

[1] In Ref. [8], both reactions were considered and used to fix parameters at the one-loop level. In that work, it is shown that the narrow near threshold states like the Z_c cannot be simply kinematical effects.

Content from this work may be used under the terms of the Creative Commons Attribution 3.0 licence. Any further distribution of this work must maintain attribution to the author(s) and the title of the work, journal citation and DOI.
Published under licence by IOP Publishing Ltd

2. Formalism

Let us denote with 1 and 2 the $J/\psi\pi$ and $\bar{D}^*D$ channels, respectively, with $I = 1$ and $J^{PC} = 1^{+-}$ (in what follows, the C-parity denotes that of the neutral member of the isospin triplet). The coupled-channel T-matrix is written as:

$$T = (\mathbb{I} - V \cdot G)^{-1} \cdot V \ , \tag{1}$$

where G is a loop function diagonal matrix, and the matrix elements of the potential V read:

$$V_{ij} = 4\sqrt{m_{i1}m_{i2}}\sqrt{m_{j1}m_{j2}}\, e^{-q_i^2/\Lambda_i^2} e^{-q_j^2/\Lambda_j^2} C_{ij} \ . \tag{2}$$

The c.m. momentum squared of the channel i is denoted by ${q_i}^2$. The details of Eq. (2) can be found in Ref. [1].

The $J/\psi\pi \to J/\psi\pi$ can be neglected [11,12], and we take $C_{11} = 0$. For the $\bar{D}^*D \to J/\psi\pi$ S-wave transition, we make the simplest possible assumption, taking it as a constant, $C_{12} \equiv \widetilde{C}$. In a momentum expansion, the lowest order contact potential for the $\bar{D}^*D \to \bar{D}^*D$ transition is simply a constant as well, denoted by $C_{22} \equiv C_{1Z}$ [13]. However, even with two coupled channels, no resonance above the highest threshold can be generated in the complex plane with only constant potentials. To allow some energy dependence for the V_{22} term, we introduce a new parameter b, and write:

$$C_{22}(E) = C_{1Z} + b\,(E - m_D - m_{D^*}) \ . \tag{3}$$

To regularize the interactions we employ a standard gaussian regulator, $e^{-q_i^2/\Lambda_i^2}$.[2] We take cutoff values $\Lambda_2 = 0.5 - 1$ GeV [13]. At the Z_c energy, the c.m. momentum of the $J/\psi\pi$ channel is $q_1 \simeq 0.7$ GeV, and hence we use $\Lambda_1 = 1.5$ GeV. The loop functions in the matrix G can be found in Ref. [1].

For the e^+e^- annihilations at the $Y(4260)$ mass, both BESIII and Belle have seen the Z_c structure in the $J/\psi\pi$ final state [2,3], but only BESIII reports data for the $\bar{D}^*D$ channel [5,6]. Hence, for consistency, we will only study the BESIII data. In particular, we will consider the most recent double-D-tag data of Ref. [6]. For definiteness, we will consider the reported spectra of the $D^{*-}D^0$ and $J/\psi\pi^-$ final states.

We denote with $\mathcal{M}_1$ ($\mathcal{M}_2$) the amplitude for the $Y \to J/\psi\pi^+\pi^-$ ($Y \to D^{*-}D^0\pi^+$) decay, and we respectively call s and t to the invariant masses squared of $J/\psi\pi^-$ and $J/\psi\pi^+$ ($D^{*-}D^0$ and $D^{*-}\pi^+$) in the first (second) decay. Up to some common irrelevant constant, both amplitudes can be written as:

$$\left|\overline{\mathcal{M}_1}(s,t)\right|^2 = |\tau(s)|^2\, q_\pi^4(s) + |\tau(t)|^2\, q_\pi^4(t) + \frac{3\cos^2\theta - 1}{4}\left(\tau(s)\tau(t)^* + \tau(s)^*\tau(t)\right) q_\pi^2(s) q_\pi^2(t) \ , \tag{4}$$

$$\tau(s) = \sqrt{2} I_3(s) T_{12}(s) + \alpha \ , \tag{5}$$

$$\left|\overline{\mathcal{M}_2}(s,t)\right|^2 = \left|\frac{1}{t - m_{D_1}^2} + I_3(s) T_{22}(s)\right|^2 q_\pi^4(s) + \left|\beta\big(1 + T_{22}(s) G_2(s)\big)\right|^2 \ , \tag{6}$$

where $q_\pi^2(s) = \lambda(M_Y^2, s, m_\pi^2)/(4M_Y^2)$, and θ denotes the relative angle between the two pions in the $Y(4260)$ rest frame. Further, $I_3(s)$ is the scalar three-meson non-relativistic loop function, and m_{D_1} is the mass of the D_1 meson (details can be found in Ref. [1]). The parameters α and β in Eqs. (5) and (6) are unknown. Details on the different contributions in Eqs. (4)– (6) can be found in Ref. [1].

[2] In principle, one expects the predictions of physical (observables) quantities to be independent of the chosen regularization procedure. In a context similar to the one presented in this work, this has been studied in Ref. [14].

Besides a contribution from the amplitudes ($\mathcal{A}_i$), the spectrum for both reactions have some background ($\mathcal{B}_i$), and thus we write them as:

$$\mathcal{N}_i(s) = K_i \left(\mathcal{A}_i(s) + \mathcal{B}_i(s)\right) \ , \tag{7}$$

$$\mathcal{A}_i(s) = \int_{t_{i,-}}^{t_{i,+}} dt \left|\overline{\mathcal{M}_i}(s,t)\right|^2 \ , \tag{8}$$

where $t_{i,\pm}(s)$ are the limits of the t Mandelstam variables for the decay mode i. To obtain a reasonable estimate of the coupling $\tilde{C}$ and the two global constants K_i, we consider a further experimental input from Ref. [5],

$$R_{\rm exp} = \frac{\Gamma\left(Z_c(3885) \to D\bar{D}^*\right)}{\Gamma\left(Z_c(3900) \to J/\psi\pi\right)} = 6.2 \pm 1.1 \pm 2.7 \ , \tag{9}$$

and estimate this ratio as

$$R_{\rm th} = \frac{\int ds \mathcal{A}_2(s)}{\int ds \mathcal{A}_1(s)} \ , \tag{10}$$

that is, as the ratio of the background subtracted areas of each physical spectrum around the Z_c mass, namely in the range $\sqrt{s} = (3900 \pm 35)$ MeV. An inspection of Fig. 1 shows that the tail of the $D^*\bar{D}$ spectrum is small, and we set $\mathcal{B}_2 = 0$. For the $J/\psi\pi$ spectrum, $\mathcal{B}_1$ is parameterized with a symmetric smooth threshold function as used in the experimental work of Ref. [2]:

$$\mathcal{B}_1(s) = B_1 \left[(\sqrt{s} - m_{1-})(m_+ - \sqrt{s})\right]^{d_1} \ , \tag{11}$$

with $m_{1-} = m_{J/\psi} + m_\pi$ and $m_+ = m_Y - m_\pi$. The parameters B_1 and d_1 are free.

3. Results

We have three free parameters directly related to our T-matrix (C_{1Z}, $\widetilde{C}$, and b), and six (B_1, d_1, α, β and $K_{1,2}$) related to the background and the overall normalization. These nine free parameters are adjusted to reproduce the data of Refs. [2, 6] (a total of 104 data points).

We perform four different fits, corresponding to the two cases of keeping the parameter b, that carries the energy dependence of the $\bar{D}^*D$ potential, free or set to zero, and for each of these, we choose Λ_2 to be 0.5 or 1 GeV [13]. The specific values of the parameters in each of these fits can be found in Ref. [1]. The description of the experimental spectra is very good in all cases, as can be seen in the top panels of Fig. 1, where the results from one of the fits (b free and $\Lambda_2 = 0.5$ GeV) are shown together with the data. The other fits lead to results similar to those shown in Fig. 1. In any case, we see that we are able to simultaneously reproduce the two available BESIII data sets related to the $Z_c^\pm(3900)/Z_c^\pm(3885)$ state with a single $\bar{D}^*D$ structure for the very first time.

We next study the pole structure of the T-matrix. Poles can be found in different Riemann sheets of the T-matrix, which are reached through analytical continuation of the loop functions. The specific definitions of the Riemann sheets can be found in Ref. [1]. We define the mass and the width of the Z_c from its pole position, $\sqrt{s} = M_{Z_c} - i\Gamma_{Z_c}/2$. For the case $b \neq 0$, we find poles on the (11) Riemann sheet, which is connected to the physical one above the $D\bar{D}^*$ threshold. In Fig. 2, we compare the pole position obtained in this work for the Z_c resonance with the experimental determinations of Refs. [2–6]. The specific values for the poles positions can be seen in Ref. [1]. As seen in Fig. 2, good agreement within errors is found. The real part of these energies is clearly above threshold, so they correspond to a resonance, which really (physically) exists as an unstable particle.

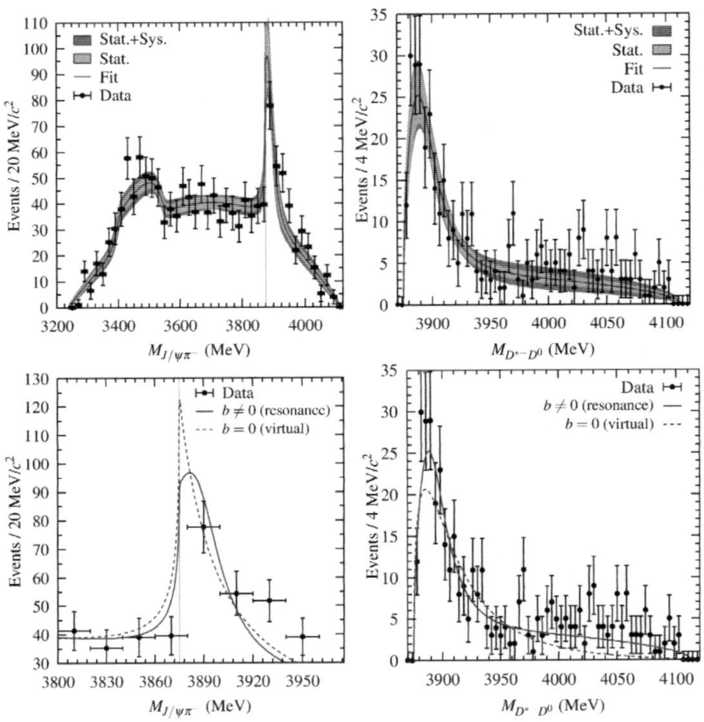

Figure 1. Invariant mass distributions for $J/\psi\pi^-$ in the decay $Y(4260) \to J/\psi\pi\pi$ (left panels) [2] and for $\bar{D}^{*-}D^0$ in the decay $Y(4260) \to \bar{D}^*D\pi$ (right panels) [6]. The top panels show the results for the fit $b \neq 0$, $\Lambda_2 = 0.5$ GeV. In the bottom panels, the two fits $b = 0$ and $b \neq 0$ are compared (without error bands) for the case $\Lambda_2 = 0.5$ GeV. In the $J/\psi\pi^-$ spectrum, the $\bar{D}^*D$ threshold is marked with a vertical black line.

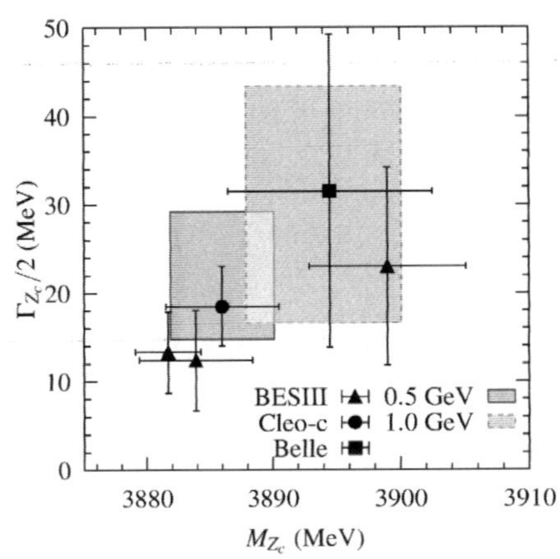

Figure 2. Comparison of the Z_c resonance pole positions determined in this work for two values of the cutoff Λ_2 with the experimental determinations of Refs. [2–6]. The shaded areas take into account our statistical and systematic uncertainties (added in quadratures). The numerical values can be found in Ref. [1].

For the case $b = 0$, however, the situation is quite different. While the description of the experimental data is still quite good, the pole in this case is located below threshold, with a small imaginary part (around 8 MeV), and in the (01) Riemann sheet. If the $J/\psi\pi$ channel were now switched off ($\widetilde{C} = 0$), this pole would move into the real axis in the unphysical Riemann sheet of the elastic amplitude T_{22}. In this sense, the obtained pole does not qualify as a resonance, and we see it as a virtual or anti-bound $D\bar{D}^*$ state. It does not correspond to a particle, but it produces observable effects at the $D\bar{D}^*$ threshold similar to those produced by a near threshold resonance or bound state [16]. The line shapes of a virtual state and a near-threshold resonance are different since the former is peaked exactly at the threshold while the latter, in principle, is above. This can be seen in the left bottom panel of Fig. 1 where the $J\psi\pi^-$ spectrum for the two fits $b = 0$ and $b \neq 0$ are shown (for the case $\Lambda_2 = 0.5$ GeV). Although the two curves are different, each one would approximately lie within the error band of the other. Very precise data with a good energy resolution and small bin size are necessary to distinguish among them.

Since both natures for the Z_c structure (resonance or virtual state) arise in fits of good quality, it must be stated that the experimental information available at this time cannot fully discriminate between both scenarios. Thus, claims about the Z_c structure should be made with caution. If it is finally shown to be a virtual state, then it cannot be a tetraquark, since it does not correspond to a normal particle, and it can only have a hadronic molecular nature, in the sense that it appears only because of the $D\bar{D}^*$ interaction.

4. Summary

Summarizing, we have reported on our recent work [1] which presents the first simultaneous study of the invariant mass distributions of the $J/\psi\pi$ and $\bar{D}^*D$ channels (where the $Z_c(3900)$ is seen) with fully unitarized amplitudes. We find that these data sets are well reproduced in two different scenarios. In the first one, in which there is an energy dependence in the $\bar{D}^*D \to \bar{D}^*D$ potential, the Z_c appears as a dynamically generated $\bar{D}^*D$ resonance. In the second one, where the aforementioned energy dependence is not allowed, it appears as a virtual state, the pole being located below the $D\bar{D}^*$ threshold. In both scenarios, it is demonstrated that both data sets can be reproduced with a single Z_c state, so that the two experimentally observed structures $Z_c^{\pm}(3900)$ and $Z_c^{\pm}(3885)$ are proven to correspond to the same state seen in different channels. It is really important to discriminate between these two scenarios, for which purpose a very precise measurement of the line shapes around (specially above) the $D\bar{D}^*$ threshold is needed.

Acknowledgments

F.-K. G. is partially supported by the Thousand Talents Plan for Young Professionals. C. H.-D. thanks the support of the JAE-CSIC Program. M. A. acknowledges financial support from the "Juan de la Cierva" program (27-13-463B-731) from the Spanish MINECO. This work is supported in part by the DFG and the NSFC through funds provided to the Sino-German CRC 110 "Symmetries and the Emergence of Structure in QCD" (NSFC Grant No. 11261130311), by the Spanish MINECO and European FEDER funds under the contract FIS2011-28853-C02-02 and FIS2014-51948-C2-1-P, FIS2014-57026-REDT and SEV-2014-0398, by Generalitat Valenciana under contract PROMETEOII/2014/0068, by the EU HadronPhysics3 project with grant agreement no. 283286.

References

[1] Albaladejo M, Guo F K, Hidalgo-Duque C and Nieves J 2016 *Phys. Lett.* B **755** 337
[2] Ablikim M *et al.* [BESIII Collaboration] 2013 *Phys. Rev. Lett.* **110** 252001
[3] Liu Z Q*et al.* [Belle Collaboration] 2013 *Phys. Rev. Lett.* **110** 252002
[4] Xiao T, Dobbs S, Tomaradze S and Seth K K 2013 *Phys. Lett.* B **727** 366
[5] Ablikim M *et al.* [BESIII Collaboration] 2014 *Phys. Rev. Lett.* **112** 022001
[6] Ablikim M *et al.* [BESIII Collaboration] 2015 *Phys. Rev.* D **92** 092006
[7] Albaladejo M, Nieves J and Fernandez-Soler P $Z_c(3900)$: Confronting theory and lattice simulations *Preprint* arXiv:1606.03008 [hep-ph]
[8] Guo F K, Hanhart C, Wang Q and Zhao Q 2015 *Phys. Rev.* D **91** 051504
[9] Wang Q, Hanhart C and Zhao Q 2013 *Phys. Rev. Lett.* **111** 132003
[10] Swanson E 2006 *Int. J. Mod. Phys.* A **21** 733); Close F, Downum C and Thomas C E 2010 *Phys. Rev.* D **81** 074033; Ding G J 2009 *Phys. Rev.* D **79** 014001
[11] Yokokawa K, Sasaki S, Hatsuda T and Hayashigaki A 2006 *Phys. Rev.* D **74** 034504; Liu L, Lin H W and Orginos K 2008 PoS LATTICE **2008** 112
[12] Liu X H, Guo F K and Epelbaum E 2013 *Eur. Phys. J.* C **73** 2284
[13] Nieves J and Valderrama M P 2012 *Phys. Rev. D* **86** 056004; Hidalgo-Duque C, Nieves J and Valderrama M P 2013 *Phys. Rev. D* **87** 076006
[14] Albaladejo M, Hidalgo-Duque C, Nieves J and Oset E 2013 *Phys. Rev. D* **88** 014510
[15] Hanhart C, Kalashnikova Y S, Kudryavtsev A E and Nefediev A V 2007 *Phys. Rev.* D **76** 034007

Journal of Physics: Conference Series **742** (2016) 012003 doi:10.1088/1742-6596/742/1/012003

Bottomonium physics at $\Upsilon(4S, 5S, 6S)$ energies with the Belle detector

Umberto Tamponi

INFN - Sezione di Torino and University of Torino, Italy

Abstract. The description of quarkonia as pure quark anti-quark bound states has been recently challenged by the observation of charged states in both the charmonium and bottomonium region and large violations of the heavy quark spin symmetry in hadronic transitions. All these effects can be ascribable to non-negligible contributions from the light quark degrees of freedom in the description of both charmonia and bottomonia. We will report the most recent experimental measurements performed by the Belle collaboration in the Y(4S), Y(5S) and Y(6S) regions, including the measurement of the ratio $\sigma[e^+e^- \to b\bar{b}]/\sigma[e^+e^- \to \mu^+\mu^-]$, the search for neutral states near the $B^0\bar{B}^0$ threshold, the first observation of the transition $\Upsilon(4S) \to \eta h_b(1P)$ and the study of the η transitions at the $\Upsilon(5S)$ energy. The contribution to the study of the structure of these states coming from the measurement of hadronic transitions will be discussed.

1. Quarkonium near the thresholds as a QCD laboratory

A new rich phenomenology consisting in new exotic states, unexpected transitions or sizable corrections to the energy levels above the threshold has been discovered in the last years, challenging the available theoretical models and triggering wide discussion in both the theoretical and experimental communities [1, 2, 3]. The common denominator of all the new, unexpected effects is the presence of light degrees of freedom not accounted by the simple $b\bar{b}$ model. The fact that thresholds located nearby a resonant state can shift its mass [4, 5] or modify its decay modes is known since long time as *coupled channel effect*. The presence in the quarkonium of light degrees of freedom, more and more important approaching the open flavor threshold, is therefore not a new or unexpected phenomenon in the quarkonium sector, and the coupled channel effects were already included in early descriptions of the dipion transitions from the $\Upsilon(3S)$ [6, 7]. What was surprising in the last years is the manifestation of light quark effects in a number of different ways, perhaps the most spectacular one being the appearance of large number of unpredicted states with exotic quantum numbers in both the bottomonium and charmonium sector. We present here the most recent results on the bottomonium states above the thresholds, $\Upsilon(4S, 5S, 6S)$, obtained using the data samples collected by the Belle detector [8] at the KEKB e^+e^- collider [9].

2. The role of hadronic transitions

The study of charmonia and bottomonia, despite their similar structure, offer an insights into different dynamic regimes: while the charmonium spectrum is known to get significant relativistic correction, the dynamic of the $b\bar{b}$ quarks in a bottomonium state lays completely in the non-relativistic regime. As a a consequence, the heavy quark spin symmetry (HQSS) is expected to

Content from this work may be used under the terms of the Creative Commons Attribution 3.0 licence. Any further distribution of this work must maintain attribution to the author(s) and the title of the work, journal citation and DOI.
Published under licence by IOP Publishing Ltd

be a good symmetry that determines the selection rules for the hadronic transition. Therefore, violations of these selection rules can lead insights on the HQSS violation mechanisms related to the presence of light quarks. Experimentally the study of the exotic charmonium states is much easier than the bottomonium ones. The former one can both be produced in e^+e^- collisions, B meson decays or complete $p\bar{p}$ annihilations. The study of charmonium in complete $p\bar{p}$ annihilation in particular, already carried on at the CERN intersecting storage rings [10] and more recently at Fermilab [11, 12] until 2000, is now looked as the most promising way to finally disentangle the ambiguities left by the observations made at the B- and tau/cram factories and to measure the width of the exotic states, which are often not precisely measurable at the e^+e^- colliders. The proposed Panda experiment [13] at the FAIR facility [14] should resume this kind of research in the future. On the other hand, the study of bottomonium is experimentally more challenging: the only feasible way to search for exotic states is the production of 1^{--} states at e^+e^- collider, and the observation of other states with different quantum numbers through hadronic and radiative transition. Therefore the study of the hadronic transitions plays, from a certain point of view, a prominent role in the bottomonium sector.

3. $\sigma[e^+e^- \to hadrons]$ in the $\Upsilon(5S, 6S)$ region

The first striking anomaly in the region above the open beauty threshold was reported by the Belle collaboration, that measured a branching fraction for the transition $\Upsilon(5S) \to \pi^+\pi^-\Upsilon(1S)$ several orders of magnitude larger than the expectations [15]. Furthermore, the $e^+e^- \to \pi^+\pi^-\Upsilon(1S)$ and $e^+e^- \to hadrons$ cross sections were reported not to be peaking at the same energy, leading to speculations about the presence of two overlapping resonances instead of one single bottomonium state [16]. Belle performed a new scan of the region during the final phase of the data taking, and re-measured more precisely the cross section for the production of both the inclusive $b\bar{b}$ pairs, $\sigma[e^+e^- \to b\bar{b}]$, and $\Upsilon(nS)$ via a dipion transition, $\sigma[e^+e^- \to \pi^+\pi^-\Upsilon(1S, 2S, 3S)]$ [17], in the form of the ratios $R_b = \frac{\sigma[e^+e^- \to b\bar{b}]}{\sigma[e^+e^- \to \mu^+\mu^-]}$ and $R_{\pi\pi} = \frac{\sigma[e^+e^- \to \pi^+\pi^-\Upsilon(1S,2S,3S)]}{\sigma[e^+e^- \to \mu^+\mu^-]}$. Thanks to the previous studies on the $\Upsilon(nS)$ and Z_b production from $\Upsilon(5S)$ [18], R_b can for the first time be analyzed into its different components. Surprisingly, the known contributions of the hadronic transitions $\Upsilon(5S) \to \pi^\pm Z_b^\mp \to \pi^\pm[B^\star\bar{B}^{(\prime)}]$, $\Upsilon(5S) \to \pi^+\pi^-h_b(1P, 2P)$ and $\Upsilon(5S) \to \pi^+\pi^-\Upsilon(nS)$ saturate the cross section at the $\Upsilon(5S)$ energy, leaving no space for direct decays of $\Upsilon(5S)$ into B meson pairs. This observation is obviously inconsistent with the copious production of B_s known to occur at the $\Upsilon(5S)$ energy. If it is not due to $\Upsilon(5S)$ annihilations as the decomposition of R_b suggests, it must come directly from the continuum. However, if the $\Upsilon(5S)$ really decays entirely into other bottomonium states and the two-body B meson production is completely due to the $b\bar{b}$ continuum, then there should be no or little interference between the $\Upsilon(5S)$ resonance and the continuum itself, in contrast with the results of the fit that show a sizable interference effect. All these evidences strongly suggest that the complex threshold structure above the open flavor threshold cannot be simply approximated by a constant distribution, and that the resonance parameters extracted with this fit could be affected by some unknown bias. This bias however can be avoided measuring the resonance parameters with $R_{\pi\pi}$ instead of R_b.

The study of $R_{\pi\pi}$ is done reconstructing the exclusive $\Upsilon(nS)$ annihilations into muon pairs. The parameters fitted in the different Υ final states are in agreement, therefore a simultaneous fit is performed to increase the overall precision. No $\pi\pi\Upsilon$ final states are expected to be produced directly by the continuum reaction, therefore the fitting function can be simplified removing the continuum-components, but introducing additional factors to account for the different phase space sizes due to the different $\Upsilon(nS)$ masses.

For the first time, we report the observation of the $\pi\pi$ transitions from the $\Upsilon(6S)$. The resonance parameters obtained in the two analyses have a 2σ level discrepancy, but this can be ascribed to the mismodelling of the continuum contribution in the fit of R_b.

4. Hadronic transitions from $\Upsilon(5S)$ and $\Upsilon(6S)$

We saw that unexpected dipion transitions from the $\Upsilon(5S)$ challenged our understanding of the $\Upsilon(5S)$ nature and lead to the discovery of the first tetraquark-like bottomonium states, the Z_b. In addition, the hadronic transitions are challenging the most common theoretical models developed for their description, in particular the QCD multipole expansion (QCDME) [19, 20, 21, 22, 23, 24, 25, 26], in which the HQSS is preserved. The HQSS violation effects in bottomonium are expected to be of the order of Λ^2_{QCD}/m_b^2. To study these effects it is required to measure both the η and the $\pi\pi$ trasitions rates between two states: the η meson transitions among vector states are HQSS violating since, in the QCDME framework, they are described by the emission of a chromo-magnetic gluon, while the corresponding di-pion transitions are HQSS preserving. Therefore the ratios between the η and $\pi\pi$ transitions among $\Upsilon(3S)$, $\Upsilon(2S)$ and $\Upsilon(1S)$, $\mathcal{R}^{3S,2S\to 1S}_{\eta,\pi\pi}$ are expected to be of the order of $\Lambda^2_{QCD}/m_b^2 \approx 4\times 10^{-3}$. This QCDME prediction meets, at least qualitatively the experimental measurements in the sector of narrow bottomonia, indicating that the HQSS is indeed a good symmetry: $\mathcal{R}^{2S\to 1S}_{\eta,\pi\pi} = (1.6\pm 0.2)\times 10^{-3}$ [27, 28, 29] and $\mathcal{R}^{3S\to 1S}_{\eta,\pi\pi} < 2.1\times 10^{-3}$ [28]. At the $\Upsilon(4S)$ energy however the situation is completely reversed. The BaBar collaboration measured $\mathcal{R}^{4S\to 1S}_{\eta,\pi\pi} < 2.4\pm 0.5$ [30], therefore the spin flipping process is even enhanced with respect to the non spin-flipping one. Soon after a similar behavior was also observed in the $\Upsilon(5S)$ decays, where the η transitions [31] are of the same order of magnitude of the $\pi\pi$ ones: $\mathcal{R}^{5S\to 1S}_{\eta,\pi\pi} = 0.13\pm 0.10$ and $\mathcal{R}^{5S\to 2S}_{\eta,\pi\pi} = 0.49\pm 0.37$. Even if the QCDME model is expected to be not reliable for wide states, such a large violation of the HQSS triggered the interest of the theoretical community.

In order to complete the exploration of the possible η an $\pi\pi$ transitions, the Belle collaboration recently studied extensively the last class of process left unexplored: the η transitions from $\Upsilon(4S)$ and $\Upsilon(5S)$ to the spin-singlets $h_b(nP)$. Since no $h_b(1,2P)$ exclusive decay modes are known to have a significant branching fraction and a clear experimental signature, the analysis is performed inclusively: the signals of $\Upsilon(4,5S)\to\eta(b\bar{b})$ transitions are expected to appear as peaks in the recoil mass spectrum of the η candidates in the samples of hadronic events collected at the $\Upsilon(4,5S)$ energies. We reconstruct the η meson in the $\gamma\gamma$ decay mode, that guarantees the lowest combinatorial background.

By this mean the $\Upsilon(4S)\to\eta h_b(1P)$ transition is observed for the first time, with a statistical significance greater than 5σ, and the corresponding branching ratio is $\mathcal{B}[\Upsilon(4S)\to \eta h_b(1P)] = (2.18\pm 0.11\pm 0.18)\times 10^{-3}$ [32], in agreement with the available theoretical predictions [33, 34]. This is by far the largest known hadronic transition from the $\Upsilon(4S)$, and offers a new, unexpected gateway to the bottomonium spin-singlet states. The same analysis performed at the $\Upsilon(5S)$ energy shows significant peaks in correspondence of the $\Upsilon(5S)\to\eta\Upsilon(2S)$, and $\Upsilon(5S)\to\eta\Upsilon(^3 1D)$, with statistical significance of 3.1σ and 4.4σ, respectively.

With these new results, we are able to improve the overall knowledge of the hadronic transitions and make a comprehensive comparison between the experimental measurements and the theoretical predictions. Figure 1 and 2 show a comparison between some theoretical prediction based on different approaches [26, 35, 36, 37, 38, 41, 40] and the experimental measurements. Each branching ratio is, individually, well predicted by at least one author. Nevertheless, no model is able to correctly reproduce the branching ratios observed below and above threshold in a single framework. W are still lacking a unified approach to all the hadron transitions from the different states due to theoretical incomprehensions, or the physical mechanism generating the transitions is strikingly different in the two mass regions. Furthermore, while the transitions from above the threshold seems to be overall well described by the introduction of coupled channel effects, loops or hybrid states as HQSS breaking mechanism in the QCDME, the theoretical landscape is surprisingly confused in the region below the

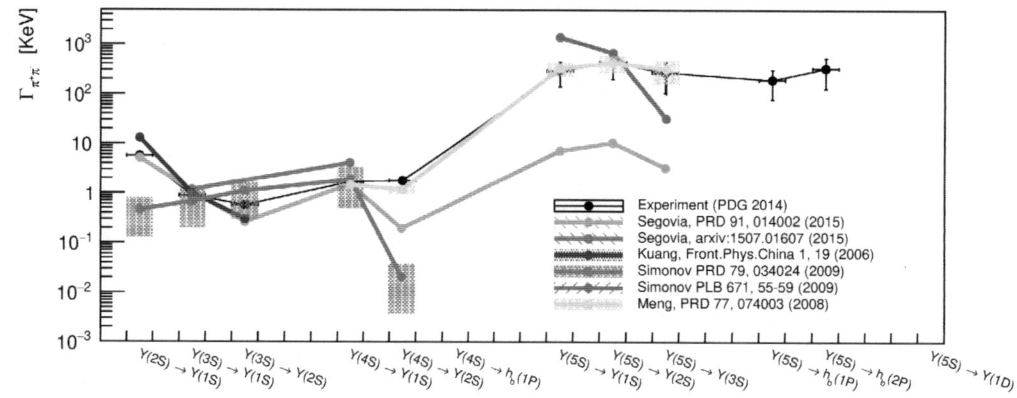

Figure 1. Theoretical predictions and experimental results for the known di-pion transitions among bottomonium states.

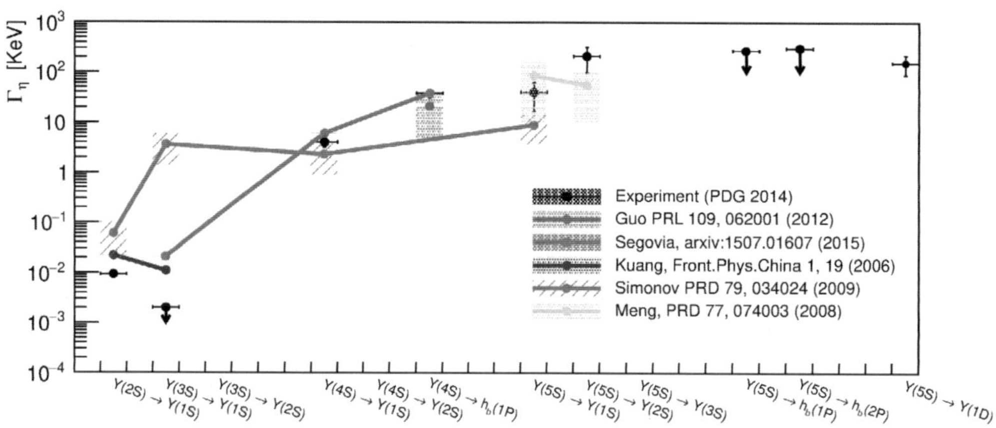

Figure 2. Theoretical predictions and experimental results for the known η transitions among bottomonium states.

threshold, where these effects are expected to be small and the HQSS should be dominant.

5. References

[1] N. Brambilla *et al.*, Eur. Phys. J. C **74**, no. 10, 2981 (2014) doi:10.1140/epjc/s10052-014-2981-5 [arXiv:1404.3723 [hep-ph]].
[2] N. Brambilla *et al.* [Quarkonium Working Group Collaboration], hep-ph/0412158.
[3] N. Brambilla *et al.*, Eur. Phys. J. C **71**, 1534 (2011) doi:10.1140/epjc/s10052-010-1534-9 [arXiv:1010.5827 [hep-ph]].
[4] N. A. Tornqvist, Phys. Rev. Lett. **53**, 878 (1984). doi:10.1103/PhysRevLett.53.878
[5] N. A. Tornqvist, Acta Phys. Polon. B **16**, 503 (1985) [Acta Phys. Polon. B **16**, 683 (1985)].
[6] P. Moxhay, Phys. Rev. D **39**, 3497 (1989). doi:10.1103/PhysRevD.39.3497
[7] H. Y. Zhou and Y. P. Kuang, Phys. Rev. D **44**, 756 (1991). doi:10.1103/PhysRevD.44.756
[8] A. Abashian *et al.*, Nucl. Instrum. Meth. A **479**, 117 (2002). doi:10.1016/S0168-9002(01)02013-7
[9] S. Kurokawa, E. Kikutani, Nucl. Instrum. Meth. A **499**, 1 (2003). and other papers included in this volume.
[10] C. Baglin *et al.* [R704 and Annecy(LAPP)-CERN-Genoa-Lyon-Oslo-Rome-Strasbourg-Turin Collaborations], Phys. Lett. B **171**, 135 (1986). doi:10.1016/0370-2693(86)91013-0

[11] T. A. Armstrong *et al.* [E760 Collaboration], Phys. Rev. Lett. **68**, 1468 (1992). doi:10.1103/PhysRevLett.68.1468
[12] C. Patrignani *et al.* [FNAL-E835 Collaboration], AIP Conf. Proc. **717**, 581 (2004) [Nucl. Phys. Proc. Suppl. **142**, 98 (2005)]. doi:10.1016/j.nuclphysbps.2005.01.017
[13] M. F. M. Lutz *et al.* [PANDA Collaboration], arXiv:0903.3905 [hep-ex].
[14] P. Spiller and G. Franchetti, Nucl. Instrum. Meth. A **561**, 305 (2006). doi:10.1016/j.nima.2006.01.043
[15] K.-F. Chen *et al.* [Belle Collaboration], Phys. Rev. D **82**, 091106 (2010) doi:10.1103/PhysRevD.82.091106 [arXiv:0810.3829 [hep-ex]].
[16] A. Ali, C. Hambrock, I. Ahmed and M. J. Aslam, Phys. Lett. B **684**, 28 (2010) doi:10.1016/j.physletb.2009.12.053 [arXiv:0911.2787 [hep-ph]].
[17] D. Santel *et al.* [Belle Collaboration], arXiv:1501.01137 [hep-ex].
[18] A. Bondar *et al.* [Belle Collaboration], Phys. Rev. Lett. **108**, 122001 (2012) doi:10.1103/PhysRevLett.108.122001 [arXiv:1110.2251 [hep-ex]].
[19] K. Gottfried, Phys. Rev. Lett. **40**, 598 (1978).
[20] G. Bhanot, W. Fischler and S. Rudaz, Nucl. Phys. B **155**, 208 (1979).
[21] M. E. Peskin, Nucl. Phys. B **156**, 365 (1979).
[22] G. Bhanot and M. E. Peskin, Nucl. Phys. B **156**, 391 (1979).
[23] M. B. Voloshin, Nucl. Phys. B **154**, 365 (1979).
[24] M. B. Voloshin and V. I. Zakharov, Phys. Rev. Lett. **45**, 688 (1980).
[25] Y. P. Kuang and T. M. Yan, Phys. Rev. D **24**, 2874 (1981). doi:10.1103/PhysRevD.24.2874
[26] Y. P. Kuang, Front. Phys. China **1**, 19 (2006) doi:10.1007/s11467-005-0012-6 [hep-ph/0601044].
[27] Q. He *et al.* (CLEO Collaboration), Phys. Rev. Lett. **101**, 192001 (2008).
[28] J. P. Lees *et al.* (BaBar Collaboration), Phys. Rev. D **84**, 092003 (2011).
[29] U. Tamponi *et al.* (Belle Collaboration), Phys. Rev. D **87**, 011104 (2013).
[30] B. Aubert *et al.* (BaBar Collaboration), Phys. Rev. D **78**, 112002 (2008).
[31] U. Tamponi, EPJ Web Conf. **70**, 00034 (2014). doi:10.1051/epjconf/20147000034
[32] U. Tamponi *et al.* [Belle Collaboration], Phys. Rev. Lett. **115**, no. 14, 142001 (2015) doi:10.1103/PhysRevLett.115.142001 [arXiv:1506.08914 [hep-ex]].
[33] F. K. Guo, C. Hanhart and U. G. Meissner, Phys. Rev. Lett. **105**, 162001 (2010) doi:10.1103/PhysRevLett.105.162001 [arXiv:1007.4682 [hep-ph]].
[34] J. Segovia, F. Fernandez and D. R. Entem, arXiv:1507.01607 [hep-ph].
[35] Y. A. Simonov and A. I. Veselov, Phys. Rev. D **79**, 034024 (2009) doi:10.1103/PhysRevD.79.034024 [arXiv:0804.4635 [hep-ph]].
[36] Y. A. Simonov and A. I. Veselov, Phys. Lett. B **671**, 55 (2009) doi:10.1016/j.physletb.2008.12.001 [arXiv:0805.4499 [hep-ph]].
[37] Y. A. Simonov and A. I. Veselov, arXiv:0806.2919 [hep-ph].
[38] Y. A. Simonov, Phys. Atom. Nucl. **71**, 1048 (2008) doi:10.1134/S1063778808060094 [arXiv:0711.3626 [hep-ph]].
[39] C. Meng and K. T. Chao, Phys. Rev. D **77**, 074003 (2008) doi:10.1103/PhysRevD.77.074003 [arXiv:0712.3595 [hep-ph]].
[40] J. Segovia, D. R. Entem and F. Fernndez, Phys. Rev. D **91**, no. 1, 014002 (2015) doi:10.1103/PhysRevD.91.014002 [arXiv:1409.7079 [hep-ph]].
[41] C. Meng and K. T. Chao, Phys. Rev. D **77**, 074003 (2008) doi:10.1103/PhysRevD.77.074003 [arXiv:0712.3595 [hep-ph]].

Charm physics and XYZ states at BESIII

Evgeny Boger for the BESIII Collaboration
Joint Institute for Nuclear Research, Joliot-Curie, 6, Dubna, Moscow region, Russia, 141980
E-mail: boger@jinr.ru

Abstract. The BESIII Experiment at the Beijing Electron Positron Collider (BEPC2) collected large data samples for electron-positron collisions with c.m. energy above 4 GeV/c^2 during 2013 and 2014. The analysis of these samples has resulted in a number of surprising discoveries, such as the discoveries of the electrically charged "Zc" structures, which, if resonant, cannot be accommodated in the traditional charm quark and anti-charm quark picture of charmonium. In this talk, we will review the current status of the analyses of the Zc structures, as well as a number of other interesting features in the new BESIII data samples. We'll also present the recent charm physics results from BESIII.

1. Introduction

BEPCII is a double ring e^+e^- collider in Beijing, China with the design luminosity of $1 \times 10^{33} cm^{-2}s^{-1}$ and the design c.m. energy $2.0 - 4.2\ GeV/c^2$. After the 2012 upgrade of BEPCII LINAC, the maximum c.m. energy was increased up to 4.6 GeV, making possible to study the XYZ states. BESIII is a general purpose 4π detector located at the BEPCII storage ring. Since 2009, the BESIII experiment has accumulated the world largest samples of J/ψ, $\psi(2S)$, ψ(3770 and $Y(4260)$, as well as large 1.3 fb^{-1} R scan data sample and XYZ data at higher energies.

2. Charm physics at BESIII

There are several objectives for studying charmonia decays above open-charm threshold at BESIII. The open-charm physics program at BESIII includes studying pure leptonic, semileptonic and hadronic charm meson decays, searches for CP violations, neutral D mixing, rare or forbidden decays. The latter are sensitive to physics beyond the Standard Model.

Studies of leptonic and semileptonic decays of D mesons are the preferred way to determine the magnitude of the Cabibbo-Kobayashi-Maskawa (CKM) matrix elements, $|V_{cs}|$ and $|V_{cd}|$, since the strong interaction binding can be described by the hadronic form factors or decay constants, which can be calculated, for instance, by Lattice QCD . Accessing these CKM matrix elements provides the way to test the unitarity of the CKM matrix. On the other hand, the CKM unitarity puts the constrains on $|V_{cs}|$ and $|V_{cd}|$ values. Using the CKM element values as an input, it's possible to access, for instance, $f_{D\to K/\pi}(q^2)$ hadronic form factors and other form factors with a great precision, providing a way to verify and calibrate the Lattice QCD calculations [7]. Although the statistics on charm physics available at the BESIII is significantly lower than that available on B- and super-B factories, the production at threshold provides extremely clean environment. Another important feature is the ability to use a tagging technique which allows to measure the absolute branching ratios.[25]

Content from this work may be used under the terms of the Creative Commons Attribution 3.0 licence. Any further distribution of this work must maintain attribution to the author(s) and the title of the work, journal citation and DOI.
Published under licence by IOP Publishing Ltd

The charmonium produced in e^+e^- collision decays into pair of charm mesons. For instance, at the open-charm threshold the $\psi(3770)$ decays into a pair of D mesons. If a decay of one D meson ("tagged decay") has been fully reconstructed in an event, then the existence of another $\bar{D}$ decay ("signal decay") in the same event is guaranteed. The tagged decays are reconstructed in the channels with larger branching fractions and lower background levels.

To improve resolution, the following two variables are used for the tagged decay selection: the beam-constrained mass $M_{\mathrm{BC}} = \sqrt{E^2_{\mathrm{beam}} - |\vec{p}_D|^2}$ and $\Delta E = E_D - E_{\mathrm{beam}}$, where E_D and $\vec{p}_D$ are the reconstructed energy and 3-momentum of the D^- candidate, and E_{beam} is the beam energy.

3. Analysis of the $D^+ \to K^-\pi^+e^+\nu_e$

The semileptonic decay $D^+ \to K^-\pi^+e^+\nu_e$ has received particular attention due to the relative simplicity of its theoretical description and the large branching fraction. The matrix element of the decay can be factorized as the product of the leptonic and hadronic currents, which makes it a natural place to study the $K\pi$ system in the absence of interactions with other hadrons, and the hadronic transition form factors. The analysis[6] was done mainly for two purposes:

i) Measure the different $K\pi$ amplitudes that contribute to this decay, including S-wave and radially excited P-wave and D-wave components. Accurate measurements of these contributions can provide helpful information for the B-meson semileptonic decays.

ii) Measure the q^2 dependent transition form factors, where q^2 is the invariant mass square of the lepton pair of the $e\nu_e$ system. They can be compared with hadronic model expectations and lattice QCD computations [1].

Using the single D-tagging technique, 18262 signal events with a background level of about 0.7% are selected. The branching fractions are measured over the full $m_{K\pi}$ (the invariant mass of $K\pi$ pair) range and in the K^*-dominated region, respectively: $\mathcal{B}(D^+ \to K^-\pi^+e^+\nu_e) = (3.71 \pm 0.03 \pm 0.08)\%$, $\mathcal{B}(D^+ \to K^-\pi^+e^+\nu_e)_{[0.8,1]} = (3.33 \pm 0.03 \pm 0.07)\%$.

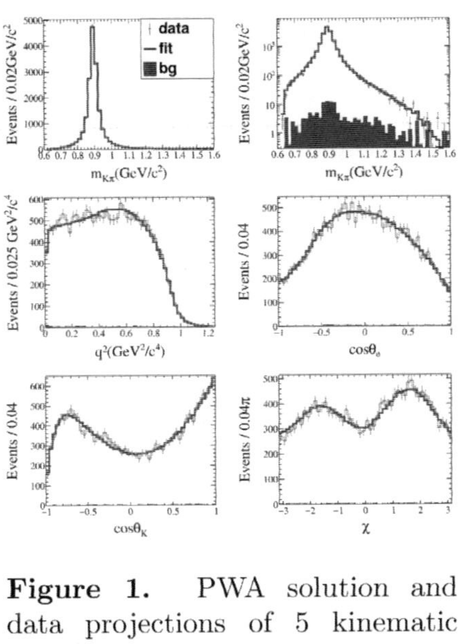

Figure 1. PWA solution and data projections of 5 kinematic variables.

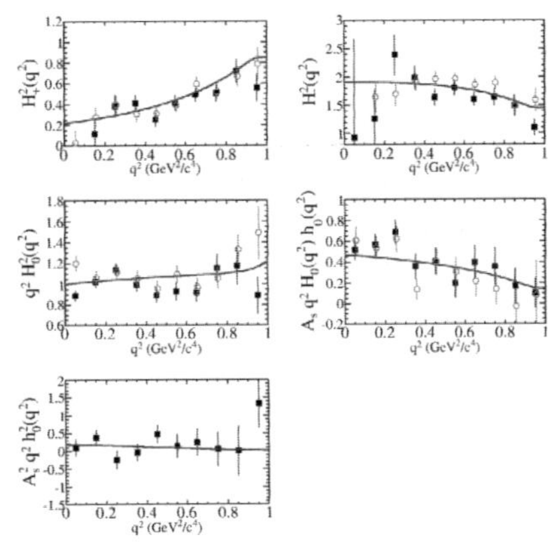

Figure 2. Model-independent helicity-basis form factor measurement results. Black dots: BESIII preliminary, red circles: CLEO-c[5], blue lines: BESIII preliminary PWA results.

A partial wave analysis (PWA) is also performed. It shows that the dominant $\bar{K}^*(892)^0$ component is accompanied by an S-wave contribution accounting for $(6.05\pm0.22\pm0.18)\%$ of the total rate and that other components are negligible. The parameters of the $\bar{K}^*(892)^0$ resonance and of the form factors based on the spectroscopic pole dominance predictions are also measured. The vector pole mass m_V in this decay is measured for the first time: $m_V = 1.81^{+0.25}_{-0.17} \pm 0.01$ GeV/c^2.

By fitting the S-wave phase separately in 12 $m_{K\pi}$ bins, the S-wave phase variation with $m_{K\pi}$ is measured in a model-independent way and found to be in agreement with the PWA solution, which is based on the parameterization in the LASS scattering experiment.

Finally, a model-independent measurement of the q^2 dependence of the helicity form factors is performed using the projective weighting technique, which was introduced in Ref. [3]. It agrees well with the CLEO-c result[5] and the PWA solution (Fig. 2).

4. Study of $D^0 \to K^-/\pi^- e^+\nu_e$

The study of dynamics of $D^0 \to K^- e^+\nu_e$ and $D^0 \to \pi^- e^+\nu_e$ decays published by BESIII last year is of particular interest. [12]

In this analysis, a careful selection procedure is performed to accurately measure both absolute and differential branching fractions of the said decays.

The absolute decay branching fractions are $\mathcal{B}(D^0 \to K^- e^+\nu_e) = (3.505 \pm 0.014 \pm 0.033)\%$ and $\mathcal{B}(D^0 \to \pi^- e^+\nu_e) = (0.295 \pm 0.004 \pm 0.003)\%$. Using the differential form factors results, the products of hadronic form factors and the CKM matrix element $f_+^K(0)|V_{cs}| = 0.7172 \pm 0.0025 \pm 0.0035$ and $f_+^\pi|V_{cd}| = 0.1435 \pm 0.0018 \pm 0.0009$ are determined.

Using the values of $|V_{cd(s)}|$ from SM constraint fit [13], the hadronic form factors f_+^K and f_+^π are extracted and their ratio is compared to LQCD and light cone sum rule (LCSR) calculations.

In turn, the values of $f_+^K(\pi)(0)$ calculated with LQCD and the measured $f_+^K(0)|V_{cd(s)}|$ products are used to extract the values of CKM matrix elements. The measurements of the said products are most precise to date and would give more precise values of $|V_{cs}|$ and $|V_{cd}|$ with their uncertainties decreasing to 0.6% and 1.4% respectively when the uncertainty in form factors calculation could be ignored (Fig. 3,4).

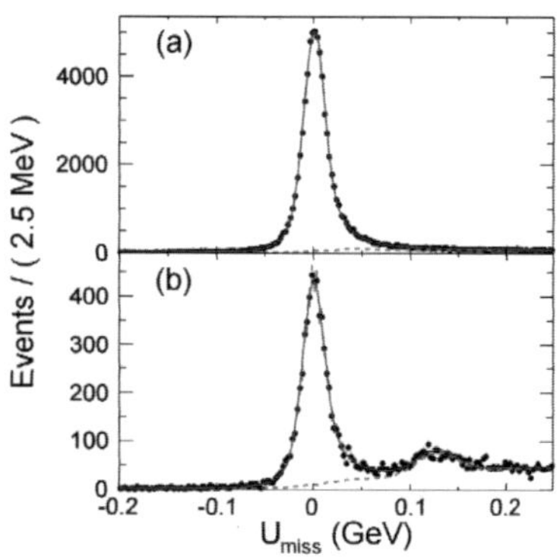

Figure 3. U_{miss} distributions of events for $D^0 \to K^- e^+\nu_e$ and $D^0 \to \pi^- e^+\nu_e$.

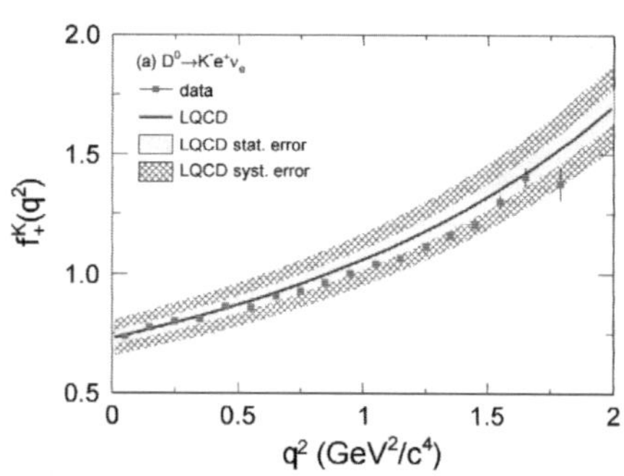

Figure 4. Comparison of the measured K form factor (squares with error bars) with the LQCD calculations [12][14].

5. Study of $D^+ \to \omega e^+\nu_e$ and search for $D^+ \to \varphi e^+\nu_e$

The amplitude analysis of $D^+ \to \omega e^+\nu_e$ four-body decay is performed for the first time [15]. The single pole (SPD) model form factor ratios are calculated: $r_V = V(0)/A_1(0) = 1.24 \pm 0.09 \pm 0.06$ and $r_2 = A_2(0)/A_1(0) = 1.06 \pm 0.15 \pm 0.05$. The absolute decay branching ratio is determined: $\mathcal{B}(D^+ \to \omega e^+\nu_e) = (1.63 \pm 0.11 \pm 0.08) \times 10^{-3}$.

In addition, the decay $D^+ \to \varphi e^+ \nu_e$ is searched for. The updated upper limit on the branching ratio is established: $\mathcal{B}(D^+ \to \varphi e^+ \nu_e) < 1.3 \times 10^{-5} (@90\% C.L.)$.

6. Observation of $D^+ \to \omega\pi^+$ and evidence for $D^0 \to \omega\pi^0$

The searches for singly Cabbibo-suppressed decays $D^+ \to \omega\pi^+$ and $D^0 \to \omega\pi^0$ are performed. Using the double D tag technique, the absolute branching fractions $\mathcal{B}(D^+ \to \omega\pi^+) = (2.79 \pm 0.57 \pm 0.16) \times 10^{-4}$ and $\mathcal{B}(D^0 \to \omega\pi^0) = (1.17 \pm 0.34 \pm 0.07) \times 10^{-4}$ are measured with the significance of 5.5σ and 4.1σ respectively. The $D^+ \to \omega\pi^+$ decay is observed for the first time [16].

7. XYZ States (Fig. 5)

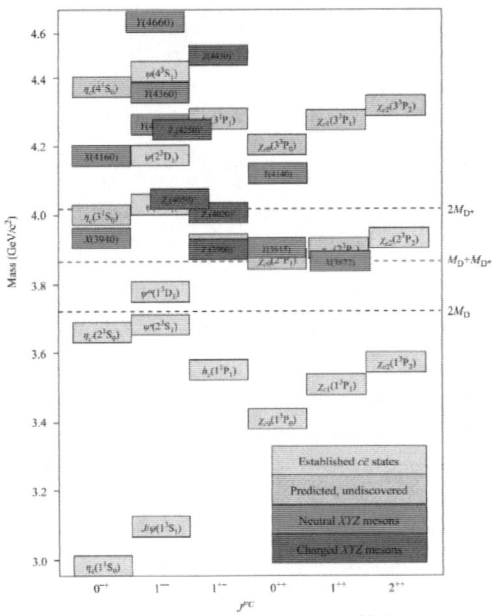

Figure 5. During the past decade strong evidence for mesons that do not fit into the $q\bar{q}$ scheme of the quark model has been accumulating. They include charmonium- (and bottomonium-) like states, called XYZ mesons, which do not fit into $c\bar{c}$ level scheme. [8] In addition to $q\bar{q}$ mesons and qqq baryons of the original quark model [10], the QCD allows color-singlet states with a more complex structure. These "exotic" states include dibaryon, pentaquark, $qq + q\bar{q}$, loosely bound dimeson molecule and $q\bar{q}g$ hybrid state as well as glueball.

In 2013, using the data collected at $\sqrt{s} = 4.26\ GeV$, BESIII Collaboration observed charged charmoniumlike state $Z_c(3900)$. [11] Following this observation, the data collection plan was changed, which led to a number of significant results. [9]

8. Z_c states at BES-III

A charged charmoniumlike particle $Z_c(3900)^+$ has been observed through its $\pi^\pm J/\psi$ decay channel on BESIII, Belle and CLEO-c [11][19][20]. The state lies very close to the $D\bar{D}^*$ production threshold. Another Z_c state called $Z_c(3885)^+$ has been discovered by BESIII in $e^+e^- \to \pi^\pm (D\bar{D})^\mp$ process.

These states are coupled to charmonium and at the same time have a non-zero electric charge, which means that they cannot be described by conventional quark model as $q\bar{q}$ mesons.

In 2015, BESIII observed the state in $e^+e^- \to \pi^+\pi^- J/\psi$ decay with a mass $M = (3894.8 \pm 2.3 \pm 3.2)\ MeV/c^2$ and significance of 10.4σ [18]. This state can be interpreted as $Z_c(3900)^0$, a neutral partner to $Z_c(3900)^+$, thus establishing an isospin triplet.

In another paper published by BESIII last year [21], the aforementioned $Z_c(3885)^+$ state is confirmed ($> 10\sigma$) using a double tag technique on larger data set at $\sqrt{s} = 4.23, 4.26\ GeV$. The polar angular distribution of the $\pi^\pm Z_c(3885)^\mp$ is found to be consistent with $J^P = 1^+$.

Finally, two neutral states were recently found by BESIII. The first one with the mass $M = (3885.7^{+4.3}_{-5.7} \pm 8.4)\ MeV/c^2$ was observed in $e^+e^- \to \pi^0(D\bar{D}^*)^0$ process with the significance of 12σ [22]. Another state with the mass $M = (4025.5^{+2.0}_{-4.7} \pm 3.1)\ MeV/c^2$ was discovered in $e^+e^- \to \pi + (D^*\bar{D}^*)$ decays (5.9σ significance) [23]. These states are attributed to the neutral partners of previously observed $Z_c(3885)^+$ and $Z_c(4025)^+$ [24] respectively, thus establishing two isospin triplets.

9. Summary

The recent progress in the charm physics and XYZ physics at BESIII are reported. A number of the analyses of a semileptonic, leptonic and rare D decays were published last year. More analyses are being performed at the moment using the high accumulated statistics for $D\bar{D}$, $D_s^+D_s^-$ and $\Lambda_c^+\Lambda_c^-$.

In 2013-2015, BESIII observed a total of four charged Z_c states: $Z_c(3900)^+$, $Z_c(3885)^+$, $Z_c(4020)^+$ and $Z_c(4025)^+$, and four neutral states, which can be associated with the charged ones. The nature of these states is yet to be discovered.

References

[1] C. W. Bernard, A. X. El-Khadra and A. Soni, Phys. Rev. D **45**, 869 (1992); V. Lubicz, G. Martinelli, M. S. McCarthy and C. T. Sachrajda, Phys. Lett. B **274**, 415 (1992); A. Abada *et al.*, Nucl. Phys. B **416**, 675 (1994); K. C. Bowler *et al.* (UKQCD Collaboration), Phys. Rev. D **51**, 4905 (1995); T. Bhattacharya and R. Gupta, Nucl. Phys. Proc. Suppl. **47**, 481 (1996); C. R. Allton *et al.* (APE Collaboration), Phys. Lett. B **345**, 513 (1995); S. Gusken, G. Siegert and K. Schilling, Prog. Theor. Phys. Suppl. **122**, 129 (1996); A. Abada *et al.* (SPQcdR Collaboration), Nucl. Phys. Proc. Suppl. **119**, 625 (2003).
[2] P. del Amo Sanchez *et al.* (BABAR Collaboration), Phys. Rev. D **83**, 072001 (2011).
[3] J.M. Link *et al.* (FOCUS Collaboration), Phys. Lett. B **633**, 183 (2006).
[4] P. del Amo Sanchez *et al.* (BABAR Collaboration), Phys. Rev. D **83**, 072001 (2011).
[5] R.A. Briere *et al.* (CLEO Collaboration), Phys. Rev. D **81**, 112001 (2010).
[6] M. Ablikim *et al.* [BESIII Collaboration], arXiv:1512.08627 [hep-ex].
[7] H. B. Li, Nucl. Phys. Proc. Suppl. **233** (2012) 185
[8] S. L. Olsen, Front. Phys. **10** (2015) 101401
[9] C. Z. Yuan [BESIII Collaboration], Front. Phys. China **10** (2015) no.6, 101401
[10] R. L. Jaffe, Phys. Rev. D **17** (1978) 1444.
[11] M. Ablikim *et al.* [BESIII Collaboration], Phys. Rev. Lett. **110** (2013) 252001
[12] M. Ablikim *et al.* [BESIII Collaboration], Phys. Rev. D **92** (2015) no.7, 072012
[13] K. A. Olive *et al.* [Particle Data Group Collaboration], Chin. Phys. C **38** (2014) 090001.
[14] C. Aubin *et al.* [Fermilab Lattice and MILC and HPQCD Collaborations], Phys. Rev. Lett. **94** (2005) 011601
[15] M. Ablikim *et al.* [BESIII Collaboration], Phys. Rev. D **92** (2015) no.7, 071101
[16] M. Ablikim *et al.* [BESIII Collaboration], Phys. Rev. Lett. **116** (2016) no.8, 082001
[17] M. Ablikim *et al.* [BESIII Collaboration], Phys. Rev. Lett. **115** (2015) no.22, 221805
[18] M. Ablikim *et al.* [BESIII Collaboration], Phys. Rev. Lett. **115** (2015) no.11, 112003
[19] Z. Q. Liu *et al.* [Belle Collaboration], Phys. Rev. Lett. **110** (2013) 252002
[20] T. Xiao, S. Dobbs, A. Tomaradze and K. K. Seth, Phys. Lett. B **727** (2013) 366
[21] M. Ablikim *et al.* [BESIII Collaboration], Phys. Rev. D **92** (2015) no.9, 092006
[22] M. Ablikim *et al.* [BESIII Collaboration], Phys. Rev. Lett. **115** (2015) no.22, 222002
[23] M. Ablikim *et al.* [BESIII Collaboration], Phys. Rev. Lett. **115** (2015) no.18, 182002
[24] M. Ablikim *et al.* [BESIII Collaboration], Phys. Rev. Lett. **112** (2014) no.13, 132001
[25] J. Adler *et al.* [MARK-III Collaboration], Phys. Rev. Lett. **60** (1988) 89.

Journal of Physics: Conference Series **742** (2016) 012005 doi:10.1088/1742-6596/742/1/012005

Effective field theory investigations of the XYZ puzzle

Jorge Segovia

Physik-Department, Technische Universität München,
James-Franck-Str. 1, 85748 Garching, Germany

E-mail: jorge.segovia@tum.de

Abstract. Quantum Chromodynamics, the theory of strong interactions, predicts several types of bound states. Among them are conventional mesons ($q\bar{q}$) and baryons (qqq), which have been the only states observed in experiments for years. However, in the last decade, many states that do not fit this picture have been observed at B-factories (BaBar, Belle and CLEO), at τ-charm facilities (CLEO-c, BESIII) and also at proton-proton colliders (CDF, D0, LHCb, ATLAS, CMS). There is growing evidence that at least some of the new charmonium- and bottomonium-like states, the so-called XYZ mesons, are new forms of matter such as quark-gluon hybrids, mesonic molecules or different arrangements of tetraquarks, pentaquarks... Effective Field Theories (EFTs) have been constructed for heavy-quark-antiquark bound states, but a general study of the XYZ mesons within the same framework has not yet been done. The scope of this conference proceedings is to discuss the possibilities we have in developing novel EFTs that, characterizing the conventional quarkonium states, facilitate also the systematic and model-independent description of the new exotic matter, in particular, the hybrid mesons.

1. Introduction

One of the basic properties of Quantum Chromodynamics (QCD) is its spectrum: the list of particles that are stable or at least sufficiently long-lived to be observed as resonances. The elementary constituents in QCD are quarks (q), antiquarks ($\bar{q}$), and gluons (g), and QCD requires them to be confined into color-singlet clusters called hadrons. The most stable hadrons are the clusters predicted by the quark model [1, 2], conventional mesons ($q\bar{q}$) and baryons (qqq), which have been the only states observed in experiments for years [3]).

A decade ago, the Belle Collaboration discovered an unexpected enhancement at 3872 MeV in the $\pi^+\pi^- J/\psi$ invariant mass spectrum while studying the reaction $B^+ \to K^+\pi^+\pi^- J/\psi$ [4]. The $X(3872)$ state was later studied by CDF, D0, and BaBar collaborations confirming that its quantum numbers, mass and decay patterns make it an unlikely conventional charmonium candidate. Therefore, the simple picture that had been so successful for 30 years was challenged leading to an explosion of related experimental activity. The number of new collectively denoted XYZ states has increased dramatically. Almost two dozen charmonium- and bottomonium-like XYZ states have forced an end to the era when heavy quarkonium was considered as a relatively well established heavy quark-antiquark bound system (see, e.g., reviews [5, 6] for more details).

The ultimate goal of theory is to describe the properties of the XYZ states from QCD's first principles. However, since this task is quite challenging, a more modest goal to start with is the development of QCD motivated phenomenological models that specify the colored constituents, how they are clustered and the forces between them. For this reason, at the same time that

Content from this work may be used under the terms of the Creative Commons Attribution 3.0 licence. Any further distribution of this work must maintain attribution to the author(s) and the title of the work, journal citation and DOI.
Published under licence by IOP Publishing Ltd

Journal of Physics: Conference Series **742** (2016) 012005 doi:10.1088/1742-6596/742/1/012005

many experimental measurements on XYZ physics were performed, theorists proposed different kind of exotic hadrons to classify them:

1) *A Quarkonium hybrid,* which has constituents $Q\bar{Q}g$, where g is a constituent gluon. The $Q\bar{Q}$-pair is in a color-octet state because the color charge of a gluon is octet.
2) *A Tetraquark,* which consists on one heavy quark (Q), one heavy antiquark ($\bar{Q}$), one light quark (q) and one light antiquark ($\bar{q}$) whose different arrangements allowed by QCD give:
   2.1) *A meson molecule* [7], which consists of color-singlet $(Q\bar{q})_1$ and $(\bar{Q}q)_1$ mesons bound by pseudoscalar-meson-exchange interactions.
   2.2) *A diquarkonium* [8], which consists of a color-antitriplet $(Qq)_{\bar{3}}$ diquark and a color-triplet $(\bar{Q}\bar{q})_3$ anti-diquark bound by gluon exchanges.
   2.3) *A hadro-quarkonium* [9], which consists of a color-singlet $(q\bar{q})_1$-pair bound through color van der Waals forces to a compact color-singlet $(Q\bar{Q})_1$-pair.
   2.4) *A quarkonium adjoint meson or Born-Oppenheimer tetraquark* [10], which consists of two bounded color-octets, $(Q\bar{Q})_8$ and $(q\bar{q})_8$, with dynamics similar to that of hybrids.
   2.5) *A compact tetraquark* [11], in which the constituents do not form any distinguishable cluster but form as a whole a color singlet state: $(Q\bar{Q}q\bar{q})_1$.
3) *A Pentaquark and so on,* which are the various multiquark configurations allowed by QCD that one can imagine and have not been already cited.

Thus far, the phenomenological descriptions specified above have produced compelling explanations for some of the XYZ states but not for all of them in an unified way. It would be desirable to have a single theoretical framework based firmly on QCD that describes all the XYZ mesons.

2. Nonrelativistic effective field theories for conventionals

Heavy quarkonium systems are characterized by their nonrelativistic nature, i.e. the heavy quark bound-state velocity, v, satisfies $v \ll 1$. This is reasonably fulfilled in bottomonium ($v^2 \sim 0.1$) and to a certain extent in charmonium ($v^2 \sim 0.3$). Moreover, at least, three widely separated scales appear: the heavy quark mass m (hard scale), the relative momentum of the bound state $p \sim mv$ (soft scale) and the binding energy $E \sim mv^2$ (ultrasoft scale). With $v \ll 1$, the following hierarchy of scales

$$m \gg p \sim 1/r \sim mv \gg E \sim mv^2 \tag{1}$$

is satisfied and this allows for a description in terms of EFTs of physical processes taking place at one of the lower scales. The integration out of modes associated with high-energy scales is performed as part of a matching procedure that enforces the equivalence between QCD and the EFT at a given order of the expansion in v. The final result is a factorization at the Lagrangian level between the high-energy modes, which are encoded in the matching coefficients, and the low-energy contributions carried by the dynamical degrees of freedom.

2.1. Physics at the scale m: NRQCD

The suitable EFT to describe heavy quarkonium annihilation and production, which take place at the scale m, is Nonrelativistic QCD (NRQCD) [12, 13]. It follows from QCD by integrating out the high energy modes of order m. As a consequence, the effective Lagrangian is organized as an expansion in $1/m$ and $\alpha_s(m)$:

$$\mathcal{L}_{\rm NRQCD} = \sum_n \frac{c_n(\alpha_s(m),\mu)}{m^n} \times \mathcal{O}_n(\mu, mv, mv^2, \ldots)\,, \tag{2}$$

where c_n are the Wilson coefficients that encode the contributions from the scale m, μ is the NRQCD factorization scale, and $\mathcal{O}_n$ are the low-energy operators constructed out of two or four heavy-quark/antiquark fields plus gluons. The matrix elements of $\mathcal{O}_n$ depend on the scales μ, mv, mv^2 and $\Lambda_{\rm QCD}$. Thus, the operators are counted in powers of v. The imaginary part of the coefficients of the four-fermion operators contains the information on heavy quarkonium annihilation and production.

The NRQCD heavy quarkonium Fock state is given by a series of terms, where the leading term is a $Q\bar{Q}$ in a color-singlet state, and the first correction, suppressed in v, comes from a $Q\bar{Q}$ in an octet state plus a gluon. Higher-order terms are subleading in increasing powers of v.

2.2. Physics at the scale mv: pNRQCD

The suitable EFT to describe heavy quarkonium formation, which takes place at the scale mv, is potential NRQCD (pNRQCD) [14, 15]. It follows from NRQCD by integrating out the modes of order $mv \sim 1/r$. The specific construction details of pNRQCD are slightly different depending upon the relative size between the soft, mv, and the confinement, $\Lambda_{\rm QCD}$, scales. Two main situations are distinguished. When $mv \gg \Lambda_{\rm QCD}$, we speak about weakly-coupled pNRQCD because the matching of NRQCD to pNRQCD may be performed in perturbation theory. When $mv \sim \Lambda_{\rm QCD}$, we speak about strongly-coupled pNRQCD because the matching of NRQCD to pNRQCD is not possible in perturbation theory.

When the soft scale is perturbative, the effective pNRQCD Lagrangian is organized as an expansion in $1/m$ and $\alpha_s(m)$, inherited from NRQCD, but also as a multipole expansion in r:

$$\mathcal{L}_{\rm pNRQCD} = \int d^3r \sum_n \sum_k \frac{c_n(\alpha_s(m),\mu)}{m^n} V_{n,k}(r,\mu',\mu)\, r^k \times \mathcal{O}_k(\mu', mv^2, \ldots)\,, \tag{3}$$

where $\mathcal{O}_k$ are the operators of pNRQCD that depend on the scales μ', mv^2 and $\Lambda_{\rm QCD}$; the pNRQCD factorization scale is μ', and $V_{n,k}$ are the Wilson coefficients that encode the contributions from the scale r.

The degrees of freedom that make up the operators $\mathcal{O}_k$ are $Q\bar{Q}$-pairs in a color-singlet or a color-octet state, and ultrasoft gluons. As one can see in Eq. (3), the Wilson coefficients $V_{n,k}$ depend on the relative distance r between the quark and the antiquark. They are potential-like terms when $k = 0$ and are the $1/m^n$ potentials that enter in a Schrödinger-type equation. Non-potential interactions, $V_{n,k\neq 0}$, account for singlet to octet transitions via ultrasoft gluons and provide loop corrections to the leading potential picture. They are typically related to nonperturbative effects.

3. Non-relativistic effective field theories for exotics

The above EFTs have proven to be very successful for the description of conventional heavy quarkonium states which are below the open-flavour threshold ($D\bar{D}$ ($B\bar{B}$) in the charmonium (bottomonium) sector). However, no EFT description has yet been constructed for those states which are close to or above threshold because the dynamical situation changes drastically [16, 17].

The threshold regions are actually the most interesting ones because many of the charmonium- and bottomonium-like XYZ states are located in such regions. Ab-initio lattice-regularized QCD computations are also in trouble when dealing with threshold regions. Moreover, calculations of excited states using this technique have been only recently pioneered and the full treatment of bottomonium on the lattice seems to be tricky. This all together explains why many of our expectations for the XYZ states still rely on phenomenological models.

Below we examine how things change when studying one particular case of the XYZ particles: the quark-gluon hybrids, and elucidate the future steps to be done towards an EFT description of them. It is important to keep in mind that the XYZ states belong to heavy quark sectors and thus the heavy quark mass is still an appropriate parameter from which begin a nonrelativistic expansion.

3.1. An example: hybrid states

Consider the case in which a system is made by a heavy quark, a heavy antiquark and gluonic excitations aiming to describe heavy quarkonium hybrids. In the static limit, at and above the $\Lambda_{\rm QCD}$ threshold, a tower of hybrid static energies (i.e. of gluonic excitations) must be considered on top of the $Q\bar{Q}$ static singlet energy [18, 19]. The spectrum has been thoroughly studied in lattice NRQCD [20] through the logarithm of large time generalized static Wilson loops divided by the interaction time:

$$E_n^{(0)}(r) = \lim_{T\to\infty} \frac{i}{T} \log \langle X_n, T/2 | X_n, -T/2 \rangle \,, \tag{4}$$

where the NRQCD initial and final states can be constructed as follows

$$|X_n\rangle = \chi(\boldsymbol{x}_2)\phi(\boldsymbol{x}_2, \boldsymbol{R}) T^a P_n^a(\boldsymbol{R}) \phi(\boldsymbol{R}, \boldsymbol{x}_1) \psi^\dagger(\boldsymbol{x}_1) |\text{vac}\rangle \,. \tag{5}$$

with $\phi(\boldsymbol{x}_2, \boldsymbol{x}_1)$ a Wilson line from $\boldsymbol{x}_1$ to $\boldsymbol{x}_2$, and P_n is some gluonic operator that generates the desired quantum numbers n to calculate the static energy. A list of possible operators P_n can be found, e.g., in Refs. [15, 21].

Berwein *et al.*, based on Refs. [15, 21, 22, 23], have set in Ref. [24] the first steps towards a pNRQCD characterization of the gluonic energies and thus of the hybrid potential. The matching between NRQCD and pNRQCD allows to establish that the spectrum of the hybrid static energies is described at short distances and in the leading multipole expansion of pNRQCD by the octet potential plus a mass scale called gluelump mass:

$$E_n^{(0)}(r) = \lim_{T\to\infty} \frac{i}{T} \log \langle X_n, T/2 | X_n, -T/2 \rangle = V_o(r) + \Lambda_H + \mathcal{O}(r^2) \,. \tag{6}$$

Equation (6) can be systematically improved by calculating higher orders in the multipole expansion. In particular, one can look at how the symmetry group of the hybrid system at short distances ($O(3)\otimes C$) is softly broken to the symmetry group that characterized the gluonic static energies in NRQCD ($D_{\infty h}$). We know that the leading correction coming from the multipole expansion is at $\mathcal{O}(r^2)$ and can be calculated in pNRQCD in terms of nonperturbative correlators to be eventually evaluated on the lattice or in QCD vacuum models. This correction has been treated phenomenologically in Ref. [24] and its computation will be one of our future goals.

Another of our goals will be going beyond the static limit and compute $1/m^n$ terms with $n = 1, 2, \ldots$ For instance, the NRQCD Hamiltonian for the one-quark-one-antiquark sector of the Fock space reads

$$H_{\rm NRQCD} = H^{(0)} + \frac{1}{m} H^{(1)} + \ldots \,, \tag{7}$$

$$H^{(0)} = \int d^3x \, \frac{1}{2} \left(\boldsymbol{E}^a \cdot \boldsymbol{E}^a + \boldsymbol{B}^a \cdot \boldsymbol{B}^a \right) - \sum_{j=1}^{n_f} \int d^3\boldsymbol{x} \, \bar{q}_j \, i\boldsymbol{D} \cdot \boldsymbol{\gamma} \, q_j \,, \tag{8}$$

$$H^{(1)} = -\frac{1}{2} \int d^3x \, \psi^\dagger \left(\boldsymbol{D}^2 + g c_F \, \boldsymbol{\sigma} \cdot \boldsymbol{B} \right) \psi + \frac{1}{2} \int d^3x \, \chi^\dagger \left(\boldsymbol{D}^2 + g c_F \, \boldsymbol{\sigma} \cdot \boldsymbol{B} \right) \chi \,, \tag{9}$$

where we have shown only terms up to order $1/m$ in the quark mass expansion, ψ ($\chi^\dagger$) is the Pauli spinor field that annihilates the heavy quark (antiquark), q_j is a massless quark of flavor j, and c_F is a matching coefficient. We are not considering the light quarks as external dynamical sources but they can still appear in the form of sea quarks.

A partial computation of the $1/m$ contributions has been performed in Ref. [24]: the kinetic part of $H^{(1)}$ was included whereas its spin dependent term was ignored. This has produced spin-multiplets which compare nicely with lattice-regularized QCD results (see, for instance, Fig. 5 of

Ref. [24]), providing the necessary support for continuing in this direction. It would be valuable to break the spin degeneracy and give a more detailed structure to the hybrid multiplets.

Finally, let us state here that our long term goal is to introduce an EFT description of heavy quarkonium hybrids without using the multipole expansion. At large distances, the dynamics of the system is nonperturbative but still reduces to a quantum mechanical problem with a number of potentials organized in powers of $1/m$ [25]. This would entail the definition of appropriate generalized Wilson loops that encode the dynamics of the nonperturbative matrix elements and obtaining in strongly-coupled pNRQCD the dynamical equations that couple them.

4. Conclusions

The search of exotic matter is a topic that has fascinated all generations of nuclear and particle physicists since the establishment of QCD as the theory of the strong interaction. In this respect, Europe is situated in a privileged position for the next decade(s) with two mayor experiments scheduled: LHCb@LHC and PANDA@FAIR. In support of the experimental effort, European institutions should also play a leading role in the theoretical counterpart. The work presented here aims to develop a state-of-the-art theoretical tool able to turn the description of exotic matter from a qualitative perspective into a quantitative one.

Acknowledgments

I would like to thank first Nora Brambilla and Antonio Vairo for insightful comments and careful reading of this manuscript; and second to some of my colleagues who attended the FAIRNESS2016 Workshop, M. Albaladejo, M. Berwein, S. Hwang, Z. Meisel, R. Navarro, C. QI, U. Tamponi and J. Tarrús-Castellà, for enjoyable physics discussions. I also want to express my gratitude to the organizers of the FAIRNESS2016 Workshop for enabling my participation, which proved very rewarding. I acknowledge financial support from the Alexander von Humboldt Foundation.

References

[1] Gell-Mann M 1964 *Phys. Lett.* **8** 214–215
[2] Zweig G 1964 *CERN Report No.8182/TH.401, CERN Report No.8419/TH.412*
[3] Olive K *et al.* (Particle Data Group) 2014 *Chin. Phys.* **C38** 090001
[4] Choi S *et al.* (Belle Collaboration) 2003 *Phys. Rev. Lett.* **91** 262001
[5] Brambilla N *et al.* 2011 *Eur. Phys. J.* **C71** 1534
[6] Brambilla N *et al.* 2014 *Eur. Phys. J.* **C74** 2981
[7] Tornqvist N A 1994 *Z. Phys.* **C61** 525–537
[8] Drenska N, Faccini R, Piccinini F, Polosa A, Renga F and Sabelli C 2010 *Riv. Nuovo Cim.* **33** 633–712
[9] Dubynskiy S and Voloshin M B 2008 *Phys. Lett.* **B666** 344–346
[10] Braaten E 2013 *Phys. Rev. Lett.* **111** 162003
[11] Vijande J, Weissman E, Valcarce A and Barnea N 2007 *Phys. Rev.* **D76** 094027
[12] Caswell W and Lepage G 1986 *Phys. Lett.* **B167** 437
[13] Bodwin G T, Braaten E and Lepage G P 1995 *Phys. Rev.* **D51** 1125–1171
[14] Pineda A and Soto J 1998 *Nucl. Phys. Proc. Suppl.* **64** 428–432
[15] Brambilla N, Pineda A, Soto J and Vairo A 2000 *Nucl. Phys.* **B566** 275
[16] Vairo A 2007 *Int. J. Mod. Phys.* **A22** 5481–5491 [,71(2006)]
[17] Brambilla N, Vairo A, Polosa A and Soto J 2008 *Nucl. Phys. Proc. Suppl.* **185** 107–117
[18] Horn D and Mandula J 1978 *Phys. Rev.* **D17** 898
[19] Hasenfratz P, Horgan R R, Kuti J and Richard J M 1980 *Phys. Lett.* **B95** 299
[20] Juge K J, Kuti J and Morningstar C 2003 *Phys. Rev. Lett.* **90** 161601
[21] Bali G S and Pineda A 2004 *Phys. Rev.* **D69** 094001
[22] Brambilla N, Pineda A, Soto J and Vairo A 2001 *Phys. Rev.* **D63** 014023
[23] Pineda A and Vairo A 2001 *Phys. Rev.* **D63** 054007 [Erratum: Phys. Rev. D64, 039902 (2001)]
[24] Berwein M, Brambilla N, Tarrús Castellà J and Vairo A 2015 *Phys. Rev.* **D92** 114019
[25] Brambilla N, Pineda A, Soto J and Vairo A 2005 *Rev. Mod. Phys.* **77** 1423

FAIRNESS 2016 IOP Publishing
Journal of Physics: Conference Series **742** (2016) 012006 doi:10.1088/1742-6596/742/1/012006

Investigation of $B\bar{B}$ four-quark systems using lattice QCD

Antje Peters[1], Pedro Bicudo[2], Krzysztof Cichy[1,3], Marc Wagner[1]
[1] Goethe-Universität Frankfurt am Main, Institut für Theoretische Physik, Max-von-Laue-Straße 1, D-60438 Frankfurt am Main, Germany
[2] CFTP, Departamento de Física, Instituto Superior Técnico, Universidade de Lisboa, Avenida Rovisco Pais, 1049-001 Lisboa, Portugal
[3] Adam Mickiewicz University, Faculty of Physics, Umultowska 85, 61-614 Poznan, Poland

E-mail: `peters@th.physik.uni-frankfurt.de`

Abstract. We investigate $B\bar{B}$ systems by computing potentials of two static quarks in the presence of two quarks of finite mass using lattice QCD. By solving the Schrödinger equation we check whether these potentials are sufficiently attractive to host bound states. Particular focus is put on the experimentally most promising bottomonium-like tetraquark candidate $Z_b^\pm$ with quantum numbers $I(J^P) = 1(1^+)$.

1. Motivation

Quite a number of mesons observed experimentally (e.g. at LHCb and Belle) are theoretically not well understood yet. One example is the charged bottomonium-like $Z_b^\pm$ with quantum numbers $I(J^P) = 1(1^+)$ (cf. e.g. [1]). It contains a $b\bar{b}$ pair, which one can infer from both its mass and decay products. Because it carries electromagnetic charge, there must be additional quarks, most likely a light quark-antiquark pair, i.e. the minimal quark content of $Z_b^\pm$ is $b\bar{b}u\bar{d}/b\bar{b}d\bar{u}$. Such four-quark states are of particular interest because most observed mesons can be described within the quark model using only a quark-antiquark pair. For this reason, four-quark states are the subject of many recent experimental and theoretical investigations.

In the following, we investigate the $Z_b^\pm$ assuming a four-quark structure $b\bar{b}u\bar{d}$ using lattice QCD. We approximate the heavy quarks b and $\bar{b}$ by infinitely heavy static quarks Q and $\bar{Q}$. We compute the potentials of the static quarks in the presence of the light quarks, which can be interpreted as the potentials of a B and a $\bar{B}$ meson. To check whether the four quarks may form a bound state, we use the Born-Oppenheimer approximation, i.e. we solve the Schrödinger equation with the computed potentials.

For similar studies of static-light BB systems within the same lattice QCD setup, cf. [2, 3, 4, 5, 6, 7, 8, 9]. Other papers studying pairs of static-light mesons with similar methods include [10, 11, 12, 13, 14, 15, 16].

2. The $b\bar{b}u\bar{d}$ four-quark system – qualitative discussion and expectations

The $b\bar{b}u\bar{d}$ four-quark system can be characterized by the separation r of the static quark $b \equiv Q$ and the antiquark $\bar{b} \equiv \bar{Q}$, by parity P, total angular momentum J and isospin I. Since we consider static b quarks, i.e. $b \equiv Q$ and $\bar{b} \equiv \bar{Q}$, the $b\bar{b}$ separation is also a quantum number.

Content from this work may be used under the terms of the Creative Commons Attribution 3.0 licence. Any further distribution of this work must maintain attribution to the author(s) and the title of the work, journal citation and DOI.
Published under licence by IOP Publishing Ltd

Moreover, since the static quark spin decouples from the Hamiltonian, both the light total angular momentum j and the static quark spin j_b ($j + j_b = J$) are separately conserved.

An experimentally very interesting case is the recently measured $Z_b^{\pm}$ with quantum numbers $I(J^P) = 1(1^+)$. Isospin $I = 1$ and positive electric charge is realised by light quark flavours $u\bar{d}$. Parity $P = +$ is consistent with a possible loosely bound $B\bar{B}$ structure, since both B and $\bar{B}$ have $P = -$ and hence in combination result in $P = +$. Numerically, we find most evidence for a four-quark bound state when total light angular momentum $j = 0$ (cf. section 3). Due to degeneracy with respect to the static quark spin j_b, the total angular momentum can either be $J = 0$ or $J = 1$, i.e. all our statements apply to a $b\bar{b}u\bar{d}$ four-quark system not only with $I(J^P) = 1(1^+)$, but also with $I(J^P) = 1(0^+)$.

States within the $I(J^P) = 1(1^+)$ sector might have different structures including the following.

- A bound four-quark state made of a loosely bound $B\bar{B}$ meson pair (a so-called mesonic molecule), referred to as $B\bar{B}$.
- A bound four-quark state made of a diquark and an anti-diquark.
- A two-meson state made of a B meson and a far separated $\bar{B}$ meson. The average distance between B and $\bar{B}$ is infinite in infinite volume in contrast to a bound four-quark state where the average distance is finite ($\mathcal{O}$(1fm)). Because the structure of the corresponding four-quark creation operator is the same as for the loosely bound $B\bar{B}$ meson pair (cf. eq. (2) below), this case is also referred to as $B\bar{B}$.
- A two-meson state made of a bottomonium state and a far separated pion π^+. In the static approximation, a bottomonium state is realized by the static quark Q and the static antiquark $\bar{Q}$ connected by a gluonic string. We refer to this case as $Q\bar{Q} + \pi$.

At small separations r, the ground state is $Q\bar{Q} + \pi$, while at large separations r the ground state is $B\bar{B}$. Using lattice QCD we investigate in the next section whether for small separations r there is a first excitation below the $B\bar{B}$ threshold that has a structure dominated by a four-quark bound state. A first excitation with such properties is an important prerequisite for a $b\bar{b}u\bar{d}$ bound state. Whether such a $b\bar{b}u\bar{d}$ bound state does indeed exist requires further analysis also discussed in the next section (using the Born-Oppenheimer approximation, solving the Schrödinger equation).

3. Lattice QCD computation of the $Q\bar{Q}$ potential in the presence $u\bar{d}$

We start by discussing suitable creation operators $\mathcal{O}_j$ that generate field excitations, which are similar to the four-quark states of interest. The correlation functions of these operators at large temporal separation provide low-lying masses, which can be interpreted as potentials $V_0(r)$, $V_1(r)$, ..., because the operators $\mathcal{O}_j$ and thus also the masses depend on the $Q\bar{Q}$ separation r. The correlation functions read

$$C_{jk}(t) = \langle\Omega|\, \mathcal{O}_j^\dagger(t)\mathcal{O}_k(0)\,|\Omega\rangle \underset{\text{large } t}{\approx} A_{jk}^0 \exp\left(-V_0(r)t\right) + A_{jk}^1 \exp\left(-V_1(r)t\right) + \ldots . \quad (1)$$

The creation operators we have currently implemented are one that excites a $B\bar{B}$ state and another that excites a $Q\bar{Q} + \pi$ state:

$$\mathcal{O}_0 \equiv \mathcal{O}_{B\bar{B}} = \Gamma_{AB}\tilde{\Gamma}_{CD}\bar{Q}_C^a(\vec{x})q_A^a(\vec{x})\bar{q}_B^b(\vec{y})Q_D^b(\vec{y}), \quad (2)$$

$$\mathcal{O}_1 \equiv \mathcal{O}_{Q\bar{Q}+\pi} = \bar{Q}_A^a(\vec{x})U^{ab}(\vec{x},t;\vec{y},t)\tilde{\Gamma}_{AB}Q_B^b(\vec{y}) \sum_{\vec{z}} \bar{q}_C^c(\vec{z})\,(\gamma_5)_{CD}\, q_D^c(\vec{z}), \quad (3)$$

with $|\vec{x} - \vec{y}| = r$. $\tilde{\Gamma}$ appearing in both operators is a combination of Dirac matrices that realizes either $j_b = 0$ or $j_b = 1$. It does not affect the potentials $V_j(r)$ since the spin of the heavy quarks

β	lattice size	$a\mu_l$	a in fm	m_π in MeV	# of configurations
3.90	$24^3 \times 48$	0.0085	0.079	480	100

Table 1. Parameters of the gauge link configurations; β: inverse gauge coupling; $a\mu_l$: bare u/d quark mass; a: lattice spacing; m_π: pion mass.

is irrelevant as explained above. The matrix Γ is a combination of Dirac matrices leading to the same quantum numbers $(j_z, P \circ C, P_x)$ as the $Q\bar{Q} + \pi$ operator (3) (for details on $(j_z$, P, C, $P_x)$, cf. e.g. [9]). Among the possible choices for Γ, the combination $\Gamma = \gamma_5 - \gamma_0\gamma_5$ yields the strongest $Q\bar{Q}$ attraction, if one takes into account only the operator $\mathcal{O}_{B\bar{B}}$. Therefore, we consider this combination as the most promising for further analysis. $U^{ab}(\vec{x}, t; \vec{y}, t)$ denotes a product of gauge links connecting the two static quarks.

At the moment, we have performed computations of the correlation functions $C_{jk}(t)$ only on a single enesemble of gauge link configurations generated by the European Twisted Mass Collaboration (ETMC) with 2 dynamical quark flavours (for details cf. Table 1). To improve statistical accuracy, we have averaged these correlation functions with respect to the symmetries: time-reversal, parity, charge conjugation, γ_5-hermiticity and cubic rotations.

In particular, the extraction of the energy of the first excited state $V_1(r)$ is a numerically rather challenging task, which we do by solving a generalized eigenvalue problem (GEP) (cf. [17] and references therein).

Figure 1 shows $Q\bar{Q}u\bar{d}$ potentials obtained via the GEP from our correlation matrix (1) as well as the ordinary static $Q\bar{Q}$ potential (obtained from standard Wilson loops). The blue curve corresponds to the ground state $V_0(r)$ (obtained from the full 2×2 matrix). By comparing with the ordinary static $Q\bar{Q}$ potential, one can clearly see that it is just shifted by the pion mass $am_\pi \approx 0.2$) indicating that the ground state of our four quark system is of the $Q\bar{Q} + \pi$ type (this is also supported by the eigenvectors obtained from the GEP). While the red curve corresponds to the first excitation $V_1(r)$ (obtained from the full 2×2 matrix), the green curve is the result obtained from a single correlation function, the correlation function of the operator $O_{B\bar{B}}$, i.e. from $C_{00}(t)$. The fact that the green curve is similar to, but somewhat below the red curve suggests that the first excitation $V_1(r)$ is predominantly of $B\bar{B}$ type (the operator $O_{B\bar{B}}$ seems to generate mostly the first excitation, but also a non-vanishing ground-state overlap). This attractive potential $V_1(r)$ is a potential candidate to host a four-quark bound state.

4. Solving the Schrödinger equation to check for a bound state

To check for a four-quark bound state and, if present, to obtain its binding energy $-E_B$, we solve the s-wave Schrödinger equation

$$\left(-\frac{1}{2\mu}\frac{d^2}{dr^2} + U(r)\right)R(r) = E_B R(r), \quad \text{with} \quad U(r) = V(r)\Big|_{V_0=0} \quad \text{and} \quad \mu = m_b/2. \tag{4}$$

V denotes an analytic parameterization of our lattice QCD result for $V_1(r)$ (red points in Figure 1) obtained by fitting

$$V(r) = V_0 - \frac{\alpha}{r}\exp(-(r/d)^2) \tag{5}$$

with respect to V_0, α and d. The existence of a four-quark bound state, i.e. a tetraquark, is indicated by $E_B < 0$. For more details on the extraction of the binding energy in this so-called Born-Oppenheimer approximation, cf. [8].

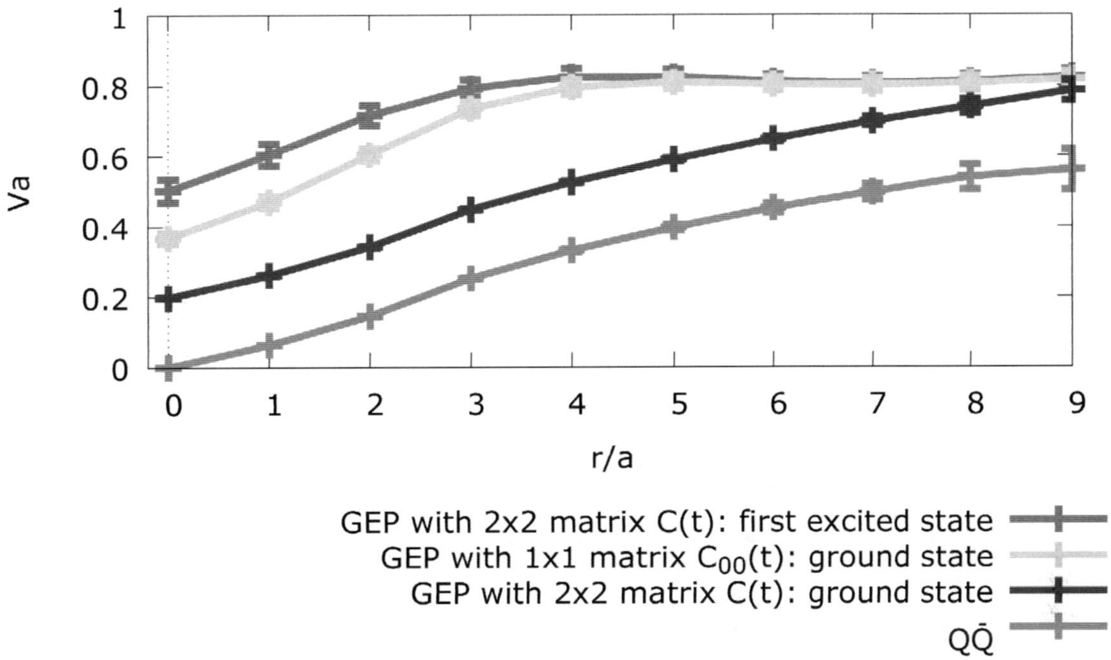

Figure 1. $Q\bar{Q}u\bar{d}$ potentials and the $Q\bar{Q}$ potential (cf. the text for details).

Determining $V_1(r)$ from the GEP using different t ranges and performing several fits of (5) with different r ranges allows to estimate both E_B and and its uncertainty,

$$E_B = (-58 \pm 71)\,\text{MeV}.$$

This negative value which has been found using lattice QCD with infinitely heavy b quarks and the Born-Oppenheimer approximation does not contradict the existence of an $I(J^P) = 1(1^+)$ $b\bar{b}u\bar{d}$ tetraquark. Our findings are consistent with the experimentally observed $Z_b^\pm$ states.

5. Outlook

The analysis presented above is very preliminary. In the near future, we plan for example to increase statistical precision and to extend our computations to lighter u/d quark masses. Moreover, it will be interesting to explore also other $I(J^P)$ sectors. Finally, the spin of the heavy quarks should be included as e.g. pioneered in [7].

Acknowledgments

P.B. thanks IFT for hospitality and CFTP, grant FCT UID/FIS/00777/2013, for support. A.P. and M.W. acknowledge support by the Emmy Noether Programme of the DFG (German Research Foundation), grant WA 3000/1-1.

This work was supported in part by the Helmholtz International Center for FAIR within the framework of the LOEWE program launched by the State of Hesse.

Calculations on the LOEWE-CSC high-performance computer of Johann Wolfgang Goethe-University Frankfurt am Main were conducted for this research. We would like to thank HPC-Hessen, funded by the State Ministry of Higher Education, Research and the Arts, for programming advice.

References

[1] A. Bondar *et al.* [Belle Collaboration], "Observation of two charged bottomonium-like resonances in $Y(5S)$ decays," Phys. Rev. Lett. **108**, 122001 (2012) [arXiv:1110.2251 [hep-ex]].
[2] M. Wagner [ETM Collaboration], "Forces between static-light mesons," PoS **LATTICE2010**, 162 (2010) [arXiv:1008.1538 [hep-lat]].
[3] M. Wagner [ETM Collaboration], "Static-static-light-light tetraquarks in lattice QCD," Acta Phys. Polon. Supp. **4**, 747 (2011) [arXiv:1103.5147 [hep-lat]].
[4] P. Bicudo and M. Wagner, "Lattice QCD signal for a bottom-bottom tetraquark," Phys. Rev. D **87**, 114511 (2013) [arXiv:1209.6274 [hep-ph]].
[5] B. Wagenbach, P. Bicudo and M. Wagner, "Lattice investigation of heavy meson interactions," J. Phys. Conf. Ser. **599**, 012006 (2015) [arXiv:1411.2453 [hep-lat]].
[6] P. Bicudo, K. Cichy, A. Peters, B. Wagenbach and M. Wagner, "Evidence for the existence of $ud\bar{b}\bar{b}$ and the non-existence of $ss\bar{b}\bar{b}$ and $cc\bar{b}\bar{b}$ tetraquarks from lattice QCD," Phys. Rev. D **92**, 014507 (2015) [arXiv:1505.00613 [hep-lat]].
[7] J. Scheunert, P. Bicudo, A. Uenver and M. Wagner, "Refined lattice/model investigation of $ud\bar{b}\bar{b}$ tetraquark candidates with heavy spin effects taken into account," Acta Phys. Polon. Supp. **8**, 363 (2015) [arXiv:1505.03496 [hep-ph]].
[8] A. Peters, P. Bicudo, K. Cichy, B. Wagenbach and M. Wagner, "Exploring possibly existing $qq\bar{b}\bar{b}$ tetraquark states with $qq = ud, ss, cc$," arXiv:1508.00343 [hep-lat].
[9] P. Bicudo, K. Cichy, A. Peters and M. Wagner, "BB interactions with static bottom quarks from Lattice QCD," Phys. Rev. D **93** (2016) 034501 [arXiv:1510.03441 [hep-lat]].
[10] C. Stewart and R. Koniuk, "Hadronic molecules in lattice QCD," Phys. Rev. D **57**, 5581 (1998) [arXiv:hep-lat/9803003].
[11] C. Michael and P. Pennanen [UKQCD Collaboration], "Two heavy-light mesons on a lattice," Phys. Rev. D **60**, 054012 (1999) [arXiv:hep-lat/9901007].
[12] M. S. Cook and H. R. Fiebig, "A lattice study of interaction mechanisms in a heavy-light meson meson system," arXiv:hep-lat/0210054.
[13] T. Doi, T. T. Takahashi and H. Suganuma, "Meson meson and meson baryon interactions in lattice QCD," AIP Conf. Proc. **842**, 246 (2006) [arXiv:hep-lat/0601008].
[14] W. Detmold, K. Orginos and M. J. Savage, "BB potentials in quenched lattice QCD," Phys. Rev. D **76**, 114503 (2007) [arXiv:hep-lat/0703009].
[15] G. Bali and M. Hetzenegger, "Static-light meson-meson potentials," PoS **LATTICE2010**, 142 (2010) [arXiv:1011.0571 [hep-lat]].
[16] Z. S. Brown and K. Orginos, "Tetraquark bound states in the heavy-light heavy-light system," Phys. Rev. D **86**, 114506 (2012) [arXiv:1210.1953 [hep-lat]].
[17] B. Blossier, M. Della Morte, G. von Hippel, T. Mendes and R. Sommer, "On the generalized eigenvalue method for energies and matrix elements in lattice field theory," JHEP **0904**, 094 (2009) [arXiv:0902.1265 [hep-lat]].

Conception and design of a control and monitoring system for the mirror alignment of the CBM RICH detector

J Bendarouach for the CBM collaboration

Justus-Liebig-University Giessen, Germany

E-mail: jordan.bendarouach@exp2.physik.uni-giessen.de

Abstract. The Compressed Baryonic Matter (CBM) experiment at the future Facility for Anti-proton and Ion Research (FAIR) complex will investigate the phase diagram of strongly interacting matter at high baryon density and moderate temperatures created in A+A collisions. For the SIS100 accelerator, the foreseen beam energy will range up to 11 AGeV for the heaviest nuclei. One of the key detector components required for the CBM physics program is the Ring Imaging CHerenkov (RICH) detector, which is developed for efficient and clean electron identification and pion suppression. An important aspect to guarantee a stable operation of the RICH detector is the alignment of the mirrors. A qualitative alignment control procedure for the mirror system has been implemented in the CBM RICH prototype detector and tested under real conditions at the CERN PS/T9 beamline. Collected data and results of image processing are reviewed and discussed. In parallel a quantitative method using recorded data has also been employed to compute mirror displacements of the RICH mirrors. Results based on simulated events and the limits of the method are presented and discussed as well. If mirror misalignment is detected, it can be subsequently included and rectified by correction routines. A first correction routine is presented and a comparison between misaligned, corrected and ideal geometries is shown.

1. Introduction

The Compressed Baryonic Matter (CBM) experiment at the future Facility for Antiproton and Ion Research (FAIR) complex will investigate the phase diagram of strongly interacting matter at high baryon density and moderate temperatures in nucleus-nucleus collisions. The beam energy will range from 2 up to 11 AGeV for the heaviest nuclei at the SIS 100 accelerator setup. In future the energy scope might be expanded to 35 AGeV beam energy, by installing the additional SIS 300 accelerator. For a characterization of the strongly interacting matter created during such collisions, electromagnetic probes appear to be very promising. Due to their penetrating nature, the information carried by these probes remains undistorted, when they reach the detector. Furthermore photons and dileptons are emitted during each stage of a heavy ion collision, therefore revealing the complete story of the collision.

For this CBM physics program to be successful, an efficient and clean electron identification (for momenta up to 8 GeV/c) together with a combined pion suppression factor of 1000 to 5000 (with the Transition Radiation Detector (TRD)) are required in a wide acceptance. One of the key detector components foreseen to satisfy these performances is the Ring Imaging CHerenkov

Content from this work may be used under the terms of the Creative Commons Attribution 3.0 licence. Any further distribution of this work must maintain attribution to the author(s) and the title of the work, journal citation and DOI.
Published under licence by IOP Publishing Ltd

(RICH) detector. In addition, 4 layers of TRD will be installed in the SIS 100 setup. The RICH detector will be made of a CO_2 gaseous radiator, Multi-Anode Photo-Multipliers for photon detection and about 80 trapezoidal glass mirror tiles, equally distributed in two half-spheres and used as focussing elements with spectral reflectivity down to the UV range [1, 2].

One specific aspect of a high quality RICH performance, in particular in the expected high track density environment of CBM, is a precise and stable mirror alignment. Each individual tile will be mounted with a three point fixation system and aligned such that their center of curvature all coincide in one virtual point. During the primary detector assembly, an initial mirror alignment will be done with a theodolite in autoreflection method, a procedure similar to what has been done for the COMPASS experiment [3]. Moreover it is crucial to control the stability in time of the initial alignment. Misalignments induce ring distortions and ring splitting, which thus harm reconstruction efficiencies and particle identification [4].

Two techniques are currently under development to provide stable alignment and will be presented consecutively. The first one is used to qualitatively control mirror alignment, using the Continuous Line Alignment Monitoring method (CLAM) [5, 6]. The second technique allows to quantify misalignments using data [7, 8]. Once misalignments have been revealed this information is included in correction routines.

2. Qualitative control of mirror misalignment

2.1. Qualitative control in the CBM RICH prototype

The alignment control method, inspired from the COMPASS experiment and employed to qualitatively monitor mirror displacements, is called Continuous Line Alignment Monitoring method (CLAM) [5, 6]. It uses retro-reflective stripes and target dots, which are glued at the entrance window of the RICH vessel. A camera surrounded by LEDs next to the photodetector plane illuminates the grid through the mirrors. For the measurement PMTs are switched off and LEDs turned on. Mirror edges are seen as black cuts on the stripes and a fast qualitative assessment regarding alignment can be made. Indeed if the stripes appear broken at mirror edges, it is obvious that the mirrors are misaligned with one another.

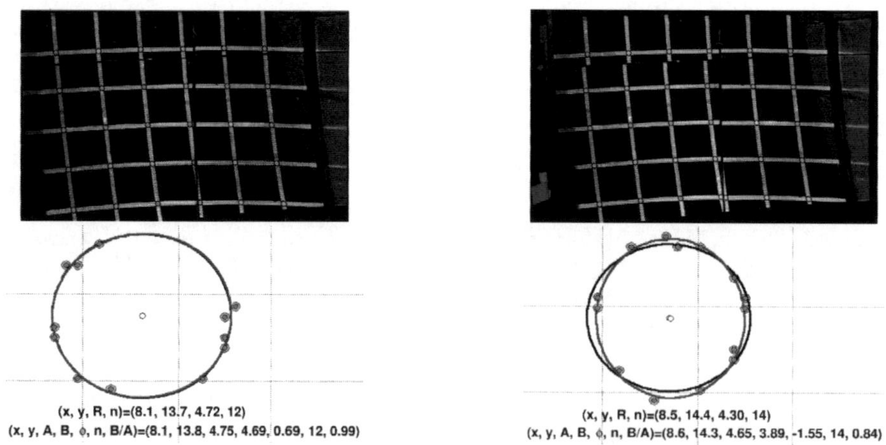

Figure 1. Pictures of the reflected grid after the initial mirror alignment (top left) and after artificial misalignment has been induced (top right). Corresponding single event rings in these configurations are shown on the bottom pictures, where the black and blue lines correspond to a ring and an ellipse shape, respectively.

This method has been implemented in a downscaled prototype of the CBM RICH detector and tested under real conditions at the CERN PS/T9 beamline. The prototype is made of a square mirror wall, made of 4 square tiles. The mirrors can be remotely controlled via actuators, allowing artificial misalignments. The vessel was placed in the beamline, such that the beam was traversing between the two lower mirrors. In Figure 1, a picture of the system after the initial alignment (top left), along with a typical reconstructed single event ring (bottom left) for this mirror configuration are depicted. The lower left mirror was then artificially misaligned by 4 milliradians [mrad] around a vertical axis going through the tile center. A picture of the misaligned system is shown in Figure 1 (top right), along with a corresponding reconstructed ring (bottom right). The vertical lines are now clearly appearing broken and the reconstructed ring is more elliptical (blue line). Indeed the induced misalignment moved the ring half reflected on the lower left mirror towards the half reflected by the right one in the photodetector plane.

2.2. *Impact of misalignment on ring parameters*

In the testbeam experiment a detailed study on the impact of misalignments on ring ellipticity has been conducted. The mirrors were in the same configuration as depicted before, i.e. the beam traversed between the two lower mirrors. To improve the beam focussing, a finger scintillator detector was used as triggering device and an additional cut was used directly on data to enhance the split ring events. Figure 2 illustrates the ring ellipticity distribution, referred to as B/A, corresponding to the minor axis of the ellipse divided by the major axis. Four different cases are presented: after reference alignment (red), with a misalignment of 1 mrad (dashed red), with a misalignment of 4 mrad (green) and with a cut enhancing the 4 mrad misaligned data sample (blue). For the last curve, one clearly sees an ellipticity drop below 0.9, with regard to the reference. Reconstruction efficiency for those rings drops from close to 100% down to 90% [9].

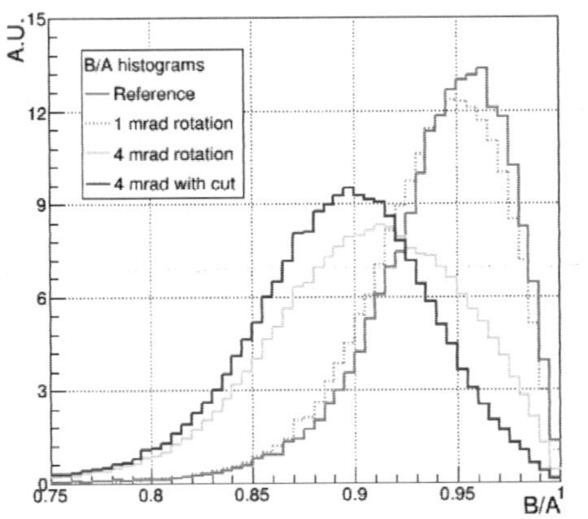

Figure 2. Dependence of ring ellipticity, B/A, with regard to mirror misalignment. After the initial mirror alignment (red), with a 1 mrad rotation (dashed red) and with a 4 mrad misalignment (green). The blue curve was obtained using a cut to enhance the 4 mrad misaligned data sample, i.e. events, for which the beam went between the two lower mirrors are selected more cleanly. For the blue curve, the mean ellipticity drops below 0.9. The distributions are normalized to the number of events used for each curve.

3. Quantitative determination of mirror misalignments

Once mirror misalignments have been located, a dedicated system is required to quantify the misalignment. The applied technique, inspired from the HERA-B experiment [7], uses data and has been tested in the simulation framework of the CBM RICH detector. The principle is illustrated in Figure 3. When a particle enters the RICH vessel, it emits Cherenkov light, which is reflected by one or several mirror tiles to the photodetector plane (red crosses). These photon hits are then reconstructed together as a ring, C' in Figure 3. The point C, represents the extrapolated particle hit to the photodetector plane, if the incident particle had been reflected by the same aligned mirror tiles. In case of misalignment, the points C and C' differ from one another. Measuring Cherenkov distances θ_{Ch} and angles Φ_{Ch} for each photon hit, a sinusoidal dependence has been demonstrated between these two parameters:

Journal of Physics: Conference Series **742** (2016) 012007 doi:10.1088/1742-6596/742/1/012007

$$\theta_{Ch} = \theta_0 + \Delta\Phi * cos(\Phi_{Ch}) + \Delta\lambda * sin(\Phi_{Ch}) \tag{1}$$

For a sufficient number of accumulated events hitting a particular mirror tile, those measurements quantify the misalignments of the considered tile, by fitting the data to equation (1) and obtaining the $\Delta\Phi$ and $\Delta\lambda$ parameters (see Figure 3 right plot) [8].

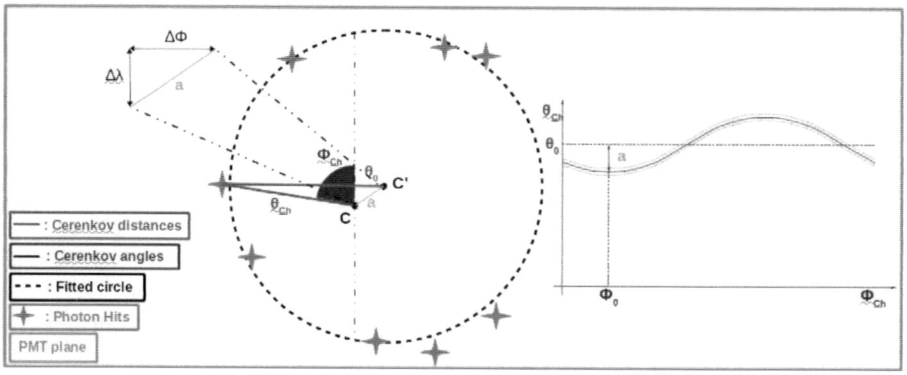

Figure 3. Principle of the method for quantifying mirror misalignment.

This technique has been adapted and tested for the CBM RICH detector with simulations in the CbmRoot framework. It produces good results for misalignments ranging from 0.1 mrad up to 12 mrad, as seen in Figure 4. For misalignments greater than 12 mrad, the point C starts lying out of the reconstructed ring, which makes the technique unstable.

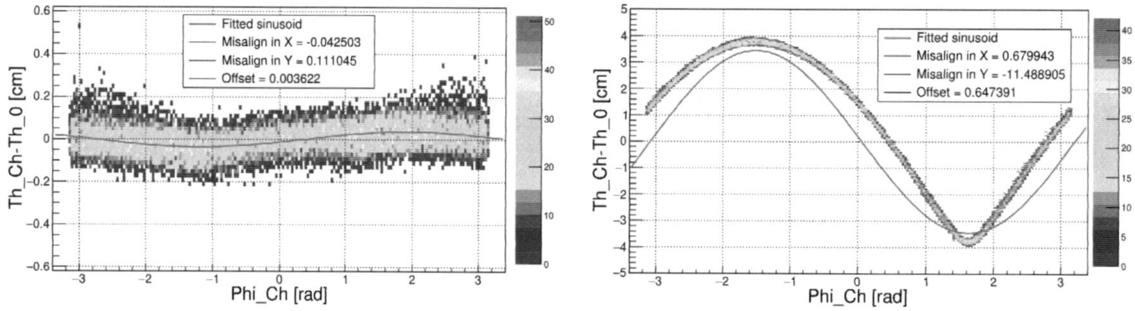

Figure 4. Performance test of the method to quantify mirror misalignments. The method yields reliable results for misalignments ranging from 0.1 mrad (left) up to 12 mrad (right). The results obtained for these misalignments using the described technique are 0.11 mrad and 11.49 mrad, respectively.

4. Correction routine

In the presented example (see Figure 5), a mirror tile has been artificially misaligned by 5 mrad around its horizontal axis and its misalignment has been evaluated using the method described in Sec. 3. The extracted misalignment is 5.26 mrad around the horizontal axis and 0.37 mrad around the vertical axis. In the correction procedure, the track projection yielding the extrapolated point (C in Figure 3) is then calculated using corrections from these misalignments. Figure 5 shows the projected distances on the horizontal (X) and vertical (Y) axes between the fitted ring center and the extrapolated track hit (C' and C in Figure 3). These distances are presented for three different cases, namely the misaligned (red), the corrected (green) and the ideal correction (blue). The latter can be directly obtained from the geometry. One sees (Figure 5 right) a clear vertical discrepancy of 1.6 cm, which is expected given the rotation applied to the tile. The slight difference in the horizontal direction (Figure 5 left), is due to the detector geometry, as the photodetector plane does not face perpendicularly the mirror wall, but with a slight tilt. For both projections the corrected distances are close to the ideal ones, indicating a proper operation of the correction routine.

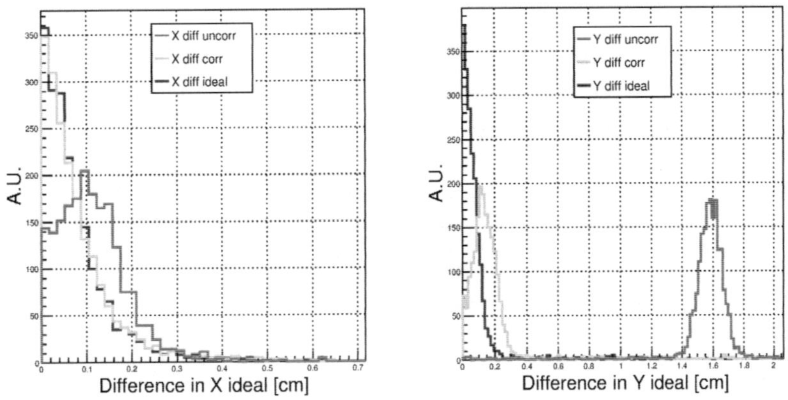

Figure 5. Distance between the points C and C' on the photodetector plane, projected in the horizontal (X) and vertical (Y) directions for the misaligned (red), corrected (green) and ideal (blue) cases.

5. Conclusion

A first alignment correction cycle for the CBM RICH detector has been presented. The system has been implemented in a RICH prototype and is able to detect misalignment qualitatively. Furthermore it is possible to quantify the detected misalignment with data. A fist correction routine has been performed and shows promising results.

Acknowledgments

This work was supported by the Helmholtz International Center for FAIR, the GSI Helmholtzzentrum *für* Schwerionenforschung F&E cooperation with Gie*β*en, the Helmholtz Graduate School for Hadron and Ion Research and the BMBF grants 05P15RGFCA and 05P12RGFCG.

6. References

[1] Höhne C *et al* 2011 *Nucl. Instr. Meth. Phys. Res.* A **639** 294-7
[2] Adamczewski-Musch J *et al* 2014 *Nucl. Instr. Meth. Phys. Res.* A **766** 101-6
[3] Albrecht E *et al* 2003 *Nucl. Instr. Meth. Phys. Res.* A **502** 236-40
[4] Adamczewski-Musch J *et al* 2014 *Nucl. Instr. Meth. Phys. Res.* A **766** 221-4
[5] Alexeev M *et al* 2008 *Nucl. Instr. Meth. Phys. Res.* A **595** 194-6
[6] Alexeev M *et al* 2011 *Nucl. Instr. Meth. Phys. Res.* A **639** 219-21
[7] Gorišek A, Križan P, Korpar S and Starič M 1999 *Nucl. Instr. Meth. Phys. Res.* A **433** 408-12
[8] Starič M and Križan P 2008 *Nucl. Instr. Meth. Phys. Res.* A **586** 174-9
[9] Höhne C *et al* 2013 *Technical Design Report for the CBM Ring Imaging Cherenkov Detector* GSI-2014-00528

Journal of Physics: Conference Series **742** (2016) 012008 doi:10.1088/1742-6596/742/1/012008

Flow harmonics of Au+Au collisions at 1.23 AGeV with HADES

Behruz Kardan for the HADES collaboration
Goethe-Universität Frankfurt, Max-von-Laue-Str. 1, 60438 Frankfurt am Main, Germany
E-mail: bkardan@ikf.uni-frankfurt.de

Abstract. Collective flow phenomena are a sensitive probe for the properties of extreme QCD matter. However, their interpretation relies on the understanding of the initial conditions e.g. the eccentricity of the nuclear overlap region. HADES [1] provides a large acceptance combined with a high mass-resolution and therefore allows to study di-electron and hadron production in heavy-ion collisions with unprecedented precision. In this contribution, the capability of HADES to study flow harmonics by utilizing multi-particle azimuthal correlation techniques is discussed. Due to the high statistics of seven billion Au+Au collisions at 1.23 AGeV collected in 2012, a systematic study of higher-order flow harmonics, the differentiation between collective and non-flow effects, and as well the multi-differential (p_t, rapidity, centrality) analysis is possible.

1. Introduction

Collective flow phenomena are a sensitive probe for the properties of extreme QCD matter. However, their interpretation relies on the understanding of the initial conditions, e.g. the eccentricity of the nuclear overlap region. To achieve a good understanding of this phenomena, event-by-event flow observables and their fluctuations are deduced and compared with model calculations and other measured data.

2. HADES

The High Acceptance DiElectron Spectrometer (HADES) is a fixed-target experiment hosted at the GSI Helmholtzzentrum für Schwerionenforschung in Darmstadt [1]. The data analyzed in this work are from the Au+Au campaign measured in 2012 [2]. This is the largest reaction system, which was measured with HADES, and requires high granularity detectors to cope with the high particle multiplicities. Due its large acceptance, which covers almost full azimuth angle, combined with its high mass-resolution and good particle-identification capability, it is well equipped to study the azimuthal flow pattern not only for protons and charged pions, but also for kaons, ϕ-mesons and electrons/positrons, as well as d, t, and light nuclear fragments.

The spectrometer is subdivided into six identical sectors, axially symmetric around the beam direction. The maximal acceptance in polar angle for charged particles, as shown in Fig.1, defined by the coverage of the magnetic field, is between 18° and 85°, which corresponds to a rapidity range for protons of $-0.6 < y - y_{cm} < 0.8$ around mid-rapidity ($y_{cm} = 0.74$). The momentum reconstruction is carried out by a tracking system consisting in total of 24 multi-wire drift-chambers (MDC), where in each sector two layers are placed in front and two behind

Content from this work may be used under the terms of the Creative Commons Attribution 3.0 licence. Any further distribution of this work must maintain attribution to the author(s) and the title of the work, journal citation and DOI.
Published under licence by IOP Publishing Ltd

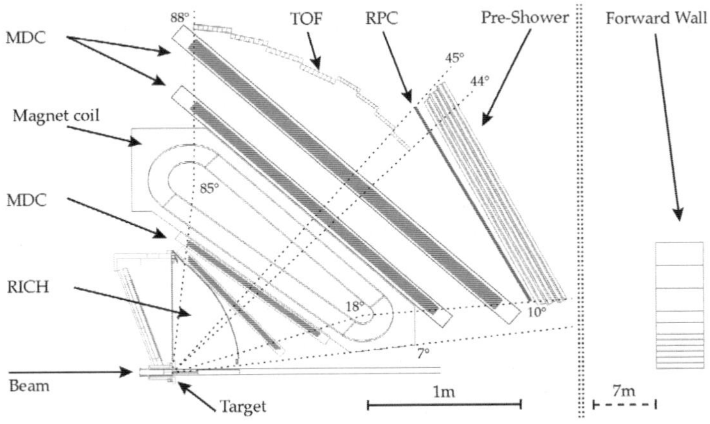

Figure 1. Cross section of one HADES sector. The segmented target irradiated by the beam, the RICH detector, the magnet spectrometer consisting of four layers of drift chambers (MDC) and the magnet coils, the two *time-of-flight* detectors (TOF and RPC) and the Forward Wall (FW) are here shown.

a toroidal magnetic field of the superconducting magnet coils (ILSE). The *Multiplicity and Trigger Array* (META) detector together with the beam detector (diamond counter) provides the time-of-flight measurements and the trigger information. For polar angles between 44° and 88° it consists of the scintillating *time-of-flight wall* (TOF), at the forward region between 18° and 45° it is instrumented with *Resistive Plate Chambers* (RPC) with a subsequent PreShower detector. The *Forward Wall* (FW), a plastic scintillator hodoscope array, is placed at a distance of 7 m behind the target at the small forward angle between 0.3° and 7.3° to detect charged projectile spectators by the time-of-flight and by the ΔE signal in the scintillator modules. The hit-position is used to reconstruct the event plane.

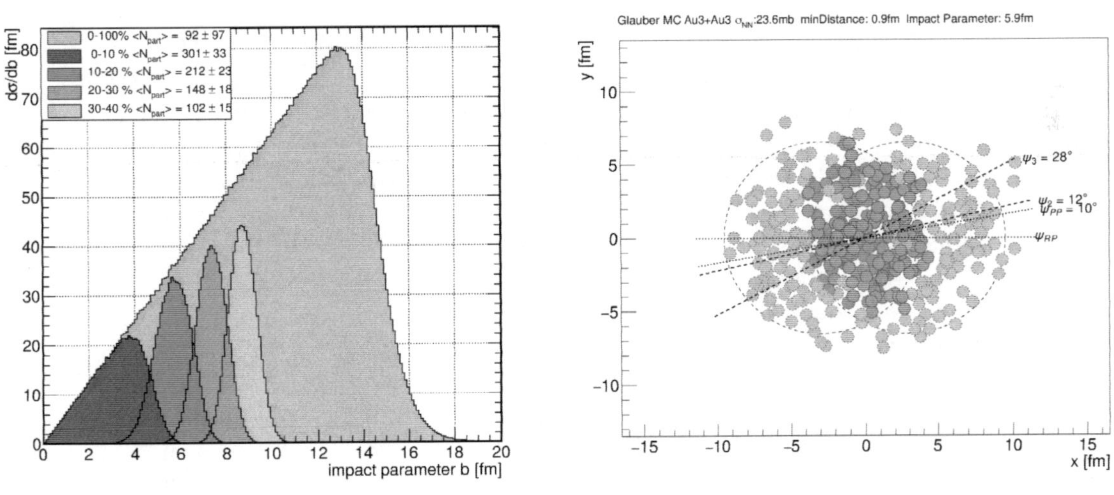

Figure 2. left: The impact parameter distribution of the total hadronic cross section of the Au+Au reaction and the estimates of the four most central centrality classes of 10% width are shown. The measured charged track and hit multiplicities are modeled by a Glauber MC simulation.
right: Glauber MC event with an impact parameter $b = 5.9$ fm showing the participating nucleons (full colored dots) and the spectators (light colored dots). The reaction plane ψ_{RP}, the participant plane ψ_{PP} and its harmonic decompositions into higher-order phase angles ψ_n are shown.

Journal of Physics: Conference Series **742** (2016) 012008 doi:10.1088/1742-6596/742/1/012008

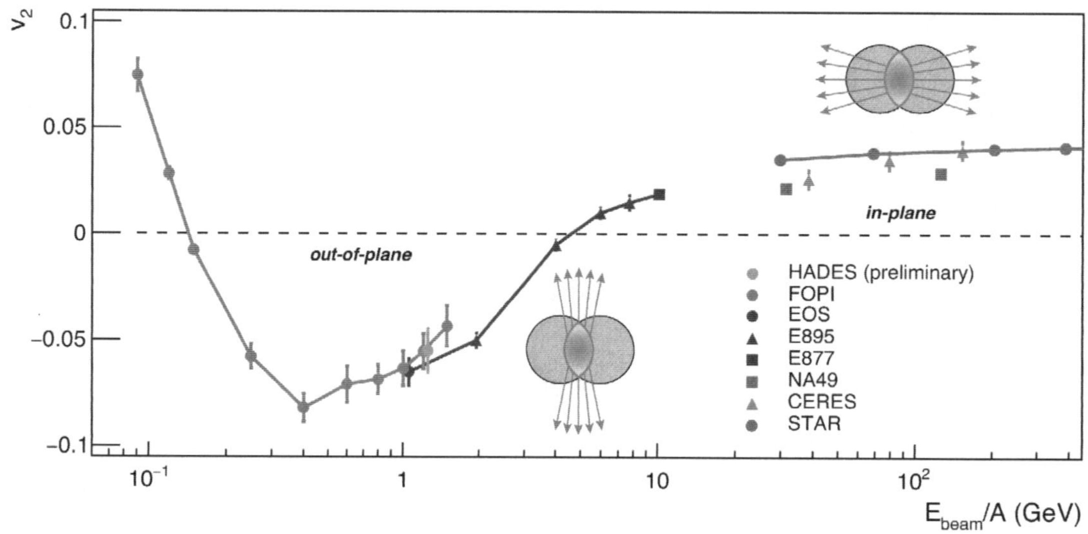

Figure 3. The excitation function of elliptic flow v_2 at mid-rapidity as a function of the beam energy. The preliminary HADES point at 1.23 AGeV, FOPI (15-29% centrality) [5], EOS and E895 [6], E877 [7], CERES [8], NA49 [9] and the STAR (10-20% centrality) [10] data points are here shown.

3. Data sample

Within the 5 weeks of Au+Au beamtime the Schwerionen-Synchrotron (SIS18) at GSI delivered 684 hours of Au^{69+} ions beam to the HADES cave [3] with an intensity of $1.2 - 2.2 \times 10^6$ ions per sec. A 15-fold segmented gold target with an interaction probability of 1.51% was used. The overall total data volume recorded on disc is 140 Tbyte, including calibration and cosmic runs. A fraction of around 80% of the total recorded events are triggered by selecting most central events with a charged hit multiplicity in the TOF detector $N_{ch} > 20$, corresponding to 5.85×10^9 events before off-line event selection. According to detailed comparison of the charged track and hit multiplicity distribution with a Glauber Model simulation, this central trigger selects about 47% of the total hadronic cross section of 6.83 ± 0.43 barn, corresponding to a maximal impact parameter of $b_{max} < 10$ fm [4], see left panel of Fig. 2.

4. Collective Flow

In a high-energy collision of nuclei a highly excited nuclear medium is created and its collective expansion produces a correlated emission of particles. In perfectly central collisions the expansion should be isotropic in the transverse plane, known as *radial flow*, which is observed in the transverse mass spectra of produces particles. Going to more non-central collisions, the overlap region is getting more anisotropically shaped. This *event-shape* characteristic is usually studied via the azimuthal anisotropy of the momentum space of identified particle w.r.t a corresponding symmetry plane and it is common to analyze this by a Fourier decomposition with the coefficients v_1, v_2, v_3 and v_4. Due to their origin induced by the collision geometry, the directed v_1 and elliptic v_2 are linked to the reaction plane which is also spanned by the spectators.

Figure 3 shows the elliptic flow of charged particles as function of beam energy (from SIS18, AGS, SPS to RHIC energies). From a minimal negative value at around 400 MeV it increases with rising energy to positive values at high energies. At low energies this *out-of-plane* emission

Journal of Physics: Conference Series **742** (2016) 012008 doi:10.1088/1742-6596/742/1/012008

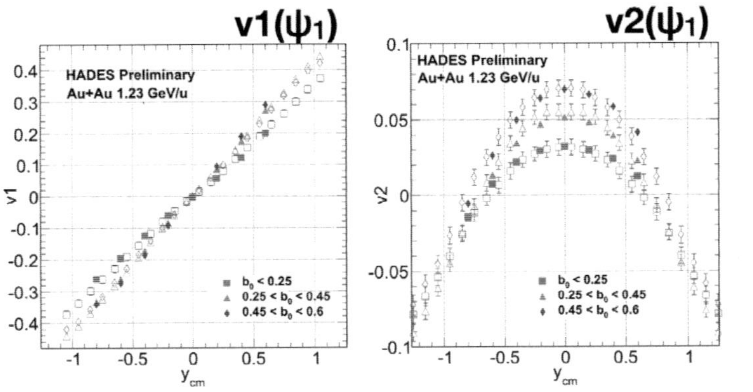

Figure 4. The preliminary HADES measurements of the directed v_1 and elliptic v_2 flow of protons for Au+Au collisions at 1.23 AGeV (full) in comparison to FOPI data [12] (open symbols) are here shown. The same procedure outlined in [12] was used to select the centrality classes $b_0 = b/b_{max}$ and the phase space region with a lower transverse momentum cut.

is caused by the long passing time of the colliding nuclei in comparison to the time of the expansion of the excited nuclear medium. The effect that the relatively slow spectators are shadowing the particle emission in the direction of the reaction plane is also called *squeeze-out*. At higher energies elliptic flow is observed in *in-plane*, made possible by a much shorter passing time of the spectators than the expansion time of the medium. A higher pressure gradient in-plane is building up due to the anisotropic shape of the overlap region. In comparison to directed and elliptic flow, originating from the dynamics of the reaction, the observation of higher-order collective flow is described by the anisotropic and fluctuating shape of the overlap region [11]. It can be shown in Glauber Monte Carlo calculations [4] that higher-order eccentricities, originating from the fluctuations of the initial configuration, are independent from the reaction plane and that each harmonic (ϵ_3, ϵ_4, ϵ_5) is aligned to its specific symmetry plane (Ψ_3, Ψ_4, Ψ_5), see right panel of Fig. 2. The approach to map *flow observables*, like elliptic v_2 or triangular flow v_3, to their corresponding *pseudo-observable*, like eccentricity ϵ_2 and ϵ_3, calculated from Glauber Monte Carlo, can reveal the response of the nuclear medium to the initial anisotropies. This enables the study of the bulk properties of the excited and extreme nuclear matter and provides access to its properties, like its viscosity.

At low energies v_1 and v_2 have been measured for pions, charged kaons, protons, neutrons and fragments, but so far higher-order harmonics have not been studied in the SIS18 energy regime. Preliminary analyses, as shown in Fig. 4, indicate a good consistency of the first two harmonic components of proton flow with FOPI data [2, 12, 13] and the unprecedented statistics not only allows the multi-differential flow analysis (p_t, rapidity, centrality) of charged particles, measurable by HADES, but also makes higher-order flow analysis possible.

5. Multi-Particle Flow-Analysis

The advantage of multi-particle azimuthal correlation techniques is that a separate determination of the event plane is not needed. This exploit the fact that if the particle flow pattern is aligned to a symmetry plane, the particles by themself are correlated to each other. Without any further knowledge of the symmetry plane the flow coefficients can be deduced by correlation functions or cumulants. Multi-particle correlation methods allow to differentiate between collective (flow) and genuine few particle correlations, also called non-flow effects, like those due to interactions (Coulomb and resonance decays) or quantum statistics effects. A general flow analysis framework based on different approaches like the event-plane or the multi-particle azimuthal correlation techniques [14] is in preparation. This will allow to implement different methods such as Cumulants, Lee-Yang Zeroes and more in future analyses.

6. Conclusions

Due to its large collected events statistics and its large acceptance HADES is able to address the measurement of higher-order flow harmonics in the low energy regime. This will allow to extend the existing measurements, as shown in Fig. 5, into so far unexplored regions and will provide new insights into the properties of strongly interacting matter at extreme densities, as e.g. its viscosity. The employment of multi-particle methods will thereby enable the disentanglement of the contributions of collective and non-flow processes.

Supported by BMBF (05P15RFFCA), HGS-HIRe and HICforFAIR.

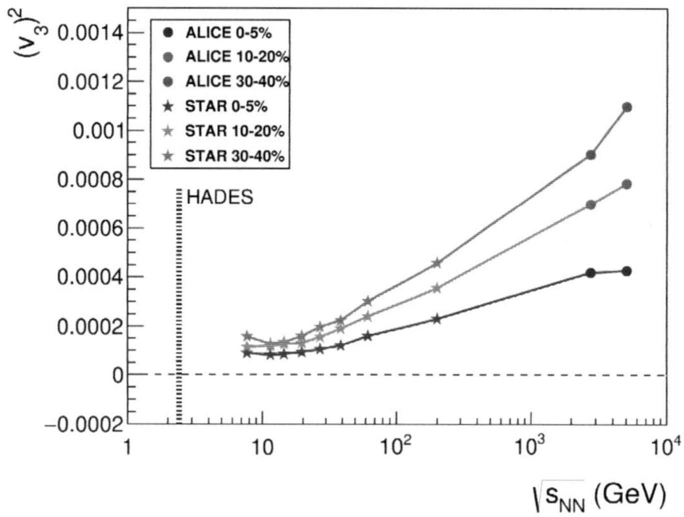

Figure 5. The energy dependence of v_3 as measured for different centralities by STAR [15] and ALICE [16, 17] are here shown. The dashed line indicates the position of the data set measured by HADES at 1.23 AGeV.

References

[1] G. Agakishiev *et al.* [HADES Collaboration], Eur. Phys. J. A **41** (2009) 243
[2] T. Galatyuk, HADES overview, Nuclear Physics A, Vol 931, November 2014, Pages 41-51.
[3] W. Bayer et al., Accelerator Operation Report, GSI Report 2013-1, pages 255-p.
[4] B. Kardan, Centrality Determination at 1.23 AGeV Gold-Gold collision and readout-electronics for the HADES electromagnetic calorimeter , Diploma thesis 2015.
[5] A. Andronic *et al.* [FOPI Collaboration], Phys. Lett. B **612** (2005) 173
[6] C. Pinkenburg *et al.* [E895 Collaboration], Phys. Rev. Lett. **83** (1999) 1295
[7] P. Braun-Munzinger and J. Stachel, Nucl. Phys. A **638** (1998) 3
[8] D. Adamova *et al.* [CERES Collaboration], Nucl. Phys. A **698** (2002) 253.
[9] C. Alt *et al.* [NA49 Collaboration], Phys. Rev. C **68** (2003) 034903
[10] L. Adamczyk *et al.* [STAR Collaboration], Phys. Rev. C **86** (2012) 054908
[11] B. Alver and G. Roland, Phys. Rev. C **81** (2010) 054905
[12] W. Reisdorf *et al.* [FOPI Collaboration], Nucl. Phys. A **876** (2012) 1
[13] A. Le Fvre, Y. Leifels, W. Reisdorf, J. Aichelin and C. Hartnack, Nucl. Phys. A **945** (2016) 112
[14] A. Bilandzic *et al.* Phys. Rev. C **89** (2014) no.6, 064904
[15] L. Adamczyk *et al.* [STAR Collaboration], Phys. Rev. Lett. **116** (2016) no.11, 112302
[16] K. Aamodt *et al.* [ALICE Collaboration], Phys. Rev. Lett. **107** (2011) 032301
[17] J. Adam *et al.* [ALICE Collaboration], Phys. Rev. Lett. **116** (2016) no.13, 132302

Journal of Physics: Conference Series **742** (2016) 012009 doi:10.1088/1742-6596/742/1/012009

Reconstructing Hyperons with the $\overline{\text{P}}$ANDA Detector at FAIR

W Ikegami Andersson
on behalf of the $\overline{\text{P}}$ANDA collaboration

Department of Physics and Astronomy, Uppsala University, Box 516, S-751 20 Uppsala, Sweden

E-mail: walter.ikegami_andersson@physics.uu.se

Abstract. Hyperon production and the study of their properties is an important part of the physics programme of the future $\overline{\text{P}}$ANDA experiment at FAIR. Antihyperon-hyperon pairs will be produced in antiproton-proton collisions through the annihilation of at least one light antiquark-quark (u, d) pair and the creation of a corresponding number of antiquark-quark $\bar{s}s$ pairs. By measuring the decay products of the hyperons, spin observables such as the polarisation can be measured.

Many hyperons have a long life-time which gives rise to final state particles originating from displaced vertices. A pattern recognition algorithm using information from the $\overline{\text{P}}$ANDA Straw Tube Tracker is extended to reconstruct not only the transversal, but also the longitudinal components of charged tracks. A Hough transform and a path finding method as tools to extract the longitudinal components are being developed.

1. Introduction

Quantum chromodynamics (QCD) is the theory of strong interaction, one of the four fundamental forces, and a part of the Standard Model. QCD describes the interaction between quarks and gluons which constitute the visible matter in the universe. In contrast to the electromagnetic and the weak interactions, the coupling constant α_s of the strong interaction diminishes over short distances and increases over long distances. At short distances, or high energy interactions, perturbation theory can be applied to perform QCD calculations. In low energy interactions, α_s grows large and perturbation theory can no longer be used. Instead, numerical methods or effective field theories, such as chiral perturbation theory, where hadrons act as degrees of freedom, can be used. The relevant degrees of freedom for an interaction are given by the energy scale. The energy scale where perturbative QCD is no longer applicable is called the QCD cut-off and is $\Lambda_{QCD} \approx 200$ MeV [1]. The energy scale of hyperon production in $\bar{p}p \rightarrow \overline{Y}Y$ reactions is given by the strange quark mass $m_s \approx 100$ MeV, which is close to the QCD cut-off. As a consequence, the relevant degrees of freedom are unclear and predictions hard to make. Instead, phenomenological models have to be developed. Various $\bar{p}p \rightarrow \overline{Y}Y$ reactions have been described in both a quark-gluon picture [2], in a hadron picture [3] as well as the combination of the two [4]. These models give different predictions for *e.g.* the polarisation. Therefore, measuring the spin observables of $\bar{p}p \rightarrow \overline{Y}Y$ reactions will help to further develop the models in the intermediate energy domain.

Content from this work may be used under the terms of the Creative Commons Attribution 3.0 licence. Any further distribution of this work must maintain attribution to the author(s) and the title of the work, journal citation and DOI.
Published under licence by IOP Publishing Ltd

The foreseen $\overline{\text{P}}$ANDA (antiProton ANnihilation at DArmstadt) detector at FAIR (Facility for Antiproton and Ion Research) provides unique opportunities for hyperon production. The experiment will provide the near 4π coverage needed to completely measure the angular distribution of the final state particles and subsequently measure spin observables. Since ground state hyperons decay weakly, they have a significantly long life-time and can travel a macroscopic distance from the interaction point.

In the following sections, spin observables in the $\bar{p}p \to \overline{Y}Y$ reaction, where Y refers to a hyperon, are discussed. The development of pattern recognition algorithms dedicated for charged particles originating from displaced vertices is also presented.

2. Spin observables in $\bar{p}p \to \overline{\Lambda}\Lambda$ reactions

The spin variables for the $\bar{p}p \to \overline{Y}Y$ reaction are derived using the density matrix formalism [5]. The density matrix for a single particle with spin j is given by

$$\rho = \frac{1}{2j+1}\mathcal{I} + \sum_{L=1}^{2j} \frac{2j}{2j+1} \sum_{M=-L}^{L} Q^L_M r^L_M, \tag{1}$$

where $\mathcal{I}$ is the identity matrix, Q^L_M are Hermitian matrices, r^L_M are the polarisation parameters, L is the angular momentum and M its third component [6]. For spin-$\frac{1}{2}$ particles, Q^L_M are the Pauli matrices and r^L_M can be interpreted as the vector polarisations P_x, P_y and P_z, and the density matrix becomes

$$\rho(1/2) = \frac{1}{2}\begin{bmatrix} 1+P_z & P_x+iP_y \\ P_x-iP_y & 1-P_z \end{bmatrix}. \tag{2}$$

Since the $\bar{p}p \to \overline{Y}Y$ reaction is a strong process, parity must be conserved. This imposes constraints on the polarisation: $P_x = P_z = 0$. The density matrix of the final state then reduces to

$$\rho(1/2) = \frac{1}{2}\begin{bmatrix} 1 & iP_y \\ -iP_y & 1 \end{bmatrix}. \tag{3}$$

Because of the parity violating weak decay of hyperons, the spin observables of the $\bar{p}p \to \overline{\Lambda}\Lambda$ reaction can be accessed by the angular distribution of the decay particles. One example is the weak $\Lambda \to p\pi^-$ decay that has a branching ratio of $\text{BR}(\Lambda \to p\pi^-) = 63.9\%$ [7]. The angular distribution of the final state particles is [6]

$$I(\theta_p) = \frac{1}{4\pi}(1 + \alpha_\Lambda P_y \cos\theta_p), \tag{4}$$

where $\alpha_\Lambda = 0.64$ [7] is the decay parameter and θ_p is the polar angle of the proton. The polar angle θ_p is measured in the Λ rest frame between the direction of the outgoing proton and the axis normal to the production plane (see Figure 1a). Thus, by measuring the scattering angle of the proton, the polarisation P_y can be deduced. In an experiment with unpolarised beam and unpolarised target such as $\overline{\text{P}}$ANDA, the polarisation of the individual hyperons as well as the spin correlation between the two are accessible.

The density matrix formalism has also been used to derive angular distributions of the $\Xi^- \to \pi^-\Lambda \to \pi^-\pi^-p$ and $\Omega^- \to K^-\Lambda \to K^-\pi^-p$ hyperon decay chains. Since the Ω^- hyperon has spin-$\frac{3}{2}$, the situation is more complicated. Instead of one non-zero polarisation parameter as in the spin-$\frac{1}{2}$ case (P_y) there are seven. These have been derived in [6].

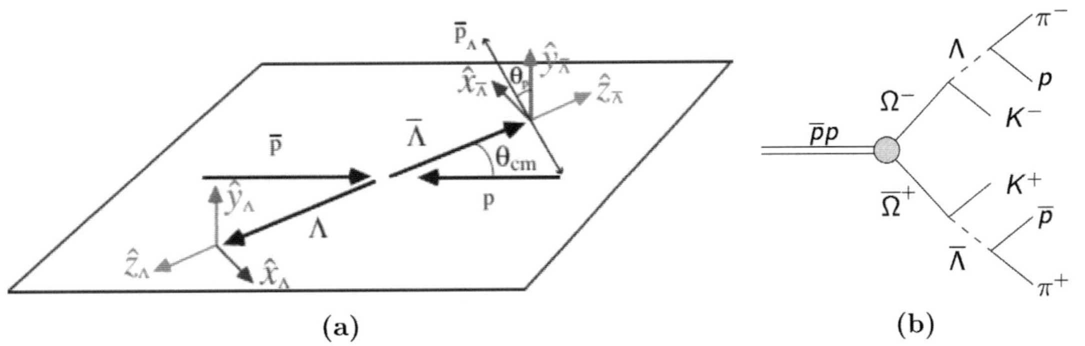

Figure 1. **(a)** Production plane of the $\bar{p}p \to \bar{\Lambda}\Lambda$ reaction. The y-axis of the Λ rest frame is perpendicular to the production plane. **(b)** The $\bar{\Omega}^{+}\Omega^{-} \to K^{+}\pi^{+}\bar{p}K^{-}\pi^{-}p$ reaction channel.

3. Pattern recognition with the $\overline{\text{P}}$ANDA Straw Tube Tracker

Simulations and measurements have shown that hyperons can travel tens of centimeters from their production vertex before decaying due to their long life-times. Furthermore, channels such as $\Omega^{-} \to K^{-}\Lambda \to K^{-}\pi^{-}p$ (see Figure 1b) produce additional displaced vertices from the production of a Λ in the first decay. This makes it challenging to reconstruct hyperon events as the information about the decay points is *a priori* unknown.

The $\overline{\text{P}}$ANDA detector will have several subdetectors for detecting charged particles. These include a Micro Vertex Detector (MVD), a Straw Tube Tracker (STT) and Gas Electron Multiplier (GEM) disks as well as a Forward Tracking System (FTS). Charged tracks originating from a hyperon decay could leave only a few hits or completely miss the MVD and/or the GEM disks, leaving only information in the STT. Here, particular attention is given to the STT.

The STT consists of 4636 gas filled straws. There are inner and outer segments which contain layers of straws parallel to the beam direction. In between, there is a skewed segment, where straws have been tilted by $\pm 2.9°$. The straws have a conductive inner layer and an anode wire in the center. A potential difference is applied between the conductor and the wire. When an ionizing particle traverses a straw, gas atoms in the particle trajectory are ionized, creating free electrons and ions. The electric field causes electrons to travel to the wire, producing an electric signal. A relation between the drift time and the radial distance to the wire of an electron is obtained through a calibration procedure. This relation is used to calculate the closest radial distance of a trajectory to the wire. The radius is called the isochrone and can be imagined as a cylinder along the wire representing all possible track positions [8].

In the $\overline{\text{P}}$ANDA software, a standard track finding algorithm exists for the detection of charged particles originating from the interaction point. In the case of hyperon decays, some tracks can only be reconstructed within the STT due to the displaced vertices. The standard track finder is not suited for the detection of such events and a different algorithm is required. A pattern recognition algorithm that can operate with only information from the STT has been developed in [9]. In this algorithm, a particle trajectory is assumed to be a perfect helix *i.e.* no energy losses and a homogeneous magnetic field are considered. A clusterization is done where parallel straws giving a signal are grouped together if the straws are adjacent to each other. A clusterization procedure is also performed to group together the skewed straws. After the clusterization is complete, a circle is fitted tangential to the isochrones of all parallel straws. This circle represents the particle track in the xy-projection. In Figure 2a, a $\bar{p}p \to \bar{\Lambda}\Lambda \to \bar{p}\pi^{+}p\pi^{-}$ event is shown, where four tracks are found. The projected isochrones (purple circles) are tangent to the tracks (black circles).

This algorithm is now being extended to reconstruct the longitudinal position z and

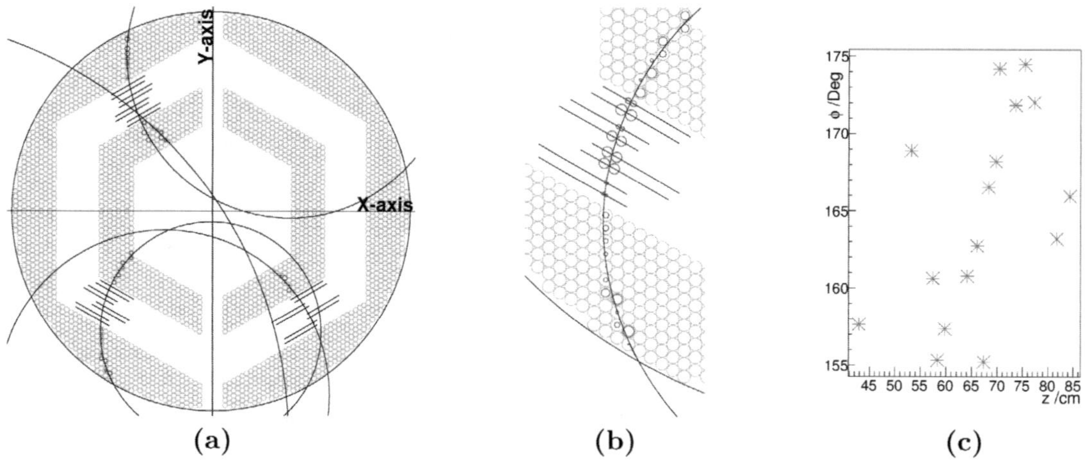

Figure 2. **(a)** xy-projection of the STT with a reconstructed $\bar{p}p \rightarrow \bar{\Lambda}\Lambda \rightarrow \bar{p}\pi^+ p\pi^-$ event at $p_{beam} = 1.64$ GeV/c is shown. **(b)** Bottom-left track from Figure 2a. The possible isochrone positions are centered in the skewed straws and tangential to the track. **(c)** The resulting (z, ϕ) coordinates of the isochrone centers from the track in Figure 2b.

momentum p_z, giving a full 3D view of a particle trajectory. The z-position where the particle traversed a skewed straw is determined by aligning the isochrone to the track. The xy-projection of the isochrone in a skewed straw is an ellipse. The ellipse is aligned such that it is tangential to the track and centered along the straw wire, shown in Figure 2b. For each skewed straw, there are two possible ways the isochrone can be aligned, analogous with a left-right ambiguity. Two possible solutions to this ambiguity are currently under development.

3.1. The Hough transform

One solution to the ambiguity problem is the *Hough transform.* It is a feature extraction method used to find geometric shapes in images [10]. The track helix is a straight line in (z, ϕ) space. The correct points in the skewed straws are found by finding the line which crosses the most points with the Hough transform. Instead of the common Cartesian parametrisation of a line, the Hesse normal form is used. A line is represented by its closest distance r to the origin and the angle θ [11]

$$r = x\cos\theta + y\sin\theta. \tag{5}$$

For each (z, ϕ) point (see Figure 2c), a set of lines in steps of 1° are generated. In Figure 3a a set of lines at a point is shown. The line parameters are stored in a histogram called the accumulator space, shown in Figure 3b. After the accumulator space has been filled, a "voting procedure" begins where a maximum is searched for in the accumulator space. The parameters at the maximum correspond to the line that crosses the most (z, ϕ) points, referred to as the *Hough line.* For each skewed straw, the distance of the two possible points to the Hough line is evaluated and the point furthest away is discarded. A straight line fit is performed with the remaining point in every skewed straw. The resulting slope parameter is used to calculate the helix angle and the longitudinal momentum and position are derived.

3.2. The Path finding method

A second method is being developed by taking a combinatorial approach. For each pair of neighbouring skewed straws, lines are calculated between all combinations of (z, ϕ) coordinates,

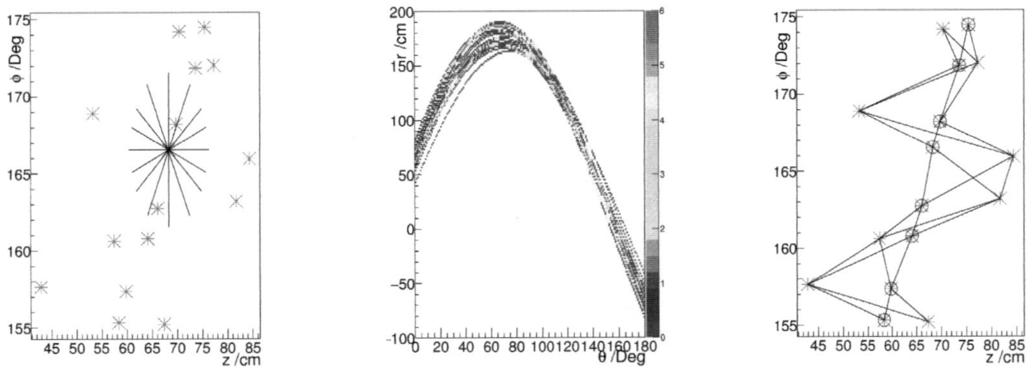

(a) (b) (c)

Figure 3. **(a)** A set of lines generated for one (z, ϕ) point. Only a couple of lines are shown. **(b)** The accumulator space after all lines have been generated for all points. **(c)** All generated lines in the path finder approach. The encircled (z, ϕ) points are chosen as correct points.

shown in Figure 3d. The angle is calculated between all pairs of lines that connect to each other at a (z, ϕ) point. The correct set of points is found by looking for a path from the innermost to the outermost skewed straw along the lines and calculating a weight

$$w = \sum_i \pi - \theta_i, \tag{6}$$

where θ_i is the angle between two lines at point (z_i, ϕ_i). To reduce the number of possible paths, a constraint requiring an angle to be above a certain value $\theta_i > \theta_{cut}$ is introduced. A path containing an angle smaller than θ_{cut} is removed. The (z, ϕ) points along the path with the smallest weight w are chosen as the correct positions and a straight line fit is performed. As before, the helix angle is obtained and the longitudinal momentum and position are calculated.

References

[1] Ellis R K, Stirling W J and Webber B R, *QCD and Collider Physics* (Cambridge Monographs on Particle Physics, Nuclear Physics and Cosmology, vol. 8, Cambridge University Press, 1996)

[2] Kohno M and Weise W, Phys. Lett. B **179**, 15 (1986); Rubinstein H R and Snellman H, Phys. Lett. B **165**, 187 (1985); Furui S and Faessler A, Nucl. Phys. A **468**, 669 (1987); Burkardt M and Dillig M, Phys. Rev. C **37**, 1362 (1988); Alberg M A et al., Z. Phys. A **331**, 207 (1988)

[3] Tabakin F and Eisenstein R A, Phys. Rev. C **31**, 1857 (1985); Kohno M and Weise W, Phys. Lett. B **179**, 15 (1986); La France P et al., Phys. Lett. B **214**, 317 (1988); Timmermans R G E et al., Phys. Rev. D **45**, 2288 (1992); Haidenbauer J et al., Phys. Rev. C **46**, 2516 (1992)

[4] Ortega P G et al., Phys. Lett. B **696**, 352 (2011)

[5] J J Sakurai, *Modern quantum mechanics*, (Addison-Wesley, 1994)

[6] Thomé E, *Multi-Strange and Charmed Antihyperon-Hyperon Physics for PANDA*, Acta Universitatis Uppsaliensis, Uppsala Dissertations from the Faculty of Science and Technology 101 (Ph. D. Thesis, Uppsala University, 2012)

[7] K A Olive *et al.* [Particle Data Group Collaboration], Chin. Phys. C **38** (2014) 090001.

[8] W. Erni *et al.* [PANDA Collaboration], Eur. Phys. J. A **49**, 25 (2013) [arXiv:1205.5441 [physics.ins-det]].

[9] Schumann J, *Beschleunigung eines Spurfindealgorithmus für den Straw Tube Tracker des PANDA-Detektors durch Parallelisierung mit CUDA C*, Master Thesis, FH Aachen (2015)

[10] Duda, R. O. and P. E. Hart, *Use of the Hough Transformation to Detect Lines and Curves in Pictures* Comm. ACM, Vol. 15, pp. 11-15 (January, 1972)

[11] Bôcher M, *Plane Analytic Geometry: With Introductory Chapters on the Differential Calculus* (H. Holt, 1915)

Journal of Physics: Conference Series **742** (2016) 012010 doi:10.1088/1742-6596/742/1/012010

The eLSM at nonzero density

Francesco Giacosa

Institute of Physics, Jan Kochanowski University, ul. Swietokrzyska 15, 25-406 Kielce, Poland and Institute for Theoretical Physics, J. W. Goethe University, Max-von-Laue-Str. 1, 60438 Frankfurt am Main, Germany

E-mail: fgiacosa@ujk.edu.pl

Abstract. The extended Linear Sigma Model (eLSM) is an effective model of QCD which includes in the mesonic sector (pseudo)scalar and (axial-)vector quarkonia mesons as well as one dilaton/glueball field and in the baryonic sector the nucleon doublet and its chiral partner in the mirror assignment. The chiral partner of the pion turns out to be the resonance $f_0(1370)$, which is then predominantly a quarkonium state. As a consequence, $f_0(500)$ is predominately not a quarkonium state but a four-quark object and is at first not part of the model. Yet, $f_0(500)$ is important in the baryonic sector and affects nuclear matter saturation, the high-density behavior, and nucleon-nucleon scattering. In these proceedings, we show how to enlarge the two-flavour version of the eLSM in order to include the four-quark field $f_0(500)$ in a chirally invariant manner. We then discuss homogeneous and inhomogeneous chiral restoration in a dense medium.

1. Introduction

Linear Sigma Models (LSMs) are effective models of QCD which contains hadrons (mesons and baryons) as degrees of freedom and which are based on the linear realization of chiral symmetry [1]. As a consequence of the spontaneous breaking of chiral symmetry, in these models the pions emerge as quasi-Goldstone bosons (a relatively small mass is present because of the explicit breaking of chiral symmetry). The chiral partner of the pion, denoted as σ_N, is also an explicit d.o.f. of such models. As various studies show, this meson corresponds to the scalar resonance $f_0(1370)$: this state is then predominantly a quark-antiquark state. Hence, the lightest scalar state listed in the PDG [2], the resonance $f_0(500)$, must be something else: its substructure corresponds to a four-quark state, either as diquark-antidiquark or pion-pion enhancement (see e.g. the recent review on the subject [3]). As such, this resonance should not be (at first) part of a LSM. Yet, as we shall describe later on, $f_0(500)$ is important at nonzero density since it describes a necessary middle-range attraction between nucleons (see also Ref. [4]).

Extensions of LSMs toward the inclusion of (axial-)vector degrees of freedom were performed in Ref. [5]. Quite recently, an as complete as possible LSM, called extended Linear Sigma Model (eLSM), was developed. The eLSM contains from the very beginning (axial-)vector fields and embodies both chiral symmetry and dilatation invariance. Spontaneous and explicit breaking of the former as well as anomalous breaking of the latter are present. As a consequence of chiral symmetry and the dilation invariance, the eLSM contains a *finite* number of terms. The eLSM was first presented for $N_f = 2$ in Refs. [6], then it was enlarged to $N_f = 3$ in Refs. [7] (this is the first version of a chiral model with $N_f = 3$ containing (axial-)vector d.o.f.), and lately it was also studied for charmed mesons ($N_f = 4$, Ref. [8]).

Content from this work may be used under the terms of the Creative Commons Attribution 3.0 licence. Any further distribution of this work must maintain attribution to the author(s) and the title of the work, journal citation and DOI.
Published under licence by IOP Publishing Ltd

In the baryonic sector, the eLSM was investigated for $N_f = 2$ in Refs. [9] (a first step toward the eLSM at $N_f = 3$ was performed in Ref. [12]). In the eLSM, the mirror assignment, which allows for chirally invariant mass term, is used [10, 11]. As mentioned in Ref. [9], a four-quark field χ corresponding to $f_0(500)$ can be coupled to the eLSM in a chirally invariant way, see Sec. 2. The eLSM with $f_0(500)$ has been investigated at nonzero density in Ref. [14], where the chiral phase transition has been investigated, and later on in Ref. [15], in which inhomogeneous condensation has been studied; these results are here summarized in Sec. 3. The role of the additional, non-conventional meson $f_0(500)$ turns out to be important: it makes a description of the properties of nuclear matter possible (both saturation and compressibility are in agreement with data) and it strongly affects the properties of nuclear matter at high density. Quite interestingly, the resonance $f_0(500)$ was recently investigated in Ref. [16] in the framework of nucleon-nucleon scattering: also in this case, its presence is necessary for a correct description of data.

2. The eLSM

2.1. eLSM without $f_0(500)$

We briefly present the eLSM for $N_f = 2$. (Pseudo)scalar mesons are contained in $\Phi = (\sigma_N + i\eta_N)t^0 + (\vec{a}_0 + i\vec{\pi}) \cdot \vec{t}$, where $t^0 = 1_2/2$, $\vec{t} = \vec{\sigma}/2$, σ_i are the Pauli matrices. In Table 1 we report the identification of the fields with resonances of the PDG [2]. [Note: $\eta_N \equiv (u\bar{u} + d\bar{d})/\sqrt{2}$ reads $\eta_N = \cos\varphi_P \eta - \sin\varphi_P \eta'$ with $\varphi_P \approx -44°$ [7]] Because of spontaneous symmetry breaking, σ_N condenses: $\sigma_N \to \sigma_N + \phi$, where ϕ is the chiral condensate. Under $U(2)_R \times U(2)_L$ chiral transformations: $\Phi \to U_L \Phi U_R^\dagger$.

The left-handed and right-handed fields L^μ and R^μ contain the vector states ω^μ and $\vec{\rho}^\mu$ and the axial–vector states f_1^μ and $\vec{a}_1^\mu$: $L^\mu = (\omega^\mu + f_1^\mu)t^0 + (\vec{\rho}^\mu + \vec{a}_1^\mu) \cdot \vec{t}$, $R^\mu = (\omega^\mu - f_1^\mu)t^0 + (\vec{\rho}^\mu - \vec{a}_1^\mu) \cdot \vec{t}$, see Table 1. Under chiral transformations: $L^\mu \to U_L \Phi U_L^\dagger$ and $R^\mu \to U_R \Phi U_R^\dagger$.

Table 1: Correspondence of eLSM $\bar{q}q$ fields to PDG [2].

Field	PDG	Quark content	I	J^{PC}	Mass (GeV)
π^+, π^-, π^0	π	$u\bar{d}, d\bar{u}, \frac{u\bar{u}-d\bar{d}}{\sqrt{2}}$	1	0^{-+}	0.13957
η	$\eta(547)$	$\frac{u\bar{u}+d\bar{d}}{\sqrt{2}} \cos\varphi_P - s\bar{s} \sin\varphi_P$	0	0^{-+}	0.54786
η'	$\eta'(958)$	$\frac{u\bar{u}+d\bar{d}}{\sqrt{2}} \sin\varphi_P + s\bar{s} \cos\varphi_P$	0	0^{-+}	0.95778
a_0^+, a_0^-, a_0^0	$a_0(1450)$	$u\bar{d}, d\bar{u}, \frac{u\bar{u}-d\bar{d}}{\sqrt{2}}$	1	0^{++}	1.474
σ_N	$f_0(1370)$	$\frac{u\bar{u}+d\bar{d}}{\sqrt{2}}$	0	0^{++}	1.350
ρ^+, ρ^-, ρ^0	$\rho(770)$	$u\bar{d}, d\bar{u}, \frac{u\bar{u}-d\bar{d}}{\sqrt{2}}$	1	1^{--}	0.77526
ω_N	$\omega(782)$	$\frac{u\bar{u}+d\bar{d}}{\sqrt{2}}$	0	1^{--}	0.78265
a_1^+, a_1^-, a_1^0	$a_1(1230)$	$u\bar{d}, d\bar{u}, \frac{u\bar{u}-d\bar{d}}{\sqrt{2}}$	1	1^{++}	1.230
$f_{1,N}$	$f_1(1285)$	$\frac{u\bar{u}+d\bar{d}}{\sqrt{2}}$	0	1^{++}	1.2819

The mesonic part of the Lagrangian reads

$$\mathcal{L}_{eLSM}^{meson} = \mathrm{Tr}\left[(D^\mu\Phi)^\dagger(D^\mu\Phi)\right] - \mu_0^2\mathrm{Tr}[\Phi^\dagger\Phi] - \lambda_2\mathrm{Tr}[(\Phi^\dagger\Phi)^2] - \frac{1}{4}\mathrm{Tr}[(L^{\mu\nu})^2 + (R^{\mu\nu})^2] +$$
$$\frac{m_1^2}{2}\mathrm{Tr}\left[L^{\mu 2} + R^{\mu 2}\right] + \mathrm{Tr}[H(\Phi + \Phi^\dagger)] + h_2\mathrm{Tr}\left[\Phi^\dagger L_\mu L^\mu \Phi + \Phi R_\mu R^\mu \Phi^\dagger\right] + 2h_3\mathrm{Tr}\left[\Phi R_\mu \Phi^\dagger L^\mu\right] \ ..., \tag{1}$$

where dots refer to large-N_c suppressed terms (including the chiral anomaly). In Refs. [6, 7] it was shown that, thanks to the inclusion of (axial-)vector d.o.f., the eLSM provides a good description of meson phenomenology. An interesting consequence is that the quark-antiquark field σ_N, which represents chiral partner of the pion, is associated to $f_0(1370)$, in agreement with previous phenomenological studies [13]. Hence, $f_0(500)$ must be something else [3].

We now turn to the baryonic sector. For $N_f = 2$ one starts from two nucleon fields Ψ_1 and Ψ_2 with opposite parity which transform mirror-like under chiral transformations: $\Psi_{1,R(L)} \to U(2)_{R(L)}\Psi_{1,R(L)}$, $\Psi_{2,R(L)} \to U(2)_{L(R)}\Psi_{2,R(L)}$. The eLSM Lagrangian is [9]:

$$\begin{aligned}\mathcal{L}_{eLSM}^{baryons} &= \bar{\Psi}_{1L} i\gamma_\mu D_{1L}^\mu \Psi_{1L} + \bar{\Psi}_{1R} i\gamma_\mu D_{1R}^\mu \Psi_{1R} + \bar{\Psi}_{2L} i\gamma_\mu D_{2R}^\mu \Psi_{2L} + \bar{\Psi}_{2R} i\gamma_\mu D_{2L}^\mu \Psi_{2R} \\ &- \hat{g}_1(\bar{\Psi}_{1L}\Phi\Psi_{1R} + \bar{\Psi}_{1R}\Phi^\dagger\Psi_{1L}) - \hat{g}_2(\bar{\Psi}_{2L}\Phi^\dagger\Psi_{2R} + \bar{\Psi}_{2R}\Phi\Psi_{2L}) \\ &- m_0(\bar{\Psi}_{1L}\Psi_{2R} - \bar{\Psi}_{1R}\Psi_{2L} - \bar{\Psi}_{2L}\Psi_{1R} + \bar{\Psi}_{2R}\Psi_{1L}) \,, \end{aligned} \tag{2}$$

where $D_{1(2)R(L)}^\mu = \partial^\mu - ic_{1(2)}R(L)^\mu$. The fields Ψ_1 and Ψ_2 mix due to the m_0-term and are related to the physical states N and its chiral partner N^* via:

$$\Psi_1 = \frac{1}{\sqrt{2\cosh\delta}}\left(Ne^{\delta/2} + \gamma_5 N^* e^{-\delta/2}\right) \,, \; \Psi_2 = \frac{1}{\sqrt{2\cosh\delta}}\left(\gamma_5 N e^{-\delta/2} - N^* e^{\delta/2}\right) \,, \tag{3}$$

where $\cosh\delta = \frac{m_N + m_{N^*}}{2m_0}$. The field N corresponds to the nucleon $N(939)$ and N^* to $N(1535)$ or $N(1650)$. For the purposes of the present work, the assignment of N^* is marginal, see however [9, 12]. The parameter m_0 represents a chirally invariant mass, which was first discussed in Ref. [10] and further investigated in Refs. [9, 11]. The masses of the nucleon N and its chiral partner N^* are given by:

$$m_{N,N^*} = \sqrt{m_0^2 + \frac{(\hat{g}_1 + \hat{g}_2)^2}{16}\phi^2} \pm \frac{1}{4}(\hat{g}_1 - \hat{g}_2)\phi \,. \tag{4}$$

In the limit $m_0 \to 0$ one obtains the result $m_N = \hat{g}_1\phi/2$, i.e., the nucleon mass is solely generated by the chiral condensate [as in the original Linear Sigma Model [1]].The parameters of the model were determined in Ref. [9], to which we refer for details.

2.2. *Inclusion of $f_0(500)$ in the eLSM*

We now introduce $\chi \equiv f_0(500)$ with quantum numbers $I(J^{PC}) = 0(0^{++})$ and mass $m_\chi = (0.475 \pm 0.25)$ GeV [2] into the eLSM. This state is regarded as an admixture of $\pi\pi$ and $[u,d][\bar{u},\bar{d}]$ configurations. For $N_f = 2$ it is a singlet under chiral transformations ($\chi \to \chi$). The coupling of χ to baryons is obtained by modifying the m_0-term as:

$$m_0(\bar{\Psi}_{1L}\Psi_{2R} - \bar{\Psi}_{1R}\Psi_{2L} - \bar{\Psi}_{2L}\Psi_{1R} + \bar{\Psi}_{2R}\Psi_{1L}) \to a\chi(\bar{\Psi}_{1L}\Psi_{2R} - \bar{\Psi}_{1R}\Psi_{2L} - \bar{\Psi}_{2L}\Psi_{1R} + \bar{\Psi}_{2R}\Psi_{1L}) \,, \tag{5}$$

where a is now a dimensionless constant. Then, the mass parameter m_0 emerges as a condensation of the four-quark field χ: $m_0 = a\chi_0$. In the mesonic sector, one has too add

$$\mathcal{L}_{eLSM}^{meson} \to \mathcal{L}_{eLSM}^{meson} + \frac{1}{2}\left((\partial_\mu\chi)^2 - m_\chi^2\chi^2\right) + g_{\chi\Phi}\chi Tr[\Phi^\dagger\Phi] + g_{\chi\Phi}\chi Tr[L^{\mu 2} + R^{\mu 2}] + \dots \,, \tag{6}$$

where dots refer to large-N_c suppressed terms. As a consequence, the condensate χ_0 takes the form $\chi_0 = g_{\chi\Phi}\phi^2/m_\chi^2$.

3. Results

3.1. *Homogenous condensation*

First, we study nuclear matter at nonzero density under the assumption that the condensates are homogenous. Two scalar fields condense: $\langle\sigma_N\rangle = \phi(\mu)$ and $\langle\chi\rangle = \bar{\chi}(\mu)$, where μ is the nuclear

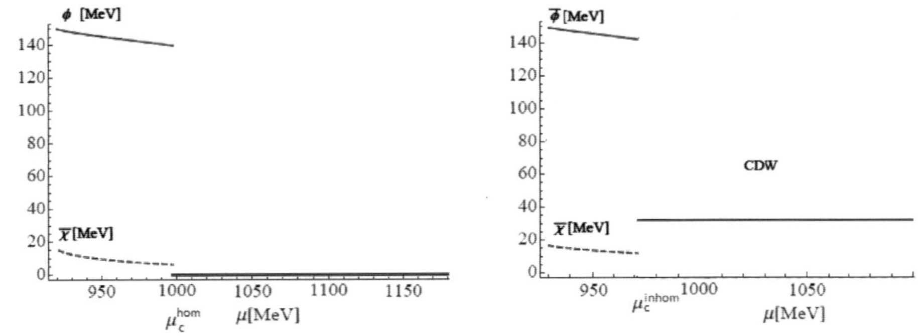

Figure 1: Left: homogenous condensation. The condensates ϕ and $\bar{\chi}$ drop to almost zero at $\mu_c^{\rm hom}$. Right: inhomogeneous condensation. At $\mu_c^{\rm inhom}$ a transition to the inhomogeneous condensation of Eq. (7) takes place. This configuration is more favorable than the homogenous case of the left panel.

chemical potential. The results are obtained by minimizing at a given μ the thermodynamical potential Ω w.r.t. ϕ, $\bar{\chi}$, as well as $\langle\omega^0\rangle$. The vacuum's relation $\chi_0 = g_{\chi\Phi}\phi^2/m_\chi^2$ holds approximately also at nonzero density: ϕ slowly decreases as function of μ together with $\bar{\chi}$. Then, at a critical $\mu_c^{\rm hom} \sim 1$ GeV (corresponding to a density $\rho/\rho_0 \sim 2.7$, where ρ_0 is the nuclear matter saturation density) a first-order phase transition takes place: ϕ and $\bar{\chi}$ drop to to very small (but nonzero) values. Chiral symmetry is restored (see Fig. 1, left panel). In terms of density, there is a long mixed phase between $2.7\rho_0$-$10\rho_0$. The compressibility K lies in the range 200-250 MeV and is therefore in good agreement with the experiment (200-300 MeV). The mass of the nucleon and that of the partner drop to almost zero in the chirally restored phase. For details, see Ref. [14].

3.2. Inhomogeneous condensation

The previous results were obtained under the assumption that the ϕ and $\bar{\chi}$ are homogenous, i.e. are not space-dependent. Yet, various studies have shown (see Ref. [17] and refs. therein) that an inhomogeneous condensation can be favoured. In Ref. [15] the so-called chiral-spiral [18] was investigated:

$$\phi(\mu, z) = \bar{\phi}\cos(2fz),\ \langle\pi^{3\equiv 0}\rangle = \bar{\phi}\sin(2fz) \tag{7}$$

with $\bar{\phi} \equiv \bar{\phi}(\mu)$ and $f \equiv f(\mu)$. The homogenous case corresponds to $f = 0$. For $f \neq 0$ one has a condensation of the neutral pion field, which corresponds to a spontaneous breaking of parity at nonzero density. At a given μ, the thermodynamical potential is now minimized for $\bar{\phi}$, f, $\bar{\chi}$, and $\langle\omega^0\rangle$. The results show that below a certain critical chemical potential $\mu_c^{\rm inhom} \lesssim 1$ GeV one has $f = 0$: homogenous condensation is realized, just as before. However, at $\mu_c^{\rm inhom}$ a phase transition takes place: f jumps to a finite value (of about 400 MeV, which then slowly increases with μ), and $\bar{\phi}$ to a lower but finite value, see Fig. 2. Interestingly, for a given parameter set, it turns out that $\mu_c^{\rm inhom} < \mu_c^{\rm hom}$: the homogenous chiral phase transition does not occur, but is only a local minimum. Chiral symmetry is only partially restored.

4. Conclusions

The resonance $f_0(500)$ is the lightest scalar listed in the PDG and hence is potentially interesting in hadron phenomenology. Its role has to be investigated case by case. While its condensate may be relevant at nonzero temperature [19], its effect on thermal models turns out to be negligible because of a very subtle and interesting cancellation with the repulsive isotensor channel: for practical purposes, $f_0(500)$ can be neglected in thermal models of heavy ion collisions [20].

At nonzero density, $f_0(500)$ plays indeed a significant role because it mediates a sizable attraction between nucleons: in these proceedings, we have incorporated $f_0(500)$ into the eLSM and reviewed the properties at finite chemical potential. We have shown that inhomogeneous condensation of the chiral-spiral type is favored w.r.t. the homogeneous one. In the future, one should go beyond the chiral-spiral Ansatz and test arbitrary types of inhomogeneous condensation by using the numerical procedure put forward in Ref. [21].

Acknowledgments
The author thanks all the members of the chiral group in Frankfurt for the many valuable cooperations and discussions.

References

[1] B.W. Lee, *Chiral Dynamics* (Gordon and Breach, New York, 1972); S. Gasiorowicz and D.A. Geffen, Rev. Mod. Phys. **41** (1969) 531; V. Koch, nucl-th/9512029.
[2] K.A. Olive *et al.* (Particle Data Group), Chin. Phys. C **38**, 090001 (2014).
[3] J.R. Pelaez, arXiv:1510.00653 [hep-ph].
[4] R. Machleidt, Phys. Rev. C **63** (2001) 024001 [nucl-th/0006014].
[5] P. Ko and S. Rudaz, Phys. Rev. D **50** (1994) 6877; M. Urban, M. Buballa and J. Wambach, Nucl. Phys. A **697** (2002) 338 [hep-ph/0102260].
[6] D. Parganlija, F. Giacosa and D.H. Rischke, Phys. Rev. D **82** (2010) 054024 [arXiv:1003.4934 [hep-ph]]; S. Janowski, D. Parganlija, F. Giacosa and D.H. Rischke, Phys. Rev. D **84** (2011) 054007 [arXiv:1103.3238 [hep-ph]].
[7] D. Parganlija, P. Kovacs, G. Wolf, F. Giacosa and D. H. Rischke, Phys. Rev. D **87**, 014011 (2013) arXiv:1208.0585 [hep-ph]]; S. Janowski, F. Giacosa and D. H. Rischke, Phys. Rev. D **90** (2014) 114005 [arXiv:1408.4921[hep-ph]].
[8] W.I. Eshraim, F. Giacosa and D.H. Rischke, Eur. Phys. J. A **51** (2015) 9, 112 [arXiv:1405.5861 [hep-ph]].
[9] S. Gallas, F. Giacosa and D.H. Rischke, Phys. Rev. D **82** (2010) 014004 [arXiv:0907.5084 [hep-ph]]; S. Gallas and F. Giacosa, Int. J. Mod. Phys. A **29** (2014) 17, 1450098 [arXiv:1308.4817 [hep-ph]].
[10] C.E. Detar and T. Kunihiro, Phys. Rev. D **39** (1989) 2805.
[11] D. Zschiesche et al, Phys. Rev. C **75** (2007) 055202 [nucl-th/0608044]; D. Jido, M. Oka, and A. Hosaka, Prog. Theor. Phys. **106** (2001) 873 [hep-ph/0110005]; D. Jido, Y. Nemoto, M. Oka and A. Hosaka, Nucl. Phys. A **671** (2000) 471 [hep-ph/9805306]; A. Hosaka, D. Jido and M. Oka, Prog. Theor. Phys. Suppl. **149** (2003) 203 [hep-ph/0305258]; M. Harada, M. Rho and C. Sasaki, Phys. Rev. D **70** (2004) 074002 [hep-ph/0312182]; C. Sasaki, I. Mishustin and K. Redlich, Phys. Rev. D **89** (2014) no.1, 014031 [arXiv:1308.3635 [hep-ph]].
[12] L. Olbrich, M. Zétényi, F. Giacosa and D. H. Rischke, Phys. Rev. D **93** (2016) no.3, 034021 [arXiv:1511.05035 [hep-ph]].
[13] C. Amsler and F. E. Close, Phys. Lett. B **353** (1995) 385; W. J. Lee and D. Weingarten, Phys. Rev. D **61** (2000) 014015 [arXiv:hep-lat/9910008]; F. E. Close and A. Kirk, Eur. Phys. J. C 21 (2001) 531 [arXiv:hep-ph/0103173]; H. Y. Cheng, C. K. Chua and K. F. Liu, Phys. Rev. D **74** (2006) 094005 [arXiv:hep-ph/0607206]; F. Giacosa, T. Gutsche, V. E. Lyubovitskij and A. Faessler, Phys. Rev. D **72**, 094006 (2005) [arXiv:hep-ph/0509247].
[14] S. Gallas, F. Giacosa and G. Pagliara, Nucl. Phys. A **872** (2011) 13 [arXiv:1105.5003 [hep-ph]].
[15] A. Heinz, F. Giacosa and D.H. Rischke, Nucl. Phys. A **933** (2015) 34 [arXiv:1312.3244 [nucl-th]].
[16] W. Deinet, K. Teilab, F. Giacosa and D. H. Rischke, arXiv:1603.04312 [nucl-th].
[17] M. Buballa and S. Carignano, Prog. Part. Nucl. Phys. **81** (2015) 39 [arXiv:1406.1367 [hep-ph]].
[18] T. Kojo, Y. Hidaka, L. McLerran and R. D. Pisarski, Nucl. Phys. A **843** (2010) 37 [arXiv:0912.3800 [hep-ph]].
[19] A. Heinz, S. Struber, F. Giacosa and D.H. Rischke, Phys. Rev. D **79** (2009) 037502 [arXiv:0805.1134 [hep-ph]].
[20] W. Broniowski, F. Giacosa and V. Begun, Phys. Rev. C **92** (2015) no.3, 034905 [arXiv:1506.01260 [nucl-th]].
[21] A. Heinz, F. Giacosa, M. Wagner and D. H. Rischke, Phys. Rev. D **93** (2016) no.1, 014007 [arXiv:1508.06057 [hep-ph]].

Journal of Physics: Conference Series **742** (2016) 012011 doi:10.1088/1742-6596/742/1/012011

Progress in Lattice QCD at nonzero density using the Complex Langevin equation

Dénes Sexty

Bergische Universität Wuppertal

E-mail: `sexty@uni-wuppertal.de`

Abstract. Recent progress in direct simulations of QCD at nonzero chemical potentials is reported upon. After a brief introduction to the sign problem in lattice QCD and a quick description of the complex Langevin equation we show recent results and discuss open questions.

1. Introduction

Lattice simulations of QCD are based on a path integral formulation using the partition sum

$$Z = \int DU e^{-S_g[U]} \det M(\mu, U), \tag{1}$$

with the gauge action of the link variables $S_g[U]$, and the fermion matrix $M(\mu, U)$, where the dependence on the quark chemical potential μ is indicated. At $\mu = 0$ the measure in this integral is positive, therefore importance sampling can be used to calculate various thermodynamical operators, propagators, etc. In contrast, at $\mu \neq 0$ the determinant is in general complex resulting in a *Sign problem*, invalidating direct simulations with importance sampling. Many methods have been proposed to evade the sign problem, see the reviews in the context of lattice QCD [1]. Most of these methods have some shortcomings, typically they are applicable at low chemical potentials ($\mu < T$ with the temperature T). While there are many proposals for different dense phases as well as a critical point, a large part of the phase diagram of QCD still remains unknown.

Recently two methods using analyticity have emerged as good candidates for an eventual solution of the QCD sign problem. They are the Complex Langevin equation (CLE)[2] and the Lefschetz thimble method [3]. Both are based on the complexification of the field manifolds, with the CLE expanding the theory to this manifold, while the thimble method shifting the original manifold into the complex plane while keeping the dimension of integrations fixed.

The Complex Langevin method was invented more than thirty years ago as a generalization of the stochastic quantization [4] (which uses the real Langevin equation), but it has had many practical problems over the years as well as unsatisfactory theoretical understanding. Recently there has been a definite progress in both grounds: On the theoretical side, there is a theorem which guarantees correct results as long as some conditions are satisfied [5]. Coupled with the gauge cooling procedure [6] this has allowed direct simulations of among others full QCD with light quarks [7, 8], at least in the high temperature regime.

The CLE is briefly introduced in section 2 and some results are presented in section 3.

Content from this work may be used under the terms of the Creative Commons Attribution 3.0 licence. Any further distribution of this work must maintain attribution to the author(s) and the title of the work, journal citation and DOI.
Published under licence by IOP Publishing Ltd

2. The complex Langevin equation

The complex Langevin equation for scalar field x and action $S(x)$ is written as

$$\partial_\tau x = -\partial_x S(x) + \eta(\tau) \tag{2}$$

with the Gaussian noise $\eta(\tau)$ satisfing $\langle \eta(\tau)\eta(\tau')\rangle = \delta(\tau - \tau')$. For a complex action the fields are complexified, and the original theory is supposed to be recovered by analytical continuation

$$\int dx P(x) O(x) = \int dx dy P(x,y) O(x+iy) \tag{3}$$

where $P(x) = \exp(-S(x))$ is the (complex) measure on the real axis, $P(x,y)$ is the real probability distribution of the complexified scalar field constructed by solving the complex Langevin equation. In practice this means that one calculates the average by collecting data along the solution of the CLE for the analytically continued observables (after thermalization). For holomorphic actions the correctness of the results can be assessed by observing the decay of $P(x,y)$ at large x and y [5], where wrong convergence is signalled by slow (power law) decay.

For gauge theories the discretised update is written as

$$U_{x,\nu}(\tau+\epsilon) = \exp\left[i \sum_a \lambda_a (-\epsilon D_{ax\nu} S[U] + \sqrt{\epsilon}\eta_{ax\nu}) \right] U_{x,\nu}(\tau), \tag{4}$$

where λ_a are the generators of the gauge group, i.e. the Gell-Mann matrices. The drift force is calculated from the action $S[U]$ with the left derivative

$$D_{ax\nu} f(U) = \partial_\alpha f(e^{i\alpha\lambda_a} U_{x,\nu})\Big|_{\alpha=0}. \tag{5}$$

Here a complex action leads to a complex drift force, and thus the original SU(3) manifold of the link variables expands to the non-compact SL(3,C), i.e. they are no longer unitary. Since the gauge symmetry also complexifies, each configuration has a gauge orbit with infinite volume. In practice this leads to the breakdown of naive CLE simulations of gauge theories. This behaviour is cured by gauge cooling [6], which uses the imaginary directions of the gauge symmetry to get the configurations closer to the original manifold, and thus keeping the system from trying to explore the non-compact directions of SL(3,C) connected to gauge symmetry.

In the case of fermionic theories, an additional problem can appear, as the measure of the theory is written as $e^{-S_g[U]}\det M(U,\mu)$. This measure can have zeroes on the complexified manifold, which is reflected by a pole in the drift term. If the distribution around the singularities is non-vanishing, incorrect results can be obtained, as first noticed in [9]. A zero of the measure outside the support of the complexified distribution can not invalidate the results, but in contrast, it can lead to incorrect results even in the cases where fast decay is observed, if the process gets close to the singularities [10]. Recently it has been proposed that a variation of the gauge cooling algorithm might be able to modify the process such that the zeros of the measure are avoided [11].

3. Numerical Results

Heavy dense QCD (HDQCD) is defined by dropping the spatial fermionic hoppings from $M(U,\mu)$, which thus corresponds to static fermionic particles. This can also be understood as the simultaneous limit $m, \mu \to \infty$ with the bare quark mass m, while the combination κe^μ is kept fixed with $\kappa = 1/(4m)$. HDQCD shares common features with QCD, while it is much simpler, in particular the fermionic determinant simplifies to a product of spatially local determinants. It has been studied with the CLE [6] as well as reweighting [12]. In the CLE treatment the gauge cooling stabilizes simulations as long as the lattice spacing is not too large (corresponding to a lower limit of the β parameter of the plaquette action). This allows to map out the phase diagram by fixing β to some fixed value, and changing the temporal extent of the lattice to perform a temperature scan. The chemical potential can be freely changed, the CLE method works fine also in the cases where the sign problem is hard. The Polyakov loop and the fermionic density is plotted in Fig. 1. The onset transition at high chemical potentials is nicely indicated by the fermionic density. One also observes that the transition gets sharper at lower densities, as required by the Silver-Blaze phenomenon. At higher chemical potentials, the fermionic density saturates, as all the available fermionic states on the lattice are filled. The deconfinement transition at zero chemical potential is shown by the nonzero value of the Polyakov loop, as well as the onset transition at high chemical potentials. The decay at large chemical potential is the consequence of the lattice artifact of saturation, therefore it has no relevance for continuum physics. So in this theory the onset transition and the deconfinement seem to be connected. The order of the transitions can be determined with e. g. finite size scaling [13]. The phase diagram of HDQCD is thus very simple compared to the expected phase diagram of full QCD, this is because quarks can't move spatially in this theory.

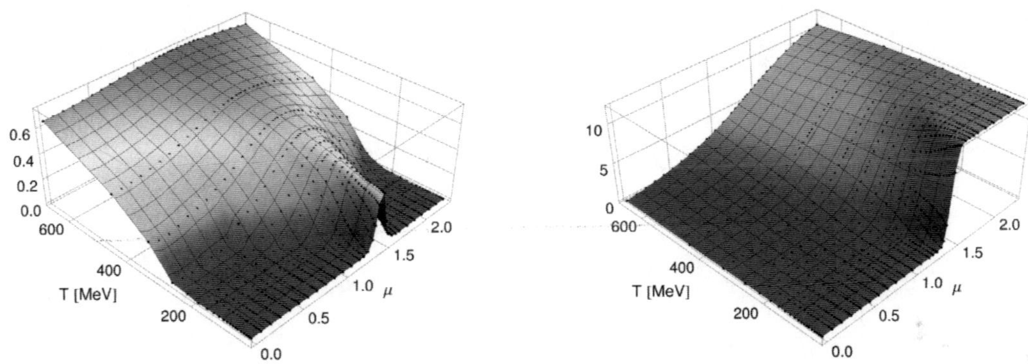

Figure 1. The Polyakov loop(left) and the fermionic density(right) as a function of the temperature and the chemical potential (in lattice units) in HDQCD at $\beta = 5.8$ and $\kappa = 0.12$.

In full QCD the fermionic contribution to the drift term is written as

$$K^F_{ax\nu} = \frac{N_F}{4}\mathrm{Tr}[M^{-1}(\mu, U)D_{ax\nu}M(\mu, U)]. \tag{6}$$

As this involves the inverse of the typically very large matrix $M(\mu, U)$, a stochastic estimator is used to calculate it using the Conjugate Gradient (CG) algorithm [7]. Here we report on simulations using the plaquette action and unimproved staggered fermions. Using gauge cooling this procedure allows simulations at not too small β values. The main cost of the simulations is the inversion of fermionic matrix with CG, similarly to Hybrid Monte Carlo (HMC) simulations. In Fig. 2 the behavior of various observables as a function of the chemical potential is shown at some fixed temperature in the deconfined phase. One observes the expected behavior with the increase and eventual saturation of the fermionic density, also signalled by the Polyakov loops and its inverse as well as the chiral condensate.

We have performed comparisons between the CLE and the multiparameter reweighting method [14], where one rewrites the measure using a positive ensemble e.g. using zero chemical potential [15]. This allows to use HMC simulations, however overlap and sign problems might

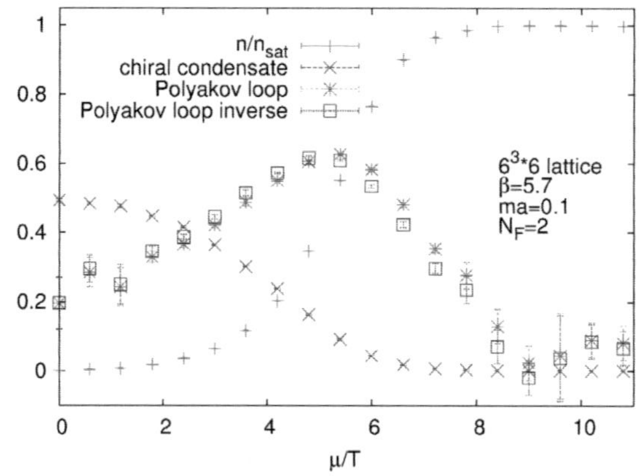

Figure 2. Various observables measured in full QCD with 2 flavours of staggered fermions as a function the chemical potential at a fixed temperatue. The quantities are plotted in lattice units.

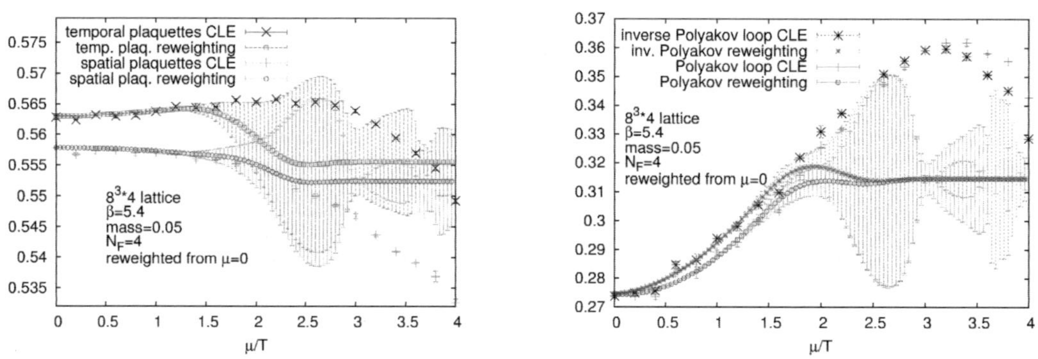

Figure 3. The spatial plaquettes (left) and the Polyakov loop (right) measured using the CLE and reweighting as a function of the chemical potential.

prevent extracting usable results if μ is large. In Fig. 3 we show the spatial plaquettes and the Polyakov loop at a fixed temperature in the deconfined phase as a function of the chemical potential. One notes that the reweighting starts to develop large errors around $\mu = 1.5T$ signalling unreliable results, while the complex Langevin results follow the expected behavior. The correctness of CLE can also be independently assessed by looking at distributions of observables, where a slow (power law) decay indicates wrong results.

In Fig. 4 the Polyakov loop is shown as a function of the β parameter of the action. We have increased the lattice size to $N_T = 6$ and 8 while keeping the pion mass over the temperature approximately fixed, and also exploring a system with larger pion mass at $N_T = 8$. As expected, at low β, the CLE simulations break down, similarly to HDQCD, as the gauge cooling can not keep the process close to the SU(3) manifold. In HDQCD, increasing the lattice size allowed to go into the confined phase already at $N_T = 8$. In contrast, full-QCD results show that $N_T = 8$ is not sufficient to reach into the confined phase. It is an open question whether a further increase in the lattice size alone will allow for exploration of the confined phase in full QCD.

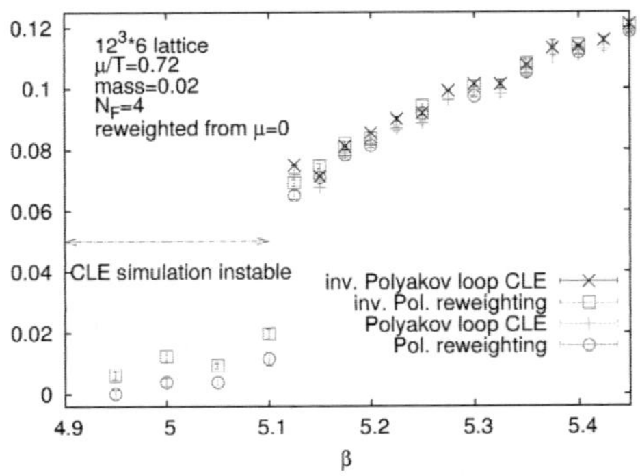

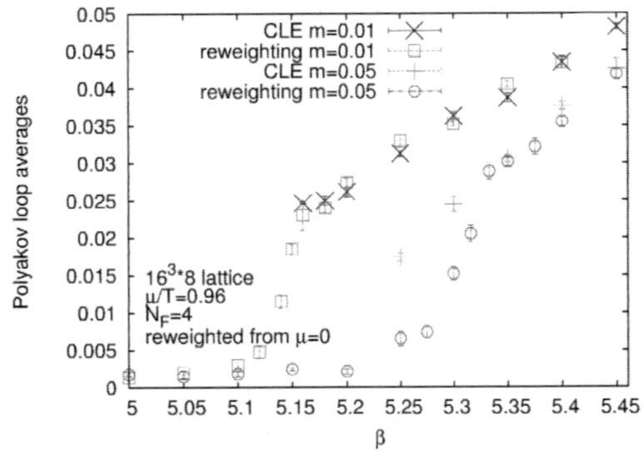

Figure 4. The Polyakov loop on $12^3 \times 6$ (left) and $16^3 \times 8$ (right) lattices measured using the CLE and reweighting as a function of the chemical potential.

Acknowledgments
I would like to thank my collaborators G. Aarts, F. Attanasio, Z. Fodor, B. Jäger, S.D. Katz, E. Seiler, I.-O. Stamatescu, Cs. Török and the Gauss Centre for Supercomputing e.V. (www.gauss-centre.eu) for providing computing time on the GCS Supercomputer SuperMUC at Leibniz Supercomputing Centre (LRZ, www.lrz.de), as well as for providing computing time through the John von Neumann Institute for Computing (NIC) on the GCS share of the supercomputer JURECA and JUQUEEN [16] at Jülich Supercomputing Centre (JSC).

References
[1] P. de Forcrand, PoS LAT **2009**, 010 (2009) [arXiv:1005.0539]; G. Aarts, PoS LATTICE **2012** (2012) 017 [arXiv:1302.3028].
[2] G. Parisi, Phys. Lett. **131 B** (1983) 393; J. Klauder, Acta Phys. Austr. Suppl. **XXXV** (1983) 251.
[3] F. Pham, Proc. Symp. Pure Math. **40**, 319 (1983); E. Witten, arXiv:1001.2933; M. Cristoforetti, F. Di Renzo and L. Scorzato, Phys. Rev. /bf D86, 074506 (2012), arXiv:1205.3996
[4] P. H. Damgaard and H. Huffel, Phys. Rept. **152** (1987) 227. doi:10.1016/0370-1573(87)90144-X
[5] G. Aarts, E. Seiler and I. -O. Stamatescu, Phys. Rev. D **81** (2010) 054508 [arXiv:0912.3360]; G. Aarts, F. A. James, E. Seiler and I. -O. Stamatescu, Eur. Phys. J. C **71** (2011) 1756 [arXiv:1101.3270].
[6] E. Seiler, D. Sexty and I. -O. Stamatescu, Phys. Lett. B **723**, 213 (2013) [arXiv:1211.3709 [hep-lat]]; G. Aarts, L. Bongiovanni, E. Seiler, D. Sexty and I. -O. Stamatescu, Eur. Phys. J. A **49** (2013) 89 doi:10.1140/epja/i2013-13089-4 [arXiv:1303.6425 [hep-lat]].
[7] D. Sexty, Phys. Lett. B **729** (2014) 108 [arXiv:1307.7748 [hep-lat]];
[8] G. Aarts, E. Seiler, D. Sexty and I. O. Stamatescu, Phys. Rev. D **90** (2014) 11, 114505; D. Sexty, PoS LATTICE **2014** (2014) 016 [arXiv:1410.8813 [hep-lat]].
[9] A. Mollgaard and K. Splittorff, Phys. Rev. D **88** (2013) 116007 [arXiv:1309.4335 [hep-lat]]; J. Greensite, Phys. Rev. D **90** (2014) no.11, 114507 [arXiv:1406.4558 [hep-lat]].
[10] G. Aarts, E. Seiler, D. Sexty, I.-O. Stamatescu, in prep.
[11] K. Nagata, J. Nishimura and S. Shimasaki, arXiv:1604.07717 [hep-lat].
[12] R. De Pietri, A. Feo, E. Seiler and I. O. Stamatescu, Phys. Rev. D **76** (2007) 114501 doi:10.1103/PhysRevD.76.114501 [arXiv:0705.3420 [hep-lat]].
[13] G. Aarts, F. Attanasio, B. Jäger, E. Seiler, D. Sexty and I. O. Stamatescu, PoS LATTICE **2014** (2014) 200 [arXiv:1411.2632 [hep-lat]]; G. Aarts, F. Attanasio, B. Jäger, E. Seiler, D. Sexty and I.-O. Stamatescu, in prep.
[14] Z. Fodor, S. D. Katz, D. Sexty and C. Török, Phys. Rev. D **92** (2015) no.9, 094516 doi:10.1103/PhysRevD.92.094516 [arXiv:1508.05260 [hep-lat]].
[15] Z. Fodor and S. D. Katz, Phys. Lett. B **534** (2002) 87 doi:10.1016/S0370-2693(02)01583-6 [hep-lat/0104001].
[16] Jülich Supercomputing Centre. (2015). JUQUEEN: IBM Blue Gene/Q Supercomputer System at the Jülich Supercomputing Centre. Journal of large-scale research facilities, 1, A1. http://dx.doi.org/10.17815/jlsrf-1-18

Journal of Physics: Conference Series **742** (2016) 012012 doi:10.1088/1742-6596/742/1/012012

Meson form factors and $P \to \gamma^*\gamma^*$ physics at BESIII

J Pettersson, For the BESIII Collaboration

Department of Physics and Astronomy, Uppsala University, Box 516, Sweden

E-mail: `joachim.pettersson@physics.uu.se`

Abstract. Using data consisting of 2.93 fb^{-1} integrated luminosity at the centre of mass energy $\sqrt{s}$ = 3773 MeV, recorded with the BESIII detector at BEPCII, the cross section and form factor $|F_\pi|^2$ was extracted in the energy range 600-900 MeV, using the method of radiative return. The cross section is used as input to calculate the leading-order vacuum polarisation contribution of the $e^+e^- \to \pi^+\pi^-$ channel to (g - 2)$_\mu$ as $a_\mu^{\pi\pi,\text{LO}}(600 - 900 \text{ MeV}) = (368.2 \pm 2.5_{\text{stat}} \pm 3.3_{\text{sys}}) \cdot 10^{-10}$. This result is compatible with corresponding values using KLOE data, but disagrees whit BaBar. The ongoing search for $e^+e^- \to \eta_c$ at BESIII is also discussed.

1. Introduction

Measurements of the process $e^+e^- \to \pi^+\pi^-$, and processes containing a light pseudoscalar meson interacting with two photons ($P \to \gamma^*\gamma^*$) are important as experimental input to Standard Model (SM) calculations of the anomalous magnetic moment of the muon a_μ. The anomalous magnetic moment of the muon has been measured to great precision by the E821 experiment at Brookhaven [1]. At present time, a deviation of 3.6σ [2] is reported between theoretical calculations and experimental measurement of a_μ. The theoretical calculation of a_μ^{theo}, is broken up into several parts $a_\mu^{theo} = a_\mu^{QED} + a_\mu^{weak} + a_\mu^{QCD}$, where the QCD part can be further broken down as $a_\mu^{QCD} = a_\mu^{VP,LO} + a_\mu^{VP,HO} + a_\mu^{LbL}$. The largest uncertainties in the calculations arises from experimental input to the leading-order (LO) hadronic vacuum polarisation ($a_\mu^{VP,LO}$) and second order light-by-light scattering (a_μ^{LbL}) contributions. The contributions to $a_\mu^{VP,LO}$ can be further broken down to the individual channels as $a_\mu^{VP,LO} = a_\mu^{\pi\pi} + a_\mu^{\pi\pi\pi} + a_\mu^{KK}$.... As can be seen in Fig. 1 the process $e^+e^- \to \pi^+\pi^-$ dominates the dispersion integral for $a_\mu^{VP,LO}$ and high precision measurements are therefore of great importance.

The $P\gamma^*\gamma^*$ vertex which appears in diagrams contributing to a_μ^{LbL}, Fig. 2, also appears in the process $\pi^0 \to e^+e^-$, see Fig. 3. Measurements of this process shows a discrepancy between theory and experiment; for the branching fraction a deviation of $2 - 3\sigma$ has been observed compared to theoretical calculations within SM [3, 4]. It has been proposed that pseudoscalar mesons decaying into lepton-antilepton pairs can provide a signal to physics beyond the Standard Model [3, 5, 6]. Within the Standard Model $P \to l^+l^-$ proceeds via a two-photon intermediate state. Therefore it is a fourth order electromagnetic process, and thus suppressed. Determinations of upper limits has previously been performed for π^0, η and η' decaying to lepton-antilepton pairs [7, 8, 9, 10]. However, until now, no attempt has been made to measure $\eta_c \to e^+e^-$.

Content from this work may be used under the terms of the Creative Commons Attribution 3.0 licence. Any further distribution of this work must maintain attribution to the author(s) and the title of the work, journal citation and DOI.
Published under licence by IOP Publishing Ltd

Journal of Physics: Conference Series **742** (2016) 012012 doi:10.1088/1742-6596/742/1/012012

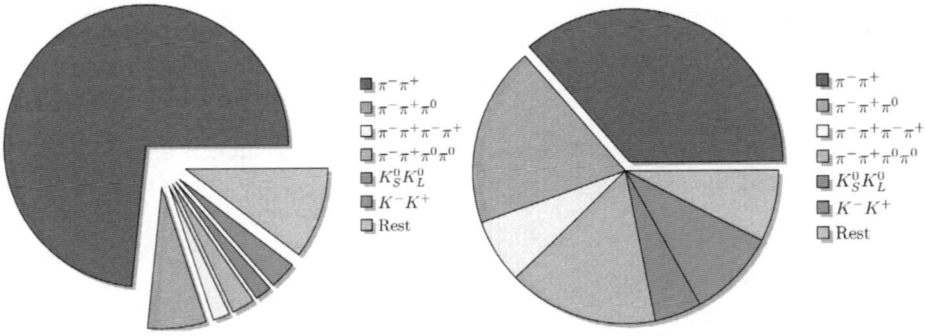

Figure 1. The left panel shows the relative contributions of the most important channels to $a_\mu^{VP,LO}$, the right panel shows the relative uncertainty in each channel [2] .

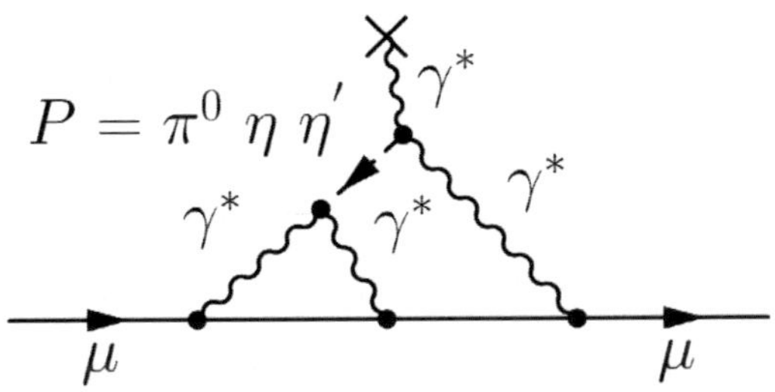

Figure 2. Light-by-Light scattering diagram describing the role of the intermediate pseudoscalar mesons.

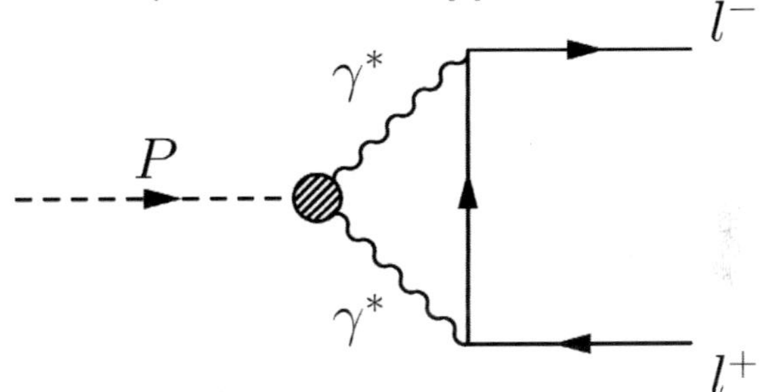

Figure 3. A pseudoscalar particle decaying into two leptons via two virtual photons.

2. The BESIII Experiment

The Beijing Electromagnetic Spectrometer III (BESIII) is a multipurpose detector located at the Beijing Electron-Positron Collider (BEPC-II), a double-ring electron-positron collider optimized for the charmonium region [11]. BESIII covers 93% of the 4π solid angle and consists of several sub-systems that provides tracking, calorimetry, particle identification, and muon detection. The main subsystems are (a) the Multilayer Drift Chamber (MDC) which contains helium gas and have 43 layers of wiring providing a 135 μm spatial resolution, 0.5% momentum resolution for charged tracks at 1 GeV/c and dE/dx resolution of < 6% in a 1 T magnetic field, provided by a superconducting solenoid. (b) A Time of flight system consisting of 176 plastic scintillators in the barrel and 96 in each endcap. Providing a 80 ps and 110 ps time resolution respectively, which gives a 2σ K/π discrimination at energies of 1 GeV/c. (c) A CsI(Tl) Electro-Magnetic Calorimeter (EMC) with a 2.5 % energy resolution in the barrel and 5% in the endcaps at energies of 1 GeV/c. (d) The Muon Chamber (MUC) consists of resistive plate chambers in nine layers in the barrel and eight in the endcaps, with a position resolution of 2 cm.

3. The $e^+e^- \to \pi^+\pi^-$ process

3.1. Measurement of $e^+e^- \to \pi^+\pi^-$

Data containing 2.93 fb^{-1} integrated luminosity, recorded at the centre of momentum energy of $\sqrt{s} = 3773$ MeV was analysed using the method of radiative return. The method of radiative return exploits Initial State Radiation (ISR) to obtain cross section measurements for a wide range of energies. ISR is when one or more photons are radiated by the initial e^+e^- pair prior to annihilation. The radiated photon(s) effectively lowers the available centrer-of-mass energy of the recoiling e^+e^- pair. High energy γ_{ISR} are likely to have a momentum pointing along the beam axis, subsequently giving the $\pi\pi$ pair a strong boost in the opposite direction, thus both the γ_{ISR} and the $\pi\pi$ are likely to escape detection. Therefore only events where the γ_{ISR} is detected in the EMC are used. The signature for the signal is $e^+e^- \to \pi^+\pi^-\gamma_{ISR}$.

The main background consists of $e^+e^- \to \mu^+\mu^-\gamma$. Using an Artificial Neural Network (ANN) from the TMVA package [12] this background was suppressed . The ANN was trained using large Monte Carlo samples containing signal $\pi^+\pi^-\gamma$ and background $\mu^+\mu^-\gamma$ events. The Monte Carlo samples were generated with the Phokhara event generator [13, 14] and were then propagated through a detector simulation based on the Geant-4 software [15, 16].

3.2. Determination of $\sigma_{e^+e^-\to\pi^+\pi^-}$ and extraction of $|F_\pi^2|$

To calculate the $a_\mu^{\pi\pi}$ contribution to $a_\mu^{VP,LO}$ the bare cross section corrected for Final State Radiation (FSR) is needed. The bare cross section is the observed cross section corrected for vacuum polarization effects. It is determined using two different normalization schemes. In the first scheme the bare cross section is extracted by correcting for vacuum polarization effects directly as

$$\sigma^{bare}_{\pi\pi(\gamma_{FSR})} = \frac{N_{\pi\pi\gamma}\cdot(1+\delta^{\pi\pi}_{FSR})}{\mathcal{L}\cdot\epsilon^{\pi\pi\gamma}_{global}\cdot H(s)\cdot\delta_{vac}}, \tag{1}$$

where $\mathcal{L}$ is the luminosity, $\epsilon^{\pi\pi\gamma}_{global}$ is the global efficiency of measuring a signal event, $H(s)$ is the ISR radiator function, δ_{vac} is the vacuum polarization correction and $(1+\delta^{\pi\pi}_{FSR})$ is the FSR correction factor. The vacuum polarization correction is obtained from the Phokhara generator [13, 14]. In the second method the number of signal events $N_{\pi\pi\gamma}$ is normalized using the measured $N_{\mu\mu\gamma}$, thus cancelling the vacuum polarization effects. Both methods agree within errors. The detection efficiency is slightly dependent on the invariant mass of the $\pi\pi$ system. To account for the energy dependence of the detector resolution, a singular value decomposition [17] was performed on the mass spectrum. The FSR correction is obtained from Phokhara [13, 14], by first simulating the spectrum including FRS effects to next-to-leading order (NLO). The same spectrum was then simulated again without the FSR effect. The two spectra were the divided to obtain the correction factor. As suggested in [18], the form factor was extracted from the cross section via

$$|F_\pi|^2(s') = \frac{3s'}{\pi\alpha^2\beta_\pi^3(s')}\sigma^{\text{dressed}}_{\pi\pi}(s')\ , \tag{2}$$

where, α is the fine-structure constant, $\beta_\pi(s')$ is the velocity of the charged pions and $\sigma^{\text{dressed}}_{\pi\pi}(s')$ is the cross section corrected for FSR but not for vacuum polarization.

3.3. Results

The fit of the extracted form factor was performed using a Gounaris-Sakurai parametrisation [18]. The cross section and form factor as a function of $\sqrt{s'}$ can be seen in Fig. 4. The relative difference between the form factor squared from BaBar [19] and KLOE [20, 21, 22] compared to the BESIII [23] fit is shown in Fig. 5. Deviations for $\sqrt{s}$ above the mass of ρ meson can be seen between BESIII and both BaBar and KLOE data, below the $\rho-\omega$ interference region the BESIII fit agrees with the KLOE data, where as a shift can be seen between the BESIII fit and the BaBar data.

To calculate the value of the contribution of the $\pi\pi$ channel in the invariant mass range 600-900 MeV/c range to $a_\mu^{\pi\pi,\text{LO}}$ the bare cross section is used as input to the integral

$$a_\mu^{\pi\pi,\text{LO}}(0.6-0.9\,\text{GeV}) = \frac{1}{4\pi^3}\int_{(0.6\text{GeV})^2}^{(0.9\text{GeV})^2} \text{ds}'\text{K}(\text{s}')\sigma^{\text{bare}}(\text{e}^+\text{e}^-\to\pi^+\pi^-(\gamma_{\text{FSR}}))\ , \tag{3}$$

Journal of Physics: Conference Series **742** (2016) 012012 doi:10.1088/1742-6596/742/1/012012

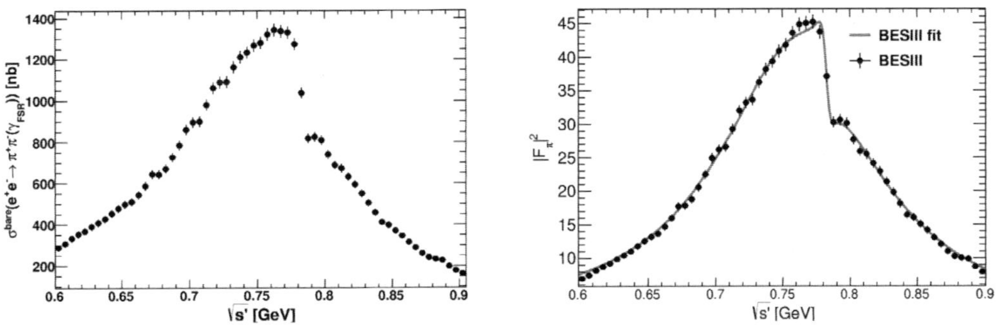

Figure 4. The left figure shows tha bare cross section with statistical errors. In the right figure the red line shows the fit of the form factor squared, using a Gounaris-Sakurai parametrization. The black dots are the measured form factor squared, with statistical errors [23].

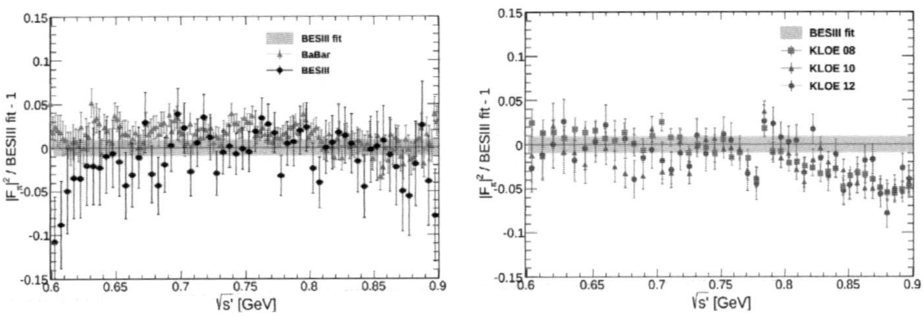

Figure 5. The left and right panel shows the relative difference between the form factor squared from BaBar [19] and KLOE [20, 21, 22], respectively, to the BESII [23] fit. For the BaBar and KLOE data points both systematic and statistical errors are included, for the BESIII fit the width of the band corresponds to the statistical error [23].

where $K(s')$ is the kernel function [24, Eq. (5)]. The integral in Eq. 3 was calculated to $a_\mu^{\pi\pi,\mathrm{LO}}(600 - 900\ \mathrm{MeV}) = (368.2 \pm 2.5_{\mathrm{stat}} \pm 3.3_{\mathrm{sys}}) \cdot 10^{-10}$. This result is compatible with corresponding values using KLOE data, but disagrees whit BaBar as can be seen in Fig. 6. The BESIII results were recently published in [23].

4. Search for $\eta_c \to e^+e^-$

BESIII have large sets of data collected around the J/Ψ peak. This data could be used to look for $\eta_c \to l^+l^-$. A possible signature for the decay would be $e^+e^- \to \gamma l^+l^-$, where $e^+e^- \to J/\Psi$ which further decays $J/\Psi \to \gamma\eta_c$ where $\eta_c \to l^+l^-$. One could in principle obtain 10^7 η_c candidates. But due to the nature of the final state, large background from electromagnetic continuum processes with the same $e^+e^- \to \gamma l^+l^-$ signature is expected. Therefore a dedicated data sample has been collected in order to study the reverse process $e^+e^- \to \eta_c$, as suggested in [25]. Using the reverse process, one also has the opportunity of choosing the final state(s) to study. Choosing a C-even final state suppresses first order C-odd background processes from the electromagnetic continuum. The dedicated data sample comprise $\approx 60\ \mathrm{pb}^{-1}$ in four points: 2950, 2981, 3000 and 3020 MeV, placing BESIII in a unique position for the search of $e^+e^- \to \eta_c$. The results will be reported during 2016.

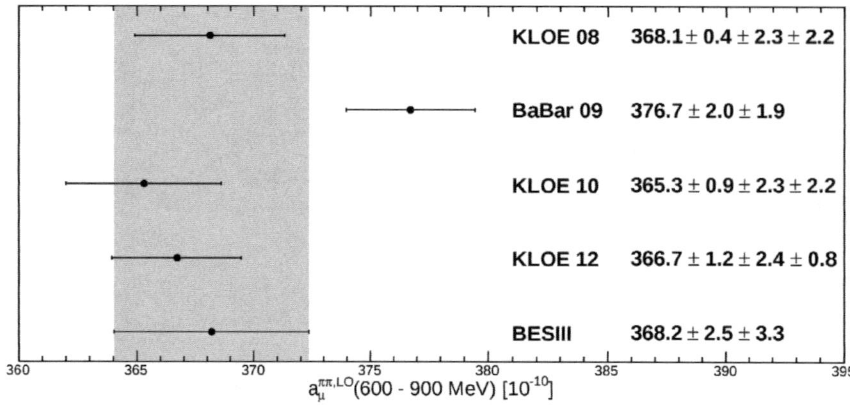

Figure 6. The recent BESIII result for the leading-order vacuum polarization $\pi\pi$ contribution compared to previous results from BaBar and KLOE with statistical and systematic errors. The coloured band shows the 1σ region of the BESIII result [23].

5. Conclusion

Precise measurements of the $\sigma^{\text{bare}}(e^+e^- \to \pi^+\pi^-(\gamma_{FSR}))$ cross section and the charged pion form factor has been performed at BESIII using the method of radiative return. The charged pion contribution to (g - 2)$_\mu$ was determined: $a_\mu^{\pi\pi,\text{LO}}(600-900\text{ MeV}) = (368.2\pm2.5_{\text{stat}}\pm3.3_{\text{sys}})\cdot10^{-10}$. The new BESIII results are in agreement with previous KLOE measurements but in disagreement with BaBar. The search for $e^+e^- \to \eta_c$ is ongoing, using data recently collected at BESIII.

References

[1] G. W. Bennett et al. *Phys. Rev.*, **D73**:072003, 2006.
[2] M. Davier, A. Hoecker, B. Malaescu, and Z. Zhang. *Eur. Phys. J.*, **C71**:1515, 2011.
[3] A. E. Dorokhov. *Phys. Part. Nucl. Lett.*, **7**:229, 2010.
[4] T. Husek, K.l Kampf, and J. Novotny. *Eur. Phys. J.*, **C74**(8):3010, 2014.
[5] L. Bergstrom. *Z.Phys.*, **C14**:129, 1982.
[6] T. Petri. PhD thesis, Julich, Forschungszentrum, 2010.
[7] R. R. Akhmetshin et al. *Phys. Lett.*, **B740**:273, 2015.
[8] E. Abouzaid et al. *Phys. Rev.*, **D75**:012004, 2007.
[9] G. Agakishiev et al. *Phys. Lett.*, **B731**:265, 2014.
[10] M. N. Achasov et al. *Phys. Rev.*, **D91**:092010, 2015.
[11] M. Ablikim et al. *Nucl. Instrum. Meth.*, **A614**:345, 2010.
[12] A. Hocker et al. *PoS*, **ACAT**:040, 2007.
[13] G. Rodrigo, H. Czyz, J. H. Kuhn, and M. Szopa. *Eur. Phys. J.*, **C24**:71, 2002.
[14] H.k Czyz, J. H. Kuhn, and A. Wapienik. *Phys. Rev.*, **D77**:114005, 2008.
[15] J. Allison et al. *IEEE Trans. Nucl. Sci.*, **53**:270, 2006.
[16] S. Agostinelli et al. *Nucl. Instrum. Meth.*, **A506**:250, 2003.
[17] A. Hocker and V. Kartvelishvili. *Nucl. Instrum. Meth.*, **A372**:469, 1996.
[18] G. J. Gounaris and J. J. Sakurai. *Phys. Rev. Lett.*, **21**:244, 1968.
[19] J. P. Lees et al. *Phys. Rev.*, **D86**:032013, 2012.
[20] D. Babusci et al. *Phys. Lett.*, **B720**:336, 2013.
[21] F. Ambrosino et al. *Phys. Lett.*, **B670**:285, 2009.
[22] F. Ambrosino et al. *Phys. Lett.*, **B700**:102, 2011.
[23] M. Ablikim et al. *Phys. Lett.*, **B753**:629, 2016.
[24] S. Eidelman and F. Jegerlehner. *Z. Phys.*, **C67**:585, 1995.
[25] N. Kivel and A. Kupsc. arXiv:1602.07460, 2016.

Journal of Physics: Conference Series **742** (2016) 012013 doi:10.1088/1742-6596/742/1/012013

Ground-state studies of charmonium via radiative transitions at BESIII

Zahra Haddadi for the BESIII collaboration

KVI-CART, University of Groningen, Zernikelaan 25, 9747 AA Groningen, The Netherlands.

E-mail: z.haddadi@rug.nl

Abstract. Charmonium states consist of a heavy charm-anticharm quark pair ($c\bar{c}$), bound by the strong interaction. Theoretically, charmonium can be analyzed based on a non-relativistic framework, since the motion of the charm quark inside the bound state is $v^2 \sim 0.3$, where v is relative velocity between the c and $\bar{c}$ with the additional of relativistic corrections. All the narrow charmonium states below the open-charm threshold have been experimentally identified and their mass spectrum can be reasonably well described by potential models that incorporate a color Coulomb term at short distances and a linear scalar confining term at large distances. Although all charmonium states below the $D\bar{D}$ mass threshold have been observed, knowledge is sparse on spin-singlet S-waves, $\eta_c(1S)$ and $\eta_c(2S)$. The BESIII/BEPCII facility in Beijing, China, has shed light on these spin-singlet states by collecting a new record of $\psi(3686)$ decays in electron-positron annihilations.

1. Introduction

The η_c state of charmonium, occupies a special place in the study of heavy quarkonia. It is the only ground state of a heavy quarkonium system which has been experimentally identified and it is the only confirmed heavy quarkonium singlet state. η_c cannot be formed directly in e^+e^- annihilation, but it can be indirectly production via radiative magnetic-dipole (M1) transitions. Radiative transitions in particular have recently been the subject of both lattice QCD calculations [1] and effective field theory techniques [2]. Among these are the most poorly understood M1 transitions $J/\psi \to \gamma\eta_c$ and $\psi(3686) \to \gamma\eta_c$. These two transitions are different ways to populate the ground state of charmonium below the open-charm threshold. The experimental challenge lies in the relatively small branching fraction involved in these M1 transitions since the process involves a charm-quark spin flip. Not only are precision measurements needed to validate our theoretical understanding, precise measurements of these M1 transitions are critical for normalizing η_c branching fractions, a key input to extracting other properties such as partial width of the η_c to two photons. The J/ψ and $\psi(3686) \to \gamma\eta_c$ transitions are also a source of information on the η_c mass and width. There is currently a 3.3σ inconsistency between previously measured J/ψ and $\psi(3686) \to \gamma\eta_c$ and $\gamma\gamma$ or $p\bar{p}$ production [3].

2. Spin singlet states: $\eta_c(1S)$, $\eta_c(2S)$

The spin-spin interaction term in potential model causes the splitting between η_c and J/ψ and the $\eta_c(2S)$ and $\psi(3686)$ as well, which are called the 1S and 2S hyperfine splitting, respectively. The mass of $\eta_c(1S)$ and $\eta_c(2S)$ is of particular interest since the hyperfine splittings provide

Content from this work may be used under the terms of the Creative Commons Attribution 3.0 licence. Any further distribution of this work must maintain attribution to the author(s) and the title of the work, journal citation and DOI.
Published under licence by IOP Publishing Ltd

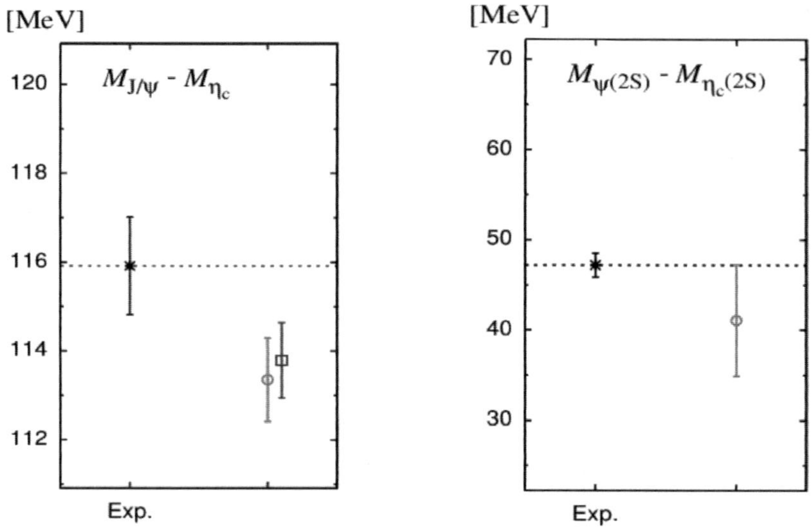

Figure 1. Mass splittings of states lying below the open-charm threshold in units of MeV. Left panel shows the mass splitting between $1S$ states and right panel between $2S$ states. In each plot, cross and circle symbols indicate the experimental data and theoretical results obtained from the non-relativistic potential model with the lattice inputs, respectively. The results of the lattice spectroscopy is shown with a square symbol [4].

valuable information about the underlying $c\bar{c}$ confinement potential. Figure 1 shows one of the most recent comparison between lattice QCD and experimental results of hyperfine mass splitting $M_{J/\psi} - M_{\eta_c(1S)}$ and $M_{\psi(3686)} - M_{\eta_c(2S)}$. In each plot, cross and circle symbols indicate the experimental data and theoretical results obtained from a non-relativistic potential model with input from lattice calculations, respectively [4]. The quoted errors indicate the sum of the statistical and systematic errors added in quadrature. Dashed lines represent the central value of the experiment. The square symbol shows the prediction of a lattice calculation for the 1S hyperfine splitting.
The mass and the width of the lowest lying charmonium state $\eta_c(1S)$, continue to have large uncertainties when compared to those other charmonium states [5]. The comparison between different experimental setups for the mass and width of η_c is shown in Figure 2. All the measurements are classified in two different groups, radiative transitions (red circles) and other process, $\gamma\gamma$ or $p\bar{p}$ production (blue circles). The data are not internally consistent considering the large spread and their uncertainties. Part of the discrepancy could be due to the interpretation of the line shape of the reconstructed $\eta_c(1S)$ which were in many cases based on a simple Breit Wigner distribution due to limited statistics in the experiment.

3. Study of $\psi(3686) \to \gamma\eta_c(1S)$

Recently, the analysis of M1 radiative transition of $\psi(3686) \to \gamma\eta_c(1S)$ was studied in BESIII based on a 106×10^6 $\psi(3686)$ data sample. The $\eta_c(1S)$ mass and width are determined from fits to the invariant mass spectra of 6 exclusive decay modes of $\eta_c(1S)$: $K_S K^\pm \pi^\mp$, $K^\pm K^\mp \pi^0$, $\eta\pi^\pm\pi^\mp$, $K_S K^\pm \pi^\mp \pi^\pm \pi^\mp$, $K^\pm K^\mp \pi^\pm \pi^\mp \pi^0$ and $3(\pi^\pm\pi^\mp)$, where the K_S is reconstructed in $\pi^+\pi^-$, and the η and π^0 in $\gamma\gamma$ decays. The solid curves in Figure 3 show the results of an unbinned simultaneous maximum likelihood fit with three components: signal, non-resonant background, and a combined background consisting of $\pi^0 X_i$ decays, continuum, and other $\psi(3686)$ decays. The mass and the width of $\eta_c(1S)$ were found to be M $= 2984.3 \pm 0.6 \pm 0.6$ MeV$/c^2$ and

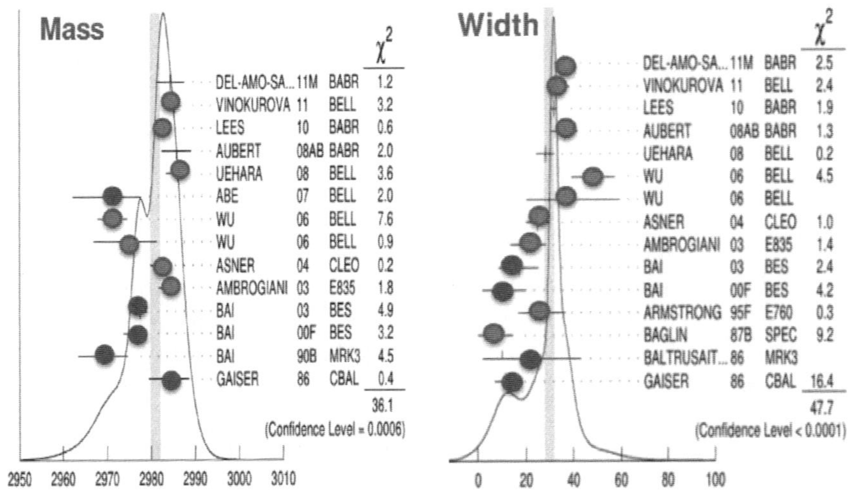

Figure 2. Comparison between different experimental measurements of the mass and the width of the η_c. Red circles indicate the results of radiative transitions and the blue circles $\gamma\gamma$ or $p\bar{p}$ production results.

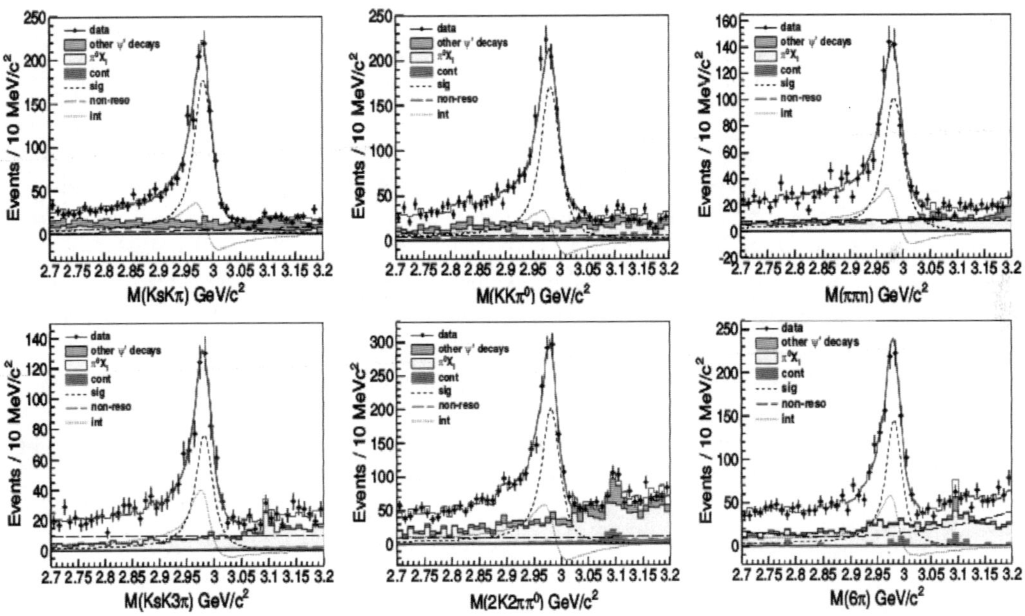

Figure 3. The $M(X_i)$ invariant mass distributions for the 6 exclusive decay modes of $\eta_c(1S)$. Points are data and the various curves are the total fit results. Signals are shown as short-dashed lines, the non-resonant components as long-dashed lines, and the interference between them as dotted lines. Shaded histograms (in red/yellow/green) are for (continuum/$\pi^0 X_i$ /other $\psi(3686)$ decays) backgrounds [6].

$\Gamma = 32.0 \pm 1.2 \pm 1.0$ MeV [6]. In this analysis, the $\eta_c(1S)$ line shapes are described successfully by using a combination of the energy dependence of the hindered-M1 transition matrix element and a full interference with non-resonant $\psi(3686)$ radiative decays. For the first time, an interference

Journal of Physics: Conference Series **742** (2016) 012013 doi:10.1088/1742-6596/742/1/012013

between $\eta_c(1S)$ and non-resonant background has been considered in $\psi(3686) \to \gamma\eta_c(1S)$ analysis. The statistical significance of the interference was found to be 15σ. This result confirms the previous CLEO-c observation of the line shape distortion [7], and may partly clarify the discrepancy found between older experiments, since the interference significantly affects the $\eta_c(1S)$ mass and width and was not always considered.
There has been a long-standing puzzle on the radiative transition rates of $\psi(3686) \to \gamma\eta_c(1S)$. The predicted partial decay width for $\psi(3686) \to \gamma\eta_c(1S)$ was nearly one order of magnitude larger than the experimental data. Such discrepancies not seem to be trivial and have initiated a lot of theoretical interests in the literature [8]. Inspired by BESIII, it has been proposed [9] to consider the intermediate meson loop (IML) corrections as an unquenched mechanism in the charmonium energy region. Such a mechanism turns out to be important for exclusive transitions especially when the mass of the initial state is close to the open-charm threshold [10]. In Ref. [9], it was shown that the J/ψ exclusive decays would experience relatively smaller open-charm effects than the $\psi(3686)$ since the latter is much closer to the $D\bar{D}$ threshold. In another word, the IML would have more important impact on the $\psi(3686)$ decays, while the J/ψ suffers less. BES collaboration is going to measure the mass and width of $\eta_c(1S)$ and branching fraction of $\psi(3686) \to \gamma\eta_c(1S)$ using four times more statistic.

4. Study of $J/\psi \to \gamma\eta_c(1S)$

The $J/\psi \to \gamma\eta_c(1S)$ decay is a M1 transition in charmonium with the most probable photon energy of about 114 MeV and a fairly large branching fraction of $(1.7 \pm 0.4)\%$ [11]. The Crystal Ball collaboration measured this branching fraction in the inclusive photon spectrum and obtained the value of $(1.27\pm0.36)\%$ [12]. There are a lot of theoretical predictions for this decay rate that are based upon QCD sum rules, lattice QCD calculations and so on, but as a rule they lead to values approximately twice as large as the Crystal Ball result. This discrepancy remained unchanged for more than twenty years. During this period no new measurements of this branching fraction were performed, and the PDG average was based on the single Crystal Ball result. Only in 2009 the CLEO Collaboration published the result of a new measurement [3], in which 12 exclusive decay modes of the $\eta_c(1S)$ were analyzed. The obtained value $B(J/\psi \to \gamma\eta_c(1S)) = (1.98 \pm 0.09 \pm 0.30)\%$ is closer to theoretical predictions. Combining the Crystal Ball and CLEO results, the PDG obtained $B(J/\psi \to \gamma\eta_c(1S)) = (1.7 \pm 0.4)\%$ [1] with a scale factor of 1.6. Recently, a new direct measurement of the partial width of $J/\psi \to \gamma\eta_c(1S)$ decay was performed by KEDR. In Figure 4 this result is compared with the Crystal Ball and CLEO measurements [13] (filled circles) as well as with theoretical predictions (open circles). This decay rate value is significantly higher compared to those experimental results, but is well consistent with the latest lattice QCD prediction [14]: $\Gamma_{\gamma\eta_c}(1S) = (2.64\pm0.11)$ keV. BESIII will be able to do the direct measurement of this decay rate by using the largest data sample of J/ψ. This is one the currenly ongoing analysis at BESIII. The expected precision of the BESIII is indicated in Figure 4 as well.

5. Mass and width of the $\eta_c(2S)$

The decay of $\psi(3686) \to \gamma\eta_c(2S)$ is experimentally challenging due to the low energy of the transition photon of about 50 MeV involved in this process. The branching fraction has been calculated by many authors, with predictions in the range $B(\psi(3686) \to \gamma\eta_c(2S)) = (0.1 - 6.2) \times 10^{-4}$. Experimentally, this transition has been searched for by Crystal Ball and CLEO. No convincing signal was observed in any of these sarches. Recently, the first observation of this M1 transition was done at BESIII through the decay processes $\psi(3686) \to \gamma K_s K^\pm \pi^\mp$ and $\gamma K^+K^-\pi^\circ$ [15]. The branching fraction measurement of the M1 transition $\psi(3686) \to \gamma\eta_c$ of $(6.8 \pm 1.1 \pm 4.5) \times 10^{-4}$ agrees with theoretical calculations and naive estimates based on the $J/\psi \to \gamma\eta_c$ transition [3]. The main systematic limitations to these measurements arise from the

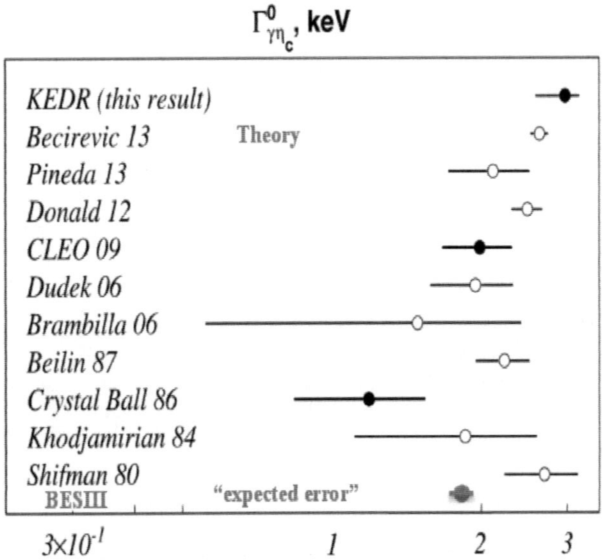

Figure 4. Results of measurements [13] (filled circles) and theoretical predictions (open circles) on $\Gamma_{\gamma\eta_c}$ [14].

choice of the functional form for the damping factor in the η_c line shape and from uncertainty in the choice of the background line shapes. In this analysis, we were able to see the signal with 106×10^6 $\psi(3686)$ data sample. This analysis is ongoing with four times more statistics. With this large data sample of $\psi(3686)$, we will be able to measure the basic properties of $\eta_c(2S)$ more precisely which might help us in understanding the line shape features of the $\eta_c(1S)$ as well.

References

[1] J. J. Dudek, R. G. Edwards, and D. G. Richards, Phys. Rev. D 73, (2006) 074507.
[2] N. Brambilla, Y. Jia, and A. Vairo, Phys. Rev. D 73, (2006) 054005.
[3] R. E. Mitchell et al. [CLEO Collaboration], Phys. Rev. Lett. 102, (2009) 011801 [arXiv:0805.0252 [hep-ex]].
[4] T. Kawanai and S. Sasaki, Phys. Rev. D 92 (2015) no.9, 094503 arXiv:1508.02178 [hep-lat].
[5] K. Nakamura et al. [Particle Data Group], J. Phys. G 37, (2010) 075021.
[6] M. Ablikim et al. (BESIII Collaboration), Phys. Rev. Lett. 108, (2012) 222002.
[7] CLEO Collaboration, R. Mitchell et al., Phys. Rev. Lett.102, (2009) 011801, arXiv:0805.0252.
[8] E. Eichten, K. Gottfried, T. Kinoshita, K. D. Lane and T. M. Yan, Phys. Rev. D 21, (1980) 203.
D. Ebert, R. N. Faustov and V. O. Galkin, Phys. Rev. D 67, (2003) 014027 [arXiv:hep-ph/0210381].
N. Brambilla, Y. Jia and A. Vairo, Phys. Rev. D 73, (2006) 054005 [arXiv:hep-ph/0512369].
N. Brambilla et al. [Quarkonium Working Group], arXiv:hep-ph/0412158.
[9] G. Li and Q. Zhao, Phys. Rev. D 84, (2011) 074005 [arXiv:1107.2037 [hep-ph]].
G. Li and Q. Zhao, Phys. Lett. B 670, (2008) 55.
[10] X. H. Liu and Q. Zhao, Phys. Rev. D 81, (2010) 014017 [arXiv:0912.1508 [hep-ph]].
X. H. Liu and Q. Zhao, J. Phys. G 38, (2011) 035007 [arXiv:1004.0496 [hep-ph]].
F. K. Guo, C. Hanhart, G. Li, U. G. Meissner and Q. Zhao, Phys. Rev. D 83, (2011) 034013 [arXiv:1008.3632 [hep-ph]].
[11] J. Beringer, et al., Phys. Rev. D 86 (2012) 010001.
[12] J. Gaiser, et al., Phys. Rev. D 34 (1986) 711.
[13] J.Z. Bai, et al., Phys. Rev. D 62 (2000) 072001.
J.Z. Bai, et al., Phys. Lett. B 555 (2003) 174.
[14] D. Becirevic, F. Sanfilippo, J. High Energy Phys. 01 (2013) 028.
[15] M. Ablikim et al. [BESIII Collaboration], Phys. Rev. Lett. 109, (2012) 042003.

Journal of Physics: Conference Series **742** (2016) 012014 doi:10.1088/1742-6596/742/1/012014

A study of the resonances $K_0^*(800)$ and $K_0^*(1430)$

Milena Sołtysiak[1,*], **Thomas Wolkanowski**[2] **and Francesco Giacosa**[1,2]

[1]Institute of Physics, Jan Kochanowski University, PL-25406 Kielce, Poland

[2]Institute for Theoretical Physics, Goethe University, D-60438 Frankfurt am Main, Germany

Abstract. We study the scalar kaonic states $K_0^*(800)$ and $K_0^*(1430)$ by using a relativistic QFT Lagrangian in which only a single kaonic field corresponding to the well-established scalar state $K_0^*(1430)$ is considered and in which both derivative and non-derivative interaction terms are taken into account. Even if the scalar spectral function shows a unique peak close to 1.4 GeV, we find two poles in the complex plane: $1.413 \pm 0.002 - i(0.127 \pm 0.003)$ GeV, which is related to the seed quark-antiquark state $K_0^*(1430)$, and $0.746 \pm 0.019 - i(262 \pm 0.014)$ GeV, which is an additional companion pole related to $K_0^*(800)$. As a further investigation for increasing N_c confirms, $K_0^*(800)$ emerges as a dynamically generated four-quark object as a consequence of pion-kaon loops.

E-mail:

* milena.soltysiak@op.pl

1. Introduction

The scalar resonance $K_0^*(1430)$ with $I(J^P) = \frac{1}{2}(0^+)$ is a well-established state listed in the summary table of the Particle Data Group (PDG) [1]. On the contrary, the scalar state $K_0^*(800)$, also with $I(J^P) = \frac{1}{2}(0^+)$ and known as κ , still needs confirmation. The acceptance of the lightest κ in the summary table of the PDG is important since it would allow to complete the nonet of the light scalar states below 1 GeV.

The main aim of this work is to understand the nature of the two resonances $K_0^*(1430)$ and $K_0^*(800)$ using the quantum field theoretical model described in Ref. [2] (on which this proceedings is based). In particular, we shall concentrate on the existence and the properties of the latter state. The key ingredient of our theoretical approach is to consider at the same time both derivative and non-derivative interaction terms, see Sec. 2. We perform a fit of the four parameters of our model to the measured data points of πK phase shift in the $I = 1/2, J = 0$ channel from Ref. [3] (See Sec. 3.1) and obtain a remarkably good agreement with the experiment. As a consequence of the fit, we determine the scalar kaonic spectral function up to 1.8 GeV, in which only one single peak corresponding to $K_0^*(1430)$ appears. Namely, there is no peak corresponding to $K_0^*(800)$ but only a small enhancement in the corresponding energy region. Next, we calculate the coordinates of the poles on the complex plane. Despite the fact that we included only one "seed" state in the effective Lagrangian, we find two poles: one corresponding to the peak in the spectral function at 1.4 GeV and thus to $K_0^*(1430)$, and one corresponding to the light κ. We get additional information about the nature of both resonances by studying the changes of the spectral function and the movement of the poles for increasing N_c: $K_0^*(1430)$ behaves as a $q\bar{q}$ state, while the light κ is a dynamically generated state and as such a non-ordinary meson (see e.g. Refs. [4, 5, 6, 7, 8, 9] and refs. therein).

Content from this work may be used under the terms of the Creative Commons Attribution 3.0 licence. Any further distribution of this work must maintain attribution to the author(s) and the title of the work, journal citation and DOI.
Published under licence by IOP Publishing Ltd

To confirm the correctness of our model, we study in Sec. 3.2 variations of it in which only the non-derivative or only the derivative interaction term is taken into account, respectively. We also studied variations of the form factor. In all cases we obtain a worse agreement with the experiment with respect to Sec. 3.1. It turns out that using both derivative and non-derivative terms is crucial to obtain a good description of experimental data. Finally, we summarize our results in Sec. 3.4.

2. The model

We use a relativistic interaction Lagrangian consisting of both derivative and nonderivative terms for a single $q\bar{q}$ scalar state denoted as K_0^*:

$$\mathcal{L}_{int} = aK_0^{*+}K^-\pi^0 + bK_0^{*+}\partial_\mu K^-\partial^\mu\pi^0 + \ldots \,, \tag{1}$$

where dots refer to complex conjugation and other members of the isospin multiplet. The model is naturally obtained as a piece of more complete chiral models, e.g. Ref. [10]. The decay width is given by:

$$\Gamma_{K_0^*}(m) = 3\frac{\left|\vec{k}_1\right|}{8\pi m^2}\left[a - b\frac{m^2 - M_K^2 - M_\pi^2}{2}\right]^2 F_\Lambda(m) \,, \tag{2}$$

where the pion and kaon masses are denoted as M_π and M_K, and $\vec{k}_1$ is the three-momentum of the pion. The form-factor $F_\Lambda(m) = e^{-2\left|\vec{k}_1\right|^2/\Lambda^2}$, where Λ stays for an energy scale, assures that calculations are finite. The propagator of K_0^* reads

$$\Delta_{K_0^*}(p^2 = m^2) = \left[m^2 - M_0^2 + \Pi(m^2) + i\varepsilon\right]^{-1} \,, \tag{3}$$

where M_0 is the bare mass of $K_0^*(1430)$ and $\Pi(m^2) = \mathrm{Re}(m^2) + i\mathrm{Im}(m^2)$ is the one-loop contribution with one kaon and one pion circulating in it. We recall that the optical theorem implies that $\mathrm{Im}\Pi(m) = m\Gamma_{K_0^*}(m)$. The spectral function is given by:

$$d_{K_0^*}(m) = \frac{2m}{\pi}\left|Im\Delta_{K_0^*}(p^2 = m^2)\right| \,. \tag{4}$$

The quantity $d_{K_0^*}(m)dm$ is the probability that the mass of the resonance is between m and $m + dm$. The normalization condition $\int_0^\infty d_{K_0^*}(m)\mathrm{dm} = 1$ must hold. For the details of the mathematical treatment used here, we refer to [11]. In the next section we investigate the form of the spectral function, the poles of the propagator, and their dependence on the number of colors N_c.

3. Results

3.1. Fit to the phase shift data

We consider the πK phase shift in the $I = 1/2, J^P = 0^-$ channel below 1.8 GeV. Under the assumption that the s-channel contribution of the field K_0^* dominates, the phase shift reads

$$\delta(m) = \frac{1}{2}\arccos\left[1 - \pi\Gamma_{K_0^*}(m)d_{K_0^*}(m)\right] \tag{5}$$

Using the data of Ref. [3] (see also Ref. [12]) we perform a fit in order to determine four parameters of our model a, b, M_0, and Λ, see Tab. 1. The result of the fit is presented in Fig. 1. We obtain a good agreement with the experimental data points ($\chi^2/d.o.f. = 1.25$).

Journal of Physics: Conference Series **742** (2016) 012014 doi:10.1088/1742-6596/742/1/012014

Table 1. The value of the parameters of our model.

Parameter	Value
a	1.60 ± 0.22 GeV
b	-11.16 ± 0.82 GeV^{-1}
m_0	1.204 ± 0.008 GeV
Λ	0.496 ± 0.008 GeV

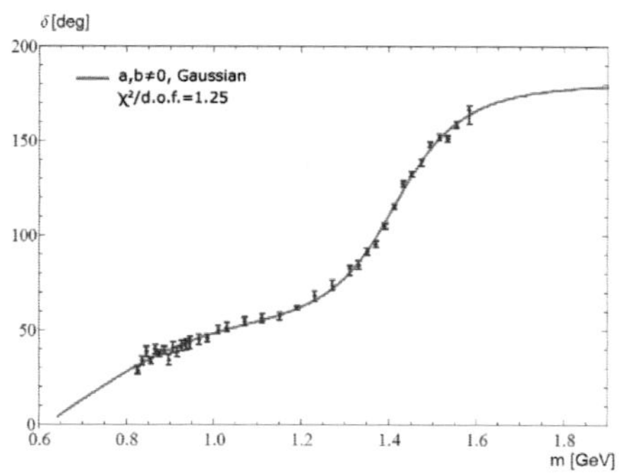

Figure 1. Result of the fit to the measured data points of the πK phase shift in $I = 1/2, J = 0$ channel.

According to Eq. (4) and using the parameters reported in Tab. 1, we report in Fig. 2 the scalar kaonic spectral function up to 1.8 GeV. We observe a single peak close to 1.4 GeV corresponding to $K_0^*(1430)$, but there is no peak for $K_0^*(800)$ (only a small enhancement in the low energy is visible). For a comparison with the spectral function of the narrow Breit-Wigner type vector state $K^*(892)$ we refer to [13].

We study the large-N_c limit of the spectral function according to the rescaling $a \longrightarrow \sqrt{3/N_c}a$, $b \longrightarrow \sqrt{3/N_c}b$, where N_c is the number of colors, see Fig. 3. (Namely, the Lagrangian (1) contains a three-leg interaction term, whose amplitude scale as $1/\sqrt{N_c}$, e.g. Ref. [14]). The peak corresponding to $K_0^*(1430)$ becomes narrower and higher with decreasing $\lambda = 3/N_c$, but the enhancement corresponding to $K_0^*(800)$ becomes smaller and finally disappears for increasing N_c.

As a next step we determine the positions of the poles in the complex plane. For the heavier state $K_0^*(1430)$ we get $1.413 \pm 0.002 - i(0.127 \pm 0.003)$ GeV. In addition, we also find the lighter state κ: $0.746 \pm 0.019 - i(262 \pm 0.014)$ GeV. In Fig. 5 we show the pole trajectories as function of N_c. The pole of $K_0^*(1430)$ moves toward to the real energy axis, while the pole of $K_0^*(800)$ moves away from it and disappears for $N_c \gtrsim 13$.

In conclusion, even if in the effective Lagrangian of Eq. (1) a single K_0^* seed state is considered, we obtain both scalar states $K_0^*(1430)$ and $\kappa \equiv K_0^*(800)$. Moreover, the behavior of the spectral function and the pole trajectories as function of N_c show that $K_0^*(1430)$ is a predominantly quark-antiquark state, while the light κ is generated dynamically, by the pion-kaon loop dressing a preformed quark-antiquark state. Accordingly, it disappears when the interaction strength is larger than a certain threshold.

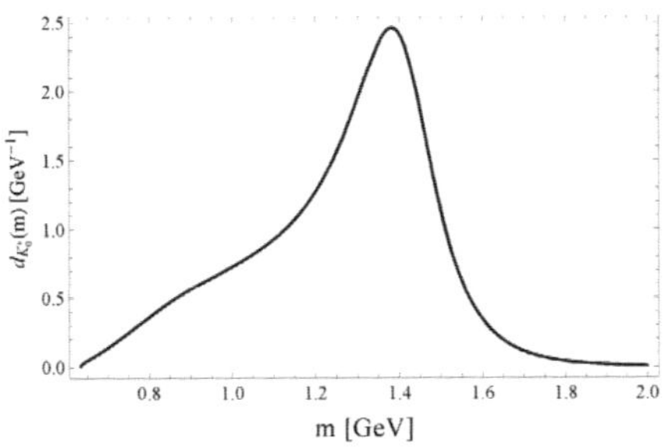

Figure 2. Scalar kaonic spectral function from Eq. (4).

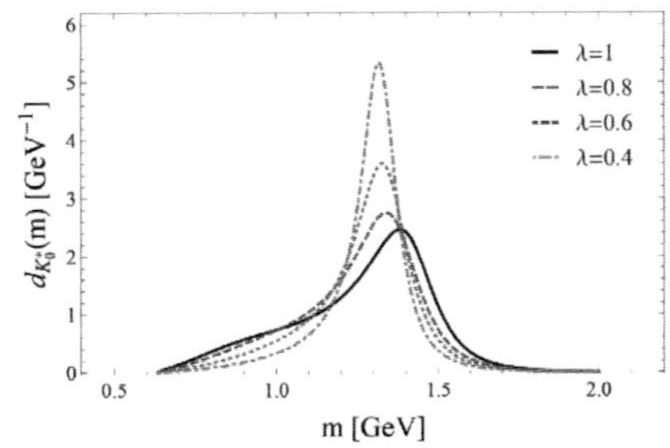

Figure 3. Scalar kaonic spectral function studied in large-N_c limit.

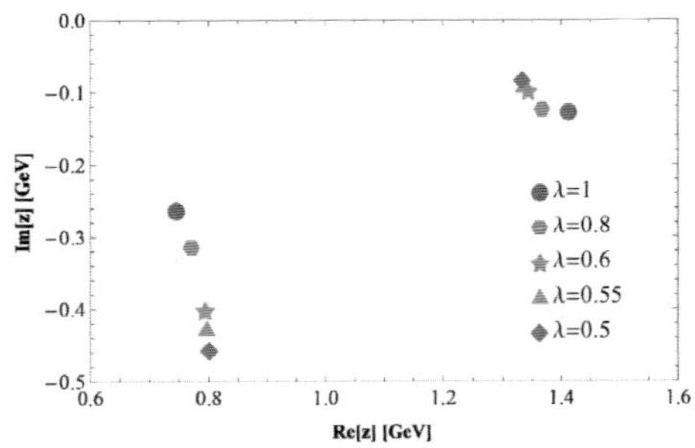

Figure 4. Positions of the poles of two resonances $K_0^*(1430)$ and $K_0^*(800)$ and their trajectories in large-N_c limit.

3.2. Modifications of the model

We study different variations of our model in order to test the robustness of the results. First, we consider two different limits: one only with the derivative term ($a \neq 0$ and $b = 0$ in Eq. (1)) and one only with the non-derivative term ($a = 0$ and $b \neq 0$ in Eq. (1)). We perform the fits to the experimental data points of the phase shift, see Fig. 5. Both results of the fits, $\chi^2/d.o.f. = 2.54$ and $\chi^2/d.o.f. = 5.41$, are worsened w.r.t. the result in Sec. 3.1 (in which $\chi^2/d.o.f. = 1.25$). Fig. 5 also shows that the derivative term is necessary to obtain a theoretical curve which is in qualitative agreement with the experiment. As a next step, we also test a different form of the form factor. Namely, although the Gaussian form used in Sec. 3.1 is typically used in such studies, there is no fundamental reason behind it. Therefore, we have repeated the fit with:

$$F_\Lambda(m) = e^{-2k^4(m)/\Lambda^4} \quad . \tag{6}$$

The result of the fit is presented in Fig. 5. It turns out that a qualitative agreement is obtained, but the χ^2 is clearly worsened: $\chi^2 = 2.86$. For the existence and positions of the poles for all these cases, we refer to [2].

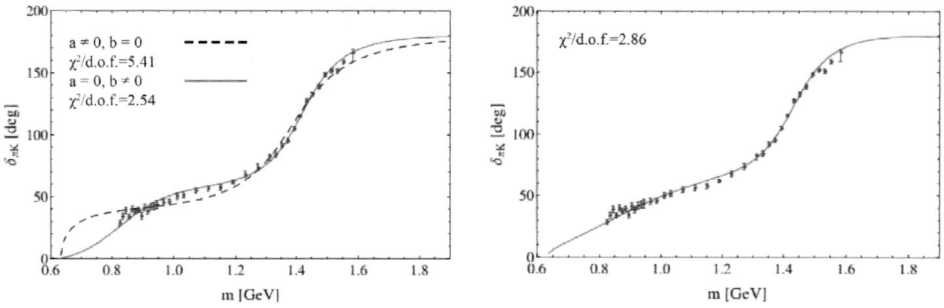

Figure 5. Fit results to the experimental points of the phase shift performed for different types of Lagrangian (left panel) and for modified form factor (right panel).

4. Conclusions

In this work we have studied the properties of the scalar kaonic resonances $K_0^*(1430)$ and $K_0^*(800)$ according to the formalism described in Ref. [2]. The original Lagrangian contains a single scalar field, but because of hadronic quantum loops two poles emerge in the complex plane once that the parameters of the model are fitted to pion-kaon scattering data: one pole corresponds to the predominantly quark-antiquark resonance $K_0^*(1430)$, while the other one is an additional (non quark-antiquark) companion pole which corresponds to $K_0^*(800)$. While the former survives in the large-N_c limit, the latter disappears. In the spectral function there is only a single peak, roughly corresponding to $K_0^*(1430)$; in the low-energy domain there is a small enhancement which is related to $K_0^*(800)$.

In the future, it will be interesting to apply these techniques to other enigmatic resonances, such as $D_{S0}^*(2317)^{\pm}$ and some of the newly discovered X, Y, and Z states [15].

References

[1] Olive K A *et al.* [Particle Data Group Collaboration] 2014 *Chin. Phys.* C **38** 090001
[2] Wolkanowski T, Soltysiak M and Giacosa F, arXiv:1512.01071 [hep-ph]
[3] Aston D *et al.* 1988 *Nucl. Phys.* B **296** 493.
[4] van Beveren E, Rijken T A, Metzger K, Dullemond C, Rupp G and Ribeiro J E 1986 *Z. Phys.* C **30** 615-620
[5] Black D, Fariborz A H, Sannino F and Schechter J 1998 *Phys. Rev.* D **58** 054012
[6] Oller J A, Oset E and Pelaez J R, 1999 *Phys. Rev.* D **59** 074001;
Pelaez J R 2004 *Phys. Rev. Lett.* **92** 102001
[7] Oller J A and Oset E, 1999 *Phys. Rev.* D **60** 074023 [hep-ph/9809337]
Jamin M, Oller J A and Pich A, 2000 *Nucl. Phys.* B **587** 331 [hep-ph/0006045]
[8] Pelaez J R 2004 *Phys. Rev. Lett.* **92** 102001
Pelaez J R 2004 *Mod. Phys. Lett.* A **19** 2879 [hep-ph/0411107]
[9] Fariborz A H, Pourjafarabadi E, Zarepour S and Zebarjad S M 2015 *Phys. Rev.* D **92** 113002 [arXiv:1511.01623 [hep-ph]]
[10] Parganlija D, Kovacs P, Wolf G, Giacosa F and Rischke D H, 2012 *Phys. Rev.* D **87** 014011;
Janowski S, Giacosa F and Rischke D H 2014 *Phys. Rev.* D **90** 11 114005 [arXiv:1408.4921 [hep-ph]]
[11] Wolkanowski T, Giacosa F and Rischke D H 2016 *Phys. Rev.* D **93** no.1, 014002 [arXiv:1508.00372 [hep-ph]]
[12] Ishida S, Ishida M, Ishida T, Takamatsu K and Tsuru T 1997 *Prog. Theor. Phys.* **98** 621 [hep-ph/9705437]
[13] Soltysiak M, Wolkanowski T and Giacosa F, arXiv:1604.01636 [hep-ph]
[14] Witten E 1979 *Nucl. Phys.* B **160** 57
[15] Chen H X, Chen W, Liu X and Zhu S L, arXiv:1601.02092 [hep-ph]

Performance of Prototypes for the Barrel Part of the $\overline{\text{P}}$ANDA Electromagnetic Calorimeter

Christoph Rosenbaum[1], S. Diehl[1], V. Dormenev[1], P. Drexler[1], M. Kavatsyuk[2], T. Kuske, S. Nazarenko[1], R. Novotny[1], P. Rosier[3], A. Ryazantsev[4], P. Wieczorek[5], A. Wilms[5] and H.-G. Zaunick[1] for the $\overline{\text{P}}$ANDA collaboration

[1]II. Physikalisches Institut, Universität Gießen, Germany
[2]KVI-CART, University of Groningen, The Netherlands
[3]IPN Orsay, France
[4]IHEP Protvino, Russia
[5]GSI, Darmstadt, Germany

E-mail: `christoph.rosenbaum@physik.uni-giessen.de`

Abstract. The performance of the most recent prototypes of the $\overline{\text{P}}$ANDA barrel electromagnetic calorimeter (EMC) will be compared. The first large scale prototype PROTO60 was designed to test the performance of the improved tapered lead tungstate crystals (PWO-II). The PROTO60 which consists of 6 × 10 crystals was tested at various accelerator facilities over the complete envisaged energy range fulfilling the requirements of the TDR of the $\overline{\text{P}}$ANDA EMC in terms of energy, position and time resolution. To realize the final barrel geometry and to test the final front end electronics, a second prototype PROTO120 has been constructed. It represents a larger section of a barrel slice, containing the most tapered crystals and the close to final components for the $\overline{\text{P}}$ANDA EMC. The performance of both prototypes will be compared with a focus on the analysis procedure including the signal extraction, noise rejection, calibration and the energy resolution. In addition, the influence of the non-uniformity of the crystal on the energy resolution will be discussed.

1. Introduction

The state-of-the-art $\overline{\text{P}}$ANDA detector at the future FAIR facility will be used to study proton - antiproton interactions. The EMC of the target spectrometer with its expected excellent performance and efficiency for electromagnetic probes over a wide energy range from 10 MeV up to 15 GeV, will be one of the central components to achieve the physics goals. The barrel part of the EMC will consist of more than 11,000 lead tungstate crystals operated at $-25°$C to achieve the required performance and efficiency stated in the TDR of the $\overline{\text{P}}$ANDA EMC [1].

2. Prototypes of the $\overline{\text{P}}$ANDA barrel EMC

2.1. Characterization of the prototypes

In order to judge the performance of the prototypes, the emphasis of the analysis is put on the obtained energy resolution and energy threshold. The energy resolution is a crucial parameter for the $\overline{\text{P}}$ANDA Barrel EMC because it has a strong impact on the accuracy of the invariant mass determination, such as of J/Ψ states. In addition, the energy resolution influences the

Content from this work may be used under the terms of the Creative Commons Attribution 3.0 licence. Any further distribution of this work must maintain attribution to the author(s) and the title of the work, journal citation and DOI.
Published under licence by IOP Publishing Ltd

determination of the E/p ratio of electrons and positrons. The following parametrization of the energy resolution was required by the TDR of the $\overline{\text{P}}$ANDA EMC:

$$\frac{\sigma}{E} = \frac{2\%}{\sqrt{E/\text{GeV}}} \oplus 1\% \quad (1)$$

The statistical term, which is dominated by the Poisson statistics of the light collection process, is crucial in particular for low energy photons due to the relatively low light yield of PWO compared to other scintillator materials. The constant term of 1% is the resolution limit at high energies affected by crystal inhomogeneities, calibration and shower leakage.
The energy threshold influences the reconstruction of low energetic photons which are a major contribution to many background channels and have to be detected completely. On the other hand, the energy threshold is limited by the electronic noise of a single channel and the distribution of the deposited energy at low energies. The TDR of the EMC [1] states an energy threshold of photons of 10 MeV which requires to a single crystal threshold of ≤ 3 MeV.

2.2. PROTO60

The PROTO60 is the first real-size prototype of the $\overline{\text{P}}$ANDA Barrel EMC and the geometry is arranged to resemble a subsection of a barrel slice. It consists of 60 PWO-II crystals which are read-out with only a single 1 cm^2 quadratic large area avalanche photodiode (LAAPD) and a low-noise low-power preamplifier [2]. The prototype is operated at $-25°$C. The performance has been studied during several beam tests over the complete energy range obtaining a sufficient energy, position and time resolution. Details can be found in [3]. The parametrization of the relative energy resolution of the PROTO60 for the total energy range from 50 MeV up to 15 GeV with an individual module threshold of 1 MeV can be described by [4]

$$\frac{\sigma}{E} = \frac{0.25\%}{E/\text{GeV}} \oplus \frac{1.86\%}{\sqrt{E/\text{GeV}}} \oplus 1.46\%, \quad (2)$$

where the first component estimates the electronic noise contribution, while the other two correspond to the statistical term and constant term.

2.3. PROTO120

The PROTO120 represents a larger section of a barrel slice, containing 120 of the most tapered crystals. It is operated at $-25°$C and contains the close to final components for the $\overline{\text{P}}$ANDA Barrel EMC. The readout is performed with two rectangular large area APDs per crystal, which are read out separately via the specially developed APFEL-ASIC preamplifier [5], providing a large dynamic range of 10,000 (1 MeV up to 10 GeV), low power consumption (55 mW/ch), a high rate capability up to 350 kHz and optimized signal shaping.

3. Results of beam tests with PROTO120

3.1. Setup

In April 2015 a first beam test with the PROTO120 was performed at the MAMI facility in Mainz. The goal was to investigate the response of the prototype to photons in the low energy regime. At MAMI, the electron beam is converted via Bremsstrahlung into a continuous and collimated photon beam using the Glasgow Photon Tagging Spectrometer which provided tagged photons up to 830 MeV. 16 crystals of type II of the PROTO120 were read out with a commercial SIS3302 sampling ADC (SADC) , but only a 3×3 sub matrix was considered for the analysis. Relatively short calibration runs were performed by directing the photon beam into each crystal and one longer run hitting the center of the 3×3 matrix.

3.2. Analysis

The SADCs are continuously taking samples of each preamplified LAAPD pulse. A trace is stored in case of a postive trigger signal, which requires a coincidence of a logical OR of the selected tagger channels and the responding PWO matrix. A typical sampled event is depicted in Fig. 1 where also the extraction of the energy information is indicated. The deposited energy in a crystal is proportional to the signal amplitude. In a basic approach, the signal amplitude is obtained by subtracting the baseline and the pulse minimum. In order to add the information of both LAAPDs a relative calibration is required using the amplitude ratio. Before summing the energy response of the whole matrix, they have to be relatively calibrated using the direct photon beam. The tagged lineshapes of the calibrated and summed 3×3 matrix with a summation threshold of 2.7 MeV are depicted in Fig. 2. In order to obtain the energy resolution, the reconstructed lineshapes are fitted with a so-called Novosibirsk function [6], a commonly used function to describe the skewed Gaussian distribution. The Novosibirsk function is defined by

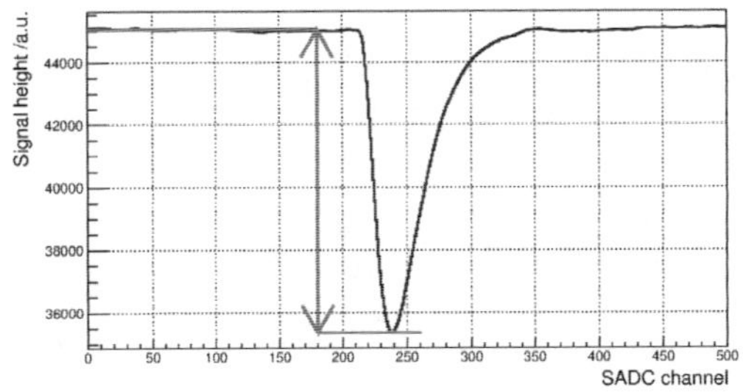

Figure 1: One sampled event (20 ns per bin) after shaping of the APFEL ASIC with indicated energy extraction.

$$f(x) = Ae^{-\frac{1}{2}\left(\frac{\ln^2(1+\Lambda\tau(x-\mu_{max}))}{\tau^2}+\tau^2\right)} \tag{3}$$

where $\Lambda = \frac{\sinh(\tau\sqrt{\ln 4})}{\sqrt{\ln 4}\tilde{\sigma}}$ is parametrized by the asymmetry τ, width $\tilde{\sigma} \equiv \frac{\text{FWHM}}{2\sqrt{\ln 4}}$ and mode μ_{max} which is the value that is most likely to be observed in a measurement. The following parametrization of the energy resolution for the beam hitting the center of the 3×3 matrix could be obtained as

$$\frac{\sigma}{E} = \frac{0.16\%}{E/\text{GeV}} \oplus \frac{2.46\%}{\sqrt{E/\text{GeV}}} \oplus 2.32\%. \tag{4}$$

The electronic noise level can be studied in more detail by obtaining the distribution of the RMS of the baseline without presence of a signal. The mean value of this distribution leads

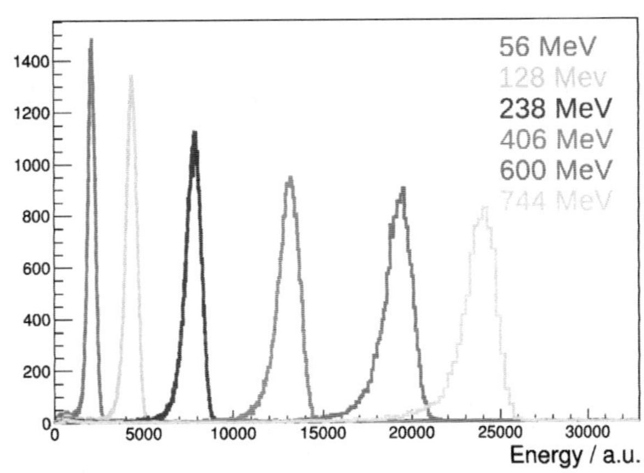

Figure 2: Tagged lineshapes of the calibrated and summed 3×3 matrix with a summation threshold of 2.7 MeV.

to an electronic noise level of 0.72 MeV. Basically, the electronic noise is the intrinsic noise of the preamplifier which comes from the input capacity of the preamplifier, in case of the PROTO120. The electronic noise can be translated directly to the summation threshold of a single crystal. The summation threshold corresponds to the level where most of the noise contribution is rejected. The noise distribution is expected to be Gaussian. Therefore, more than 99.7% of the events are within a $3 \cdot \sigma$-range. This range is usually set as the summation threshold. Hence, for the beamtime test a summation threshold was set to 2.7 MeV.

4. Influence of light collection non-uniformity on the energy resolution

To realize the barrel EMC geometry, crystals with different degree of tapering are used. The distribution of the light yield of differently tapered crystals, depicted in Fig. 3, is determined within the quality control with low energy gamma sources. Therefore, the crystals of the barrel EMC show a non-uniformity in light collection, which is a result from an interplay between the focusing and the absorption of light within the crystal. Since the focusing effect strongly depends on the degree of tapering, the resulting NUF curve varies with the type of crystal. While crystals with an average degree of tapering show a difference in light yield of $\sim$ 20% when created in the front or the rear part of the crystal, this value increases for the most tapered crystals up to 40% with a slope in the front part of almost 3 %/cm [7]. Furthermore, the non-uniformity in light collection causes a smearing of the energy response, resulting in a significant increase of the constant term of the parameterization of the energy resolution. This effect is dominant for energies above a few hundred MeV and can be reduced by de-polishing one side face of the crystals as it has been investigated by the CMS-ECAL collaboration [8]. However, since the reconstruction of low energy photons down to 10 MeV is essential for $\overline{\text{P}}$ANDA and due to the low light yield of PWO, de-polished crystals have not been considered so far. Nevertheless, an investigation has been started to study a 3×3 matrix with reduced non-uniformity. The analysis was performed in a similar manner. The obtained relative energy resolution for the 3 × 3 linearized crystal array can be parametrized by [9]

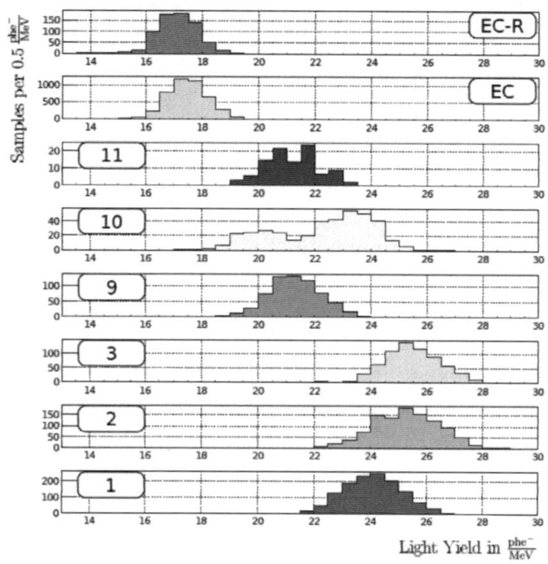

Figure 3: The larger the level of tapering, the higher the light yield output. The crystals of both end caps are almost rectangular (EC-R, EC). The type 1 has the most tapered geometry [1].

$$\frac{\sigma}{E} = \frac{0.27\%}{E/\text{GeV}} \oplus \frac{2.30\%}{\sqrt{E/\text{GeV}}} \oplus 0.50\%. \tag{5}$$

A comparison of both energies resolutions, depicted in Fig. 4, shows a significant improvement for higher energies and similar results for lower energies.

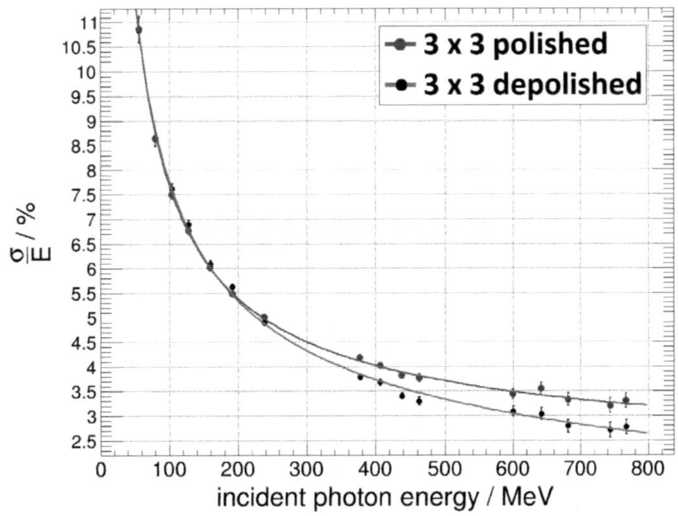

Figure 4: Energy resolution of the 3 × 3 matrix with polished and de-polished crystals, respectively, for photons impinging on the central crystal [9].

5. Conclusion

The response of both $\overline{\text{P}}$ANDA EMC prototypes fulfills the requirements of the TDR [1]. The study of the influence of the non-uniformity of the light collection on the energy resolution has shown that the performance of the $\overline{\text{P}}$ANDA barrel EMC could gain from a reduction of the non-uniformity. However, the de-polishing of the crystals should be done preferentially in the forward region of the barrel where higher photon energies are expected. Nevertheless, the effective noise of the PROTO120 is higher than of the PROTO60 and requires a further reduction of the noise level in particular due to pick-up and cross-talk.

6. Acknowledgments

This work was supported by BMBF, GSI and HIC4FAIR. The authors like to thank the staff of the MAMI facility for providing the beam.

References

[1] W. Erni *et al.* [$\overline{\text{P}}$ANDA Collaboration], "Technical Design Report for $\overline{\text{P}}$ANDA Electromagnetic Calorimeter (EMC)," arXiv:0810.1216 [physics.ins-det].

[2] W. Erni, "Technical Design Report for discrete Preamplifier for APD Readout, LNP Preamplifier Version SP 883A02", http://panda-wiki.gsi.de/cgi-bin/view/DCS/SupportedHardware, 2008

[3] M. Kavatsyuk, "Performance of the prototype of the electromagnetic calorimeter for $\overline{\text{P}}$ANDA", doi:10.1016/j.nima.2011.06.044

[4] M. Moritz, PhD Thesis, Giessen 2013

[5] P. Wieczorek, H. Flemming, "Low noise preamplifier ASIC for the $\overline{\text{P}}$ANDA EMC", *Nuclear Science Symposium Conference Record (NSS/MIC)*, 2010 IEEE, DOI: 10.1109/NSSMIC.2010.5873982

[6] A. Bevan, F. Wilson, "A Fit User Guide", http://pprc.qmul.ac.uk/~bevan/afit/afit.pdf

[7] D. Bremer, PhD Thesis, Giessen 2013

[8] [CMS collaboration], "The Electromagnetic Calorimeter Project: Technical Design Report", CERN-LHCC-97-033, 1997.

[9] S. Diehl, PhD Thesis, Giessen 2016

(Anti-)strangeness in heavy-ion collisions

P. Moreau[1], W. Cassing[2], A. Palmese[2], E. L. Bratkovskaya[1,3,4]

[1]Frankfurt Institute for Advanced Studies, Johann Wolfgang Goethe Universität, Frankfurt am Main, Germany

[2]Institut für Theoretische Physik, Universität Gießen, Germany

[3]Institut für Theoretische Physik, Johann Wolfgang Goethe Universität, Frankfurt am Main, Germany

[4]GSI Helmholtzzentrum für Schwerionenforschung GmbH, Darmstadt, Germany

E-mail: moreau@fias.uni-frankfurt.de

Abstract. We study the production of hadrons in nucleus-nucleus collisions within the Parton-Hadron-String Dynamics (PHSD) transport approach that is extended to incorporate essentials aspects of chiral symmetry restoration (CSR) in the hadronic sector (via the Schwinger mechanism) on top of the deconfinement phase transition as implemented in PHSD before. The essential impact of CSR is found in the Schwinger mechanism (for string decay) which fixes the ratio of strange to light quark production in the hadronic medium. Our studies suggest a microscopic explanation for the maximum in the K^+/π^+ and $(\Lambda + \Sigma^0)/\pi^-$ ratios at about 30 A GeV which only shows up if in addition to CSR a deconfinement transition to partonic degrees-of-freedom is incorporated in the reaction dynamics.

1. Introduction

In this contribution we summarize the ideas and results from the study in Ref. [1] that investigates chiral symmetry restoration in heavy-ion reactions and addresses phenomena such as strangeness enhancement [2, 3] or the 'horn' in the K^+/π^+ ratio [4, 5]. Both phenomena have been attributed to a deconfinement transition before. Indeed, the actual experimental observation could not be described within conventional hadronic transport theory [6, 7, 8] and remained a major challenge for microscopic approaches.

2. Extensions in PHSD3.3

Our studies are performed within the PHSD transport approach that has been described in Refs. [9, 10]. The PHSD incorporates explicit partonic degrees-of-freedom in terms of strongly interacting quasiparticles (quarks and gluons) in line with an equation-of-state from lattice QCD (lQCD) as well as dynamical hadronization and hadronic elastic and inelastic collisions in the final reaction phase. For a recent review we refer the reader to Ref. [11].

2.1. Strings in (P)HSD

We recall that in the PHSD/HSD, the high energy inelastic hadron-hadron collisions in the hadronic phase are described by the FRITIOF model [12], where two incoming nucleons emerge the reaction as two excited color singlet states, i.e. 'strings'. The production probability P of

Content from this work may be used under the terms of the Creative Commons Attribution 3.0 licence. Any further distribution of this work must maintain attribution to the author(s) and the title of the work, journal citation and DOI.
Published under licence by IOP Publishing Ltd

massive $s\bar{s}$ or $qq\bar{q}\bar{q}$ pairs is suppressed in comparison to light flavor production ($u\bar{u}$, $d\bar{d}$) according to the Schwinger-like formula [13], i.e.

$$\frac{P(s\bar{s})}{P(u\bar{u})} = \frac{P(s\bar{s})}{P(d\bar{d})} = \gamma_s = \exp\left(-\pi\frac{m_s^2 - m_q^2}{2\kappa}\right), \tag{1}$$

with $\kappa \approx 0.176$ GeV^2 denoting the string tension and $m_s, m_q = m_u = m_d$ the appropriate (dressed) strange and light quark masses. Inserting the constituent (dressed) quark masses $m_u \approx 0.33$ GeV and $m_s \approx 0.5$ GeV in the vacuum a value of $\gamma_s \approx 0.3$ is obtained from Eq.(1). This ratio is expected to be different in a nuclear medium and actually should depend on the in-medium scalar quark condensate $< \bar{q}q >$.

2.2. The scalar quark condensate

The well known scalar quark condensate $< \bar{q}q >$ is viewed as an order parameter for the restoration of chiral symmetry at high baryon density and temperature. A reasonable estimate for the quark scalar condensate in dynamical calculations has been suggested by Friman et al. [14],

$$\frac{< \bar{q}q >}{< \bar{q}q >_V} = 1 - \frac{\Sigma_\pi}{f_\pi^2 m_\pi^2}\rho_S - \sum_h \frac{\sigma_h \rho_S^h}{f_\pi^2 m_\pi^2}, \tag{2}$$

where σ_h denotes the σ-commutator of the relevant mesons h and ρ_S^h the scalar density of meson h [1]. Furthermore, $< \bar{q}q >_V$ denotes the vacuum condensate, $\Sigma_\pi \approx 45$ MeV is the pion-nucleon Σ-term, f_π and m_π the pion decay constant and pion mass, respectively. The last (and essential) quantity in the relation (2), that still has to be determined for an evaluation of the quark condensate, is the nucleon scalar density ρ_S. The latter quantity can be calculated in relativistic mean-field theory by replacing the mass and momentum by the effective quantities

$$m_N^*(x) = m_N^v - g_s\sigma(x) \tag{3}$$

with m_N^v denoting the nucleon mass in vacuum. In Eq. (3) the scalar field $\sigma(x)$ mediates the scalar interaction with the surrounding medium while g_s is a coupling. When including self-interactions of the σ-field up to 4th order [15] the scalar field is determined locally by the nonlinear gap equation [15, 16]

$$m_s^2\sigma(x) + B\sigma^2(x) + C\sigma^3(x) = g_s\rho_S = g_s d \int \frac{d^3p}{(2\pi)^3} \frac{m_N^*(x)}{\sqrt{p^2 + m_N^{*2}}} f_N(x, \mathbf{p}) \tag{4}$$

with $d = 4$ in case of isospin symmetric nuclear matter. The parameters g_s, m_s, B, C are fixed in the non-linear $\sigma - \omega$ model for nuclear matter [16] and are displayed in Table I of Ref. [1]. We will use the well-known NL3 parameter set in the following with a compressibility $K = 380$ MeV and effective mass $m^*/m = 0.7$ at normal nuclear matter density. We recall that in the non-linear $\sigma - \omega$ model the energy density for symmetric nuclear matter is given by [16]

$$\epsilon = U(\sigma) + \frac{g_v^2}{2m_v^2}\rho_N^2 + d\int \frac{d^3p}{(2\pi)^3} E^*(\mathbf{p})\left(N_c(\mathbf{p}) + N_d(\mathbf{p})\right), \tag{5}$$

with

$$E^*(\mathbf{p}) = \sqrt{\mathbf{p}^2 + m_B^{*2}} \quad ; \quad U(\sigma) = \frac{m_s^2}{2}\sigma^2 + \frac{B}{3}\sigma^3 + \frac{C}{4}\sigma^4,$$

while ρ_N denotes the baryon density and $N_c(\mathbf{p})$ and $N_d(\mathbf{p})$ the particle/antiparticle numbers at fixed momentum $\mathbf{p}$.

Journal of Physics: Conference Series **742** (2016) 012016 doi:10.1088/1742-6596/742/1/012016

2.3. The string decay in a hot and dense medium

The basic assumption now is that the strange and light quark masses in the hadronic medium drop both in line with the ratio (2),

$$m_s^* = m_s^0 + (m_s^v - m_s^0)\left|\frac{<q\bar{q}>}{<q\bar{q}>_V}\right| \quad , \quad m_q^* = m_q^0 + (m_q^v - m_q^0)\left|\frac{<q\bar{q}>}{<q\bar{q}>_V}\right|, \tag{6}$$

using $m_s^0 \approx 100$ MeV and $m_q^0 \approx 7$ MeV for the bare quark masses while the vacuum (dressed) masses are $m_s^v \approx 500$ MeV and $m_q^v \approx 330$ MeV, respectively.

In order to illustrate our main conjecture we show in Fig. 1 the ratio s/u as a function of the energy density ϵ at temperature T=0 as evaluated by Eq. (5). In this case only nucleons contribute to the scalar density in Eq. (2) and the hadronic energy density which is (apart from slight corrections due to the interaction energy) roughly given by $\epsilon \approx m_N \rho_N$, where m_N is the vacuum nucleon mass and ρ_N the nuclear density. It is seen that the ratio s/u rises with energy density up to the limiting values given by bare masses $m_s = m_s^0$ and $m_q = m_q^0$ in Eq. (1). In fact, this illustration corresponds to the case of HSD; in PHSD the ratio s/u increases in the same way up to $\epsilon_c \approx 0.5$ GeV/fm^3 but then drops down to a value of ~1/3 in the deconfined phase. Accordingly, the approach to CSR occurs in the hadronic phase and should be seen experimentally for local energy densities below $\epsilon_c \approx 0.5$ GeV/fm^3 since there is no more any 'string decay' in the partonic phase above ϵ_c due to the direct conversion of energy and momentum to the massive quasiparticles of the QGP.

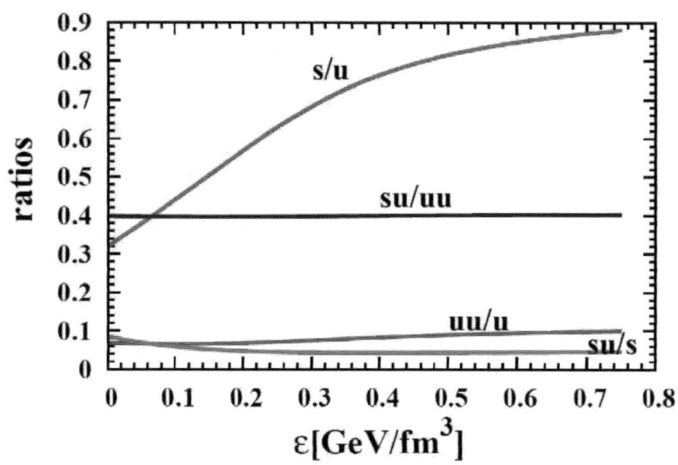

Figure 1. The ratios of strange to light quarks (s/u) according to Eq. (1) and the various ratios for diquarks as a function of the nuclear energy density ϵ for $T = 0$.

In order to implement this scenario of 'chiral symmetry restoration' in the PHSD code we solve the gap equation (4) for each cell in space-time in order to determine the scalar nucleon density ρ_S, the scalar quark condensate by Eq. (2) and the strangeness ratio by Eq. (1) which drives the string decay in each local cell. Since in the case of HSD (without any deconfined partonic phase) the ratio s/u increases further with energy density ϵ as in Fig. 1, one has to expect an overestimation of strangeness production in the HSD when implementing CSR via (1) at high bombarding energies where the scalar quark condensate is vanishing in the overlap zone of the reaction.

3. Comparison of PHSD3.3 results to A+A data

For an illustration we present here the ratio (2) in Fig. 2 as a function of x and z (for $y = 0$) at different times t for a central Au+Au collisions at 30 A GeV. Whereas in the approach phase the ratio drops to about 2/3 inside the impinging nuclei (cf. Fig. 2) the scalar quark condensate practically vanishes in the full overlap phase from about 3 to 7 fm/c and regains the vacuum value only in the late expansion phase as noted before. However, for all times from contact ($\sim$ 2.5 fm/c) to about 8 fm/c the ratio of the scalar quark condensate remains far below unity, which implies that the decay of strings in the hadronic environment is modified substantially in the hot and dense medium.

The white borderlines in Fig. 2 separate the space-time regions of deconfined matter to hadronic matter. It is clearly seen that although the chiral symmetry is approximately restored in the full overlap phase from 2.8 to 6 fm/c some region is occupied by deconfined partons in the PHSD. In these regions, however, an enhanced production of strangeness should not occur since the Schwinger mechanism (1) no longer applies due to a vanishing string tension and a transformation of energy and momentum to massive partonic degrees-of-freedom.

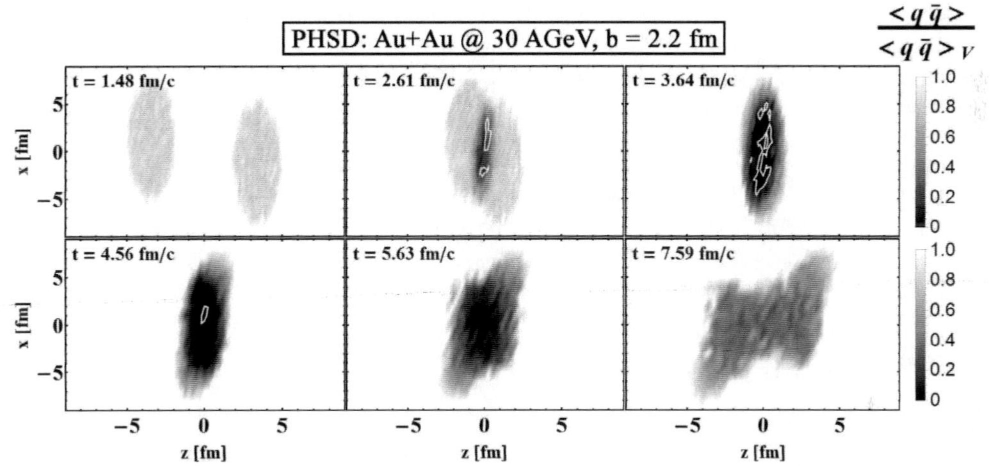

Figure 2. The ratio (2) for the scalar quark condensate as a function of x and z (for $y = 0$) at different times t for central Au+Au collisions at 30 AGeV. The white borderline separates the space-time regions of deconfined matter and hadronic matter.

Incorporating the effective masses (6) into the probability (1), we can determine the effects of CSR in the production of hadrons by string fragmentation. In order to illustrate our findings we show the ratios K^+/π^+ and $(\Lambda + \Sigma^0)/\pi^-$ at midrapidity from 5% central A+A collisions in Fig. 3 as a function of the invariant energy $\sqrt{s_{NN}}$ in comparison to the experimental data available [17]. It is clearly seen from Fig. 3 that the results in the conventional scenario (without incorporating the CSR) clearly underestimate the ratios at low $\sqrt{s_{NN}}$ for both HSD and PHSD – as found earlier in Refs. [7, 8] – while the inclusion of CSR leads to results significantly closer to the data. Especially, the rise of the K^+/π^+ ratio at low invariant energy follows closely the experimental excitation function when incorporating 'chiral symmetry restoration'. Then a drop is observed in the PHSD case due to the appearance of a deconfined medium at higher bombarding energies contrary to the HSD case where an overestimation was expected.

4. Conclusions

When comparing the results from the extended PHSD approach for the ratios K^+/π^+ and $(\Lambda + \Sigma^0)/\pi^-$ from the different scenarios we see in Fig. 3 that the results from PHSD fail to

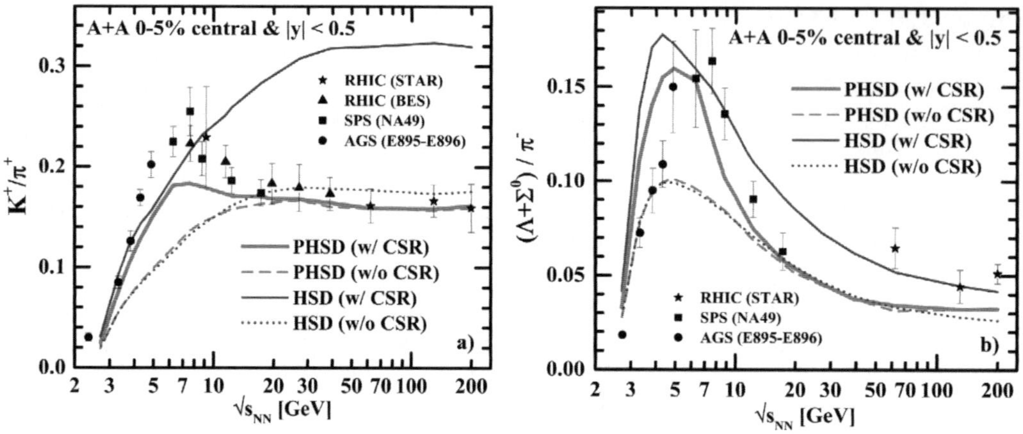

Figure 3. The ratios K^+/π^+ and $(\Lambda + \Sigma^0)/\pi^-$ at midrapidity from 5% central A+A collisions as a function of the invariant energy $\sqrt{s_{NN}}$ in comparison to the experimental data from [17]. The results from PHSD3.3 and HSD3.3 (with CSR) are displayed in terms of the full lines while the dashed lines show the default results without CSR.

describe the data in the conventional scenario [18] without incorporating the CSR. Especially, the rise of the K^+/π^+ ratio at low invariant energies follows closely the experimental excitation function when including 'chiral symmetry restoration' in the string decay. Nevertheless, the drop in this ratio again is due to 'deconfinement' since there is no longer any hadronic string decay in a partonic medium at higher energies. Accordingly, the experimental 'horn' in the excitation function is caused by chiral symmetry restoration but also deconfinement is essential to observe a maximum in the K^+/π^+ ratio.

Acknowledgements

The authors acknowledge the support by BMBF, HIC for FAIR and the HGS-HIRe for FAIR. The computational resources were provided by the LOEWE-CSC.

References

[1] W. Cassing *et al.*, Phys. Rev. C **93** (2016), 014902
[2] J. Rafelski and B. Müller, Phys. Rev. Lett. **48** (1982) 1066.
[3] R. Stock, J. Phys. G **28** (2002) 1517.
[4] M. Gazdzicki and M. I. Gorenstein, Acta Phys. Polon. B **30** (1999) 2705.
[5] M. Gazdzicki *et al.*, Acta Phys. Polon. B **42** (2011) 307.
[6] J. Geiss, W. Cassing, and C. Greiner, Nucl. Phys. A **644** (1998) 107.
[7] E. L. Bratkovskaya *et al.*, Phys. Rev. C **69** (2004) 054907.
[8] H. Weber *et al.*, Phys. Rev. C **67** (2003) 014904.
[9] W. Cassing and E.L. Bratkovskaya, Nucl. Phys. A **831** (2009) 215.
[10] E. L. Bratkovskaya *et al.*, Nucl. Phys. A **856** (2011) 162.
[11] O. Linnyk *et al.*, Prog. Part. Nucl. Phys. **87** (2016) 50.
[12] B. Nilsson-Almqvist and E. Stenlund, Comp. Phys. Comm. **43** (1987) 387; B. Andersson, G. Gustafson, and H. Pi, Z. Phys. C **57** (1993) 485.
[13] J. Schwinger, Phys. Rev. **83** (1951) 664.
[14] B. Friman, W. Nörenberg and V. D. Toneev, Eur. Phys. J. A **3** (1998) 165.
[15] J. Boguta and A. R. Bodmer, Nucl. Phys. A **292** (1977) 413.
[16] A. Lang et al., Z. Phys. A **340** (1991) 287.
[17] B. I. Abelev *et al.* [STAR Collaboration], Phys. Rev. C **81** (2010) 024911; M. M. Aggarwal *et al.* [STAR Collaboration], Phys. Rev. C **83** (2011) 024901.
[18] P. Moreau *et al.*, J. Phys. Conf. Ser. **668** (2016), 012072

Quench calculations for the superconducting dipole magnet of CBM experiment at FAIR

P. Kurilkin[1,4], P. Akishin[1], A. Bychkov[1], E. Floch[2], Yu. Gusakov[1], V. Ladygin[1], A. Malakhov[1], G. Moritz[2], H. Ramakers[2], P. Senger[2], A. Shabunov[1], P. Szwangruber[2] and F. Toral[3]

1 JINR, Joliot-Curie 6, 141980, Dubna, Russia.
2 GSI, Planckstraße 1, 64291 Darmstadt, Germany.
3 CIEMAT, 28040, Madrid, Spain.
4 FSBI "SSC RF ITEP" of NRC "Kurchatov Institute", 117218, Moscow, Russia.

E-mail: pkurilkin@jinr.ru

Abstract. The scientific mission of the Compressed Baryonic Matter (CBM) experiment is the study of the nuclear matter properties at the high baryon densities in heavy ion collisions at the Facility of Antiproton and Ion Research (FAIR) in Darmstadt. The 5.15 MJ superconducting dipole magnet will be used in the silicon tracking system of the CBM detector. It will provide a magnetic field integral of 1 Tm which is required to obtain a momentum resolution of 1% for the track reconstruction. This paper presents quench modeling and evaluation of candidate protection schemes for the CBM dipole magnet. Two quench programs based on finite-difference method were used in simulation. One of them is currently used at GSI, and the other based on CIEMAT (Madrid, Spain) was modified to perform quench calculation for the CBM magnet.

1. Introduction

The Compressed Baryonic Matter (CBM) will be one of the four major scientific experiments of the future Facility for Antiproton and Ion Research (FAIR) in Darmstadt[1]. The goal of the CBM research program is to explore the QCD phase diagram in the region of high baryon densities using high-energy nucleus-nucleus collisions. This includes the study of the equation-of-state of nuclear matter at high densities, and the search for the deconfinement and chiral phase transitions.
The geometrical acceptance of the CBM detector systems is full 2π for azimuth (ϕ) and from 2.5° to 25° for polar (Θ) angles. Muon and electron pairs created during particle decays will be measured during separate runs. In electron configuration the following detectors will be used: Micro-vertex Detector (MVD), Silicon Tracking System (STS), then Ring Imaging Cherenkov Detector, Transition Radiation Detectors, Resistive Plate Chambers for time-of-flight measurements, Electromagnetic Calorimeter (ECAL) and Projectile Spectator Detector. In muon configuration the RICH detector will be replaced by the Muon Detection System and ECAL will be removed.

The superconducting magnet [2] is a central component of the CBM detector setup. The target, MVD and STS with a polar angle acceptance of $\pm 25^\circ$ and a horizontal acceptance of $\pm 30^\circ$ will be placed in the magnet gap. For design of any superconducting magnet, an analysis of its stability is necessary. An important phenomenon in superconducting magnets is quenching.

Content from this work may be used under the terms of the Creative Commons Attribution 3.0 licence. Any further distribution of this work must maintain attribution to the author(s) and the title of the work, journal citation and DOI.
Published under licence by IOP Publishing Ltd

Journal of Physics: Conference Series **742** (2016) 012017 doi:10.1088/1742-6596/742/1/012017

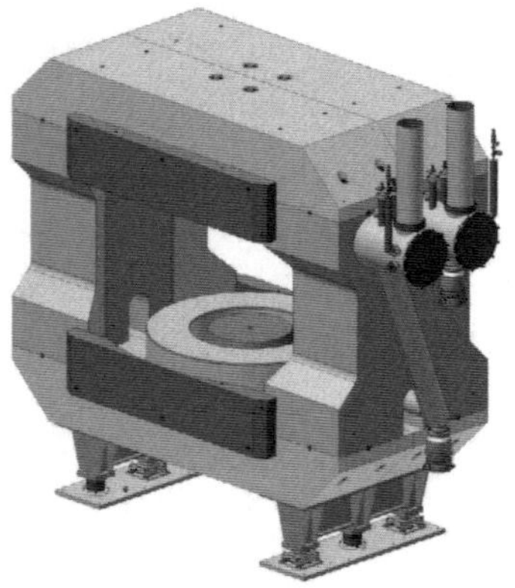

Figure 1. Perspective view of the CBM dipole magnet with support. The pole-gap width is 1.05 m.

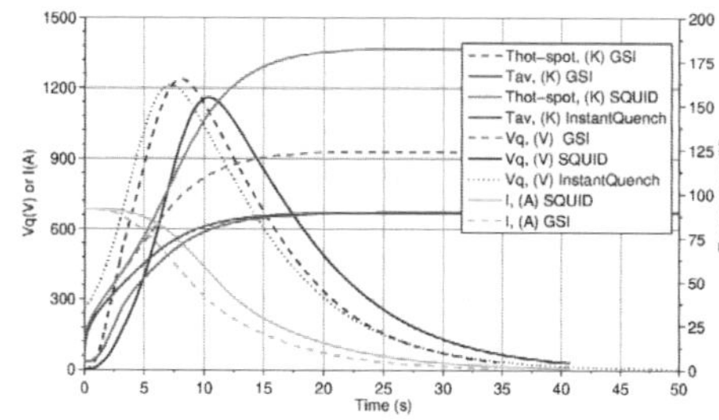

Figure 2. 3D Quench modelling results for the short-circuited CBM magnet: magnet current, coil voltage, hot spot and average coil temperatures.

Quench is a process of a resistive transition in a superconducting magnet. There are many factors that cause a superconducting coil of a magnet to a quench, such as mechanical disturbance, heat disturbance, nuclear radiation, flux jump, etc.[3, 4, 5]. During the quench, the energy stored in the magnet, and power provided by the power supply can cause a high temperature rise in the region where quench started, as well as high voltages in the coil and between the coil and ground. These factors may char the insulation or even melt the conductor and destroy the magnet. Thus the quench must be detected fast and some actions should lead to a shutdown of the power supply and the discharge of the magnet by dissipation of its energy via a dump resistor, or onto the own thermal mass of the magnet.

During last decades, several approaches to model the quench propagation in the superconducting magnets have been proposed [6]. In this paper, the results of the numerical quench modeling and evaluation of candidate protection schemes for the CBM dipole magnet are presented.

2. CBM dipole magnet

The H-type superconducting magnet of the CBM experiment is designed to provide a vertical magnetic field with bending power of 1 Tm over a length of 1 m from the target. A perspective view of the magnet with the support is presented in Fig. 1. The magnet yoke consists of two conical poles and two horizontal and two vertical parts to close a flux return. Two cylindrical superconducting coils will be installed in their respective cryostats and will be cooled separately by helium bath based cooling system like the SAMURAI magnet at RIKEN[7]. The maximum energy storage is evaluated as 5.15 MJ when the operating current of the CBM magnet is rated at 686 A. The design of coil for the CBM magnet is based on the design of the Super-FRS dipole[8]. Each coil has 1749 turns and 53 layers. The superconducting wire is a Nb/Ti wire. The conductor insulation consists of 0.1 mm polyimide type and 0.2 mm glass fiber (G10) material. Each layer is additionally insulated from the others by three layers of 0.1 mm fiber glass fabric with epoxy resin. The detailed description of the CBM magnet can be found in [2].

3. Quench calculations for the CBM magnet

In the magnet design it is necessary to calculate the temperature, voltage and current in case of a quench. There are several levels of quench simulation. The simplest level of modeling is

Journal of Physics: Conference Series **742** (2016) 012017 doi:10.1088/1742-6596/742/1/012017

performed within the framework of adiabatic model. This model is based on implementation of the well-known balance equation between Joule heating due to the current and temperature increase through specific heats. The heat propagation through the coil on this level is ignored. One of such approximation is "instantaneous quench"[2, 9]. The "instantaneous quench" means that the whole coil is heated up instantaneously above the critical temperature. The calculation is performed for short-circuit magnet and gives good estimation of the maximum quench voltage. The value of average temperature will be lower than real "hot spot" (the location of initial quench in the coil) temperature.

The second level of approximation includes the heat propagation within the coil. This implies having a multidimensional mesh of the coil and the code on this level incorporates the magnetic distribution in the coil and also the model of critical surface. The heat exchange between the coil and helium bath is not taken into account on this level. The 3D quench calculations for the CBM magnet coil have been carried out by two numerical programs based on the finite-difference method. One of them is used currently at GSI and its detailed description can be found in [10].The other (SQUID) based on CIEMAT numerical code[11] was adapted to perform the quench simulation for the CBM magnet. The coil is represented as a straight slab with a length equal to the average turn length. A regular mesh is applied to the slab so that the cross-section of the one mesh element equal to the cross-section of the insulated conductor. Longitudinally, the coil is divided into a number of slices which define a mesh element size. The material properties are homogeneous within each mesh element.

The electrical equation for the short-circuit magnet is predicted by the following differential equation:

$$L_d(I)\frac{dI}{dt} + R_q(T_{av}) \cdot I = 0 \tag{1}$$

where L_d, I, R_q and T_{av} are the differential inductance, magnet current, quench resistance and average temperature, respectively.

The heat balance equation per unit volume of the coil can be presented as:

$$\rho(T) \cdot J^2 = C(T)\frac{dT}{dt} - \nabla \cdot (K(T)\nabla T) \tag{2}$$

where $C(T)$ is specific heat per unit volume ($Jm^{-3}K^{-1}$), $K(T)$ is thermal conductivity ($Wm^{-1}K^{-1}$), $\rho(T)$ is resistivity ($\Omega \cdot m$). It means that the energy deposited due to Joule effect in the unit volume is stored in the material increasing its temperature and it is evacuated to the nearby material by conduction. This equation is solved by means of the finite-difference method. The quench programs couple the thermal (Eq.2) and electrical (Eq.1) equations at each time step. The material properties (NbTi, Cu, insulation) are recalculated for each iteration. They depend on the temperature and magnetic field. The quench origin is simulated by the set of temperature of one (or several, depending on the minimal propagation zone) element slightly above its critical value. The quenched element is located in the inner layer of the coil where the field is higher. In the simulation one can get the "hot spot" temperature, coil voltage, current and also quench propagation velocities.

The results of quench modeling for the CBM coil at 686 A are presented in Fig.2. The results of the "hot spot" temperatures calculated by GSI and SQUID 3D quench program are shown by red solid and red dashed lines, respectively. The average coil temperature is presented by the green and brown lines for GSI and "Instantaneous quench" calculations, respectively. The maximum temperature during the quench is about 180 K. All models give that the maximum quench voltage is about of 1200 V. The magnetic field distribution in the coil of the CBM magnet taken from designers has been used in the SQUID computations, while the GSI quench simulations have been performed for the uniform distribution of the magnetic field. It should be noted that the SQUID modeling has been performed for the double insulated conductor

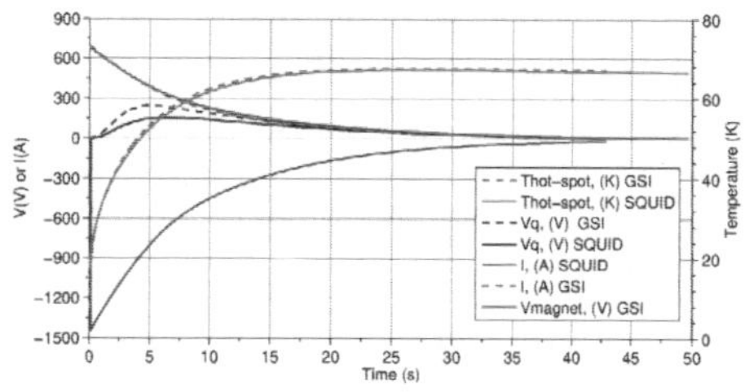

Figure 3. 3D quench simulation for the CBM magnet: magnet current, coil voltage, magnet voltage and "hot spot" temperature. $R_d = 2.1\Omega$.

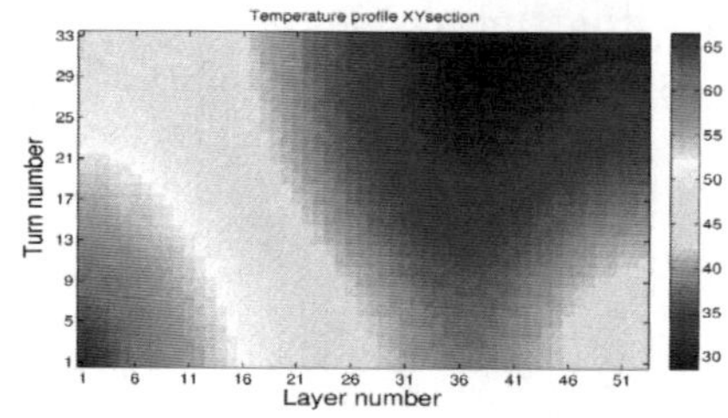

Figure 4. 3D quench simulation for the CBM magnet: themperature distribution in the coil cross section. $R_d = 2.1\Omega$.

(0.2 mm G10 and 0.1mm polyimide) while the GSI computation has been done for the single insulated (0.3 mm G10) one. There is also the difference in the material data bases used in the calculations. The protection system is necessary to reduce the maximum "hot spot" temperature during the quench bellow 100 K. Two possibile quench protection systems for the CBM magnet were considered.

The first one is based on the extraction of the energy stored in the magnet via a dump resistor. The details can be found in [2]. The dump resistor is connected in parallel to the magnet and it is always on. In case of the quench the magnet will be disconnected from the power supply and the magnet current will be dumped via the dump resistor. The dump resistor with resistance of $1.9-2.1\Omega$ will be used. The quench simulation showed that in this case around $80-86\%$ of the energy stored in the magnet (5 MJ) will be dissipated outside of the cryostat. Fig.3 presents the quench calculation results using the 2.1 Ohm dump resistor for the energy extraction. Simulation shows that the maximal "hot spot" temperature in the coil in this case is about 70 K. The maximum voltage across the magnet (1441 V) occurs when the dump resistor turns on. Fig.4 shows the temperature profile distribution in the coil cross section when the "hot spot" temperature has a maximum value.

The second quench protection system is similar that is used for the SAMURAI dipole magnet at RIKEN (JAPAN)[7]. This system is based on the heating of the outer layer of the coil. In this case, the transition of the winding of the coil to the normal state occurs faster than in the previous case. In this scheme, the superconducting coil is connected in parallel with the low-temperature ("cold") diodes and a copper heater wire that are connected in series. The quench detector monitors the difference of the voltage between the upper and the lower coils. In case of a quench, the coils are cut off from the power supply, and the current flows in the closed circuit consisting of the superconducting coil, the diodes and the heater wire. The outermost layers of both coils are forced to quench simultaneously due to the heat generated by the heater wire. Then the normal zone in the coils propagates from the outer to the inner layer. The simulations of the quench process at different start condition and resistance of the coil heater have been performed[12]. The maximum quench voltage is about 1200 V. Fig.5 and Fig.6 present the temperature distribution in the cross section of the CBM magnet coil in case when the quench appears in the inner and outer layer of the coil, respectively. For the heating of the outermost layer of the coil, the copper heater with 33 turns of wire with the cross section of $2.02 \times 3.25mm^2$ was used. The maximum "hot spot" temperature is about 140 K. The obtained results show also the big temperature gradient in the CBM magnet coil cross section in case of the heating of the coil. The use of the copper heater with the cross section of $2.02 \times 3.25mm^2$ in the quench simulations gives the similar results.

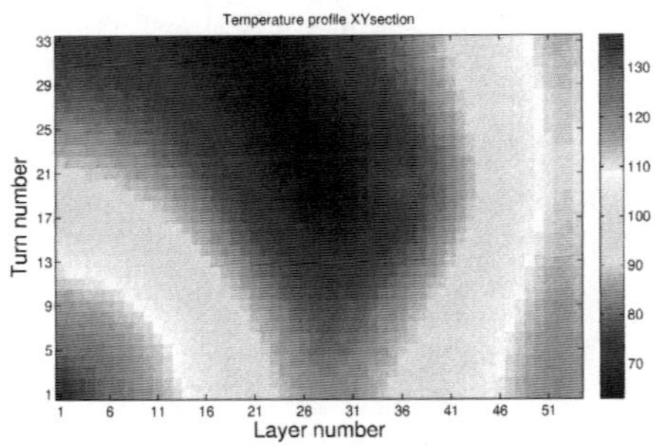

Figure 5. 3D quench simulation for the CBM magnet: temperature distribution in the coil cross section. Quench appears in the inner layer of the coil winding. Heater parameters are described in the text.

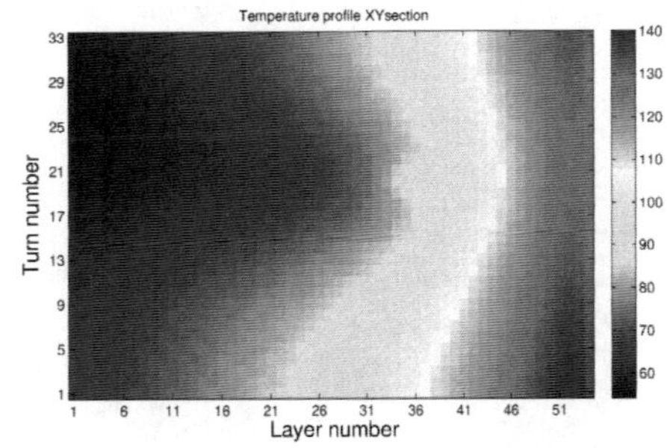

Figure 6. 3D quench simulation for the CBM magnet: temperature distribution in the coil cross section.Quench appears in the outer layer of the coil winding.Heater parameters are described in the text.

4. Conclusion

The research program of the Compressed Baryonic Matter experiment aims at the exploration of the QCD phase diagram in the regime of high densities and moderate temperatures. The superconducting dipole magnet for the CBM experiment is designed to have a large pole gap and wide horizontal opening. It will be a unique component of the CBM detector setup. The SQUID and GSI[10] quench programs have been used to the quench simulation of the CBM dipole magnet. Both programs properly take the dependence of strong variation of the matter properties from the temperature properties into account. They also use the adaptive time stepping which significantly reduce calculation time. Using both programs, we computed that the energy extraction resistor needs to extract 85% of the energy stored in the magnet.

The quench protection system based on the energy evacuation from the dump resistor of $1.9-2.1\Omega$ is chosen for the CBM magnet due to the lower "hot spot" temperature and the lower temperature gradient in the coil winding.

Acknowledgments

The work has been supported in part by the BMBF-JINR and FRCC grants.

References

[1] Compressed Baryonic Matter (CBM) http://www.fair-center.eu/public/experiment-program/nuclear-matter-physics/cbm.html
[2] The CBM collaboration, 2014, Technical Design Report for the CBM Superconducting Dipole Magnet.
[3] Y. Wang, 2013, *Fundamental elements of Applied superconductivity in electrical Engineering*, (Science press: Wiley) 114-135
[4] M.Wilson, 1983, *Superconducting magnet*, (London U.K.: Oxford Univ. Press), 200-232
[5] Yu.Iwasa, 2009, *Case studies in superconducting magnet: design and operational issues* (Springer)
[6] H.Felice et al., *CERN Yellow Report CERN-2013-006* 17-20; arXiv:1401.3934v1[physics.acc-ph]
[7] H.Sato et al., 2013, *IEEE Trans. Appl. Supercond.*, **23**, No3, 4500308
[8] L. Z. Ma et al., June 2012, *IEEE Trans. Appl. Supercond.*, **22** No3, 9502104
[9] E.Floch, private communication
[10] P. Szwangruber et al., 2013, *IEEE Transactions on Applied Superconductivity*, **23** No.3 4701704
[11] F. Toral, 2001, "Design and Calculation Procedure for Particle AcceleratorSuperconducting Magnets: Application to an LHC Superconducting Quadrupole", *Ph. D. Thesis.* (Madrid)
[12] P.Kurilkin et al., 2016, *CBM progress report 2015*, (Darmstadt) 6-7

Three-flavor chiral effective model with four baryonic multiplets within the mirror assignment

L Olbrich[1], M Zétényi[2], and F Giacosa[1,3]

Abstract. We study three-flavor octet baryons by using the so-called extended Linear Sigma Model (eLSM). Within a quark-diquark picture, the requirement of a mirror assignment naturally leads to the consideration of four spin-$\frac{1}{2}$ baryon multiplets. A reduction of the Lagrangian to the two-flavor case leaves four doublets of nucleonic states which mix to form the experimentally observed states $N(939)$, $N(1440)$, $N(1535)$ and $N(1650)$. We determine the parameters of the nucleonic part of the Lagrangian from a fit to masses and decay properties of the aforementioned states. By tracing their masses when chiral symmetry is restored, we conclude that the pairs $N(939)$, $N(1535)$ and $N(1440)$, $N(1650)$ form chiral partners.

[1] Institute for Theoretical Physics, Goethe University, Frankfurt am Main, Germany
[2] Wigner Research Center for Physics, Budapest, Hungary
[3] Institute of Physics, Jan Kochanowski University, Kielce, Poland

E-mail: olbrich@th.physik.uni-frankfurt.de, zetenyi.miklos@wigner.mta.hu, fgiacosa@ujk.edu.pl

1. Introduction

In these proceedings, which are based on the results of Ref. [1], we focus our attention on baryons consisting of the light quarks u, d, and s and with quantum numbers $J^P = \frac{1}{2}^+$ and $J^P = \frac{1}{2}^-$. First and foremost, this includes the nucleon $N(939)$ and the resonances $N(1440)$, $N(1535)$, $N(1650)$, but also Λ, Σ, and Ξ resonances [2].

The fundamental force of nature describing baryons (and hadrons in general) is Quantum Chromodynamics (QCD), whose Lagrangian is given in terms of quarks and gluons by (e.g. Ref. [3]):

$$\mathcal{L}_{\text{QCD}} = \bar{q}\big(i\gamma_\mu D^\mu - m\big)q - \frac{1}{2}\,\text{Tr}(\mathcal{G}_{\mu\nu}\mathcal{G}^{\mu\nu}) \ , \tag{1}$$

where $D^\mu = \partial_\mu + ig\mathcal{A}_\mu$ is the covariant derivative and g is the coupling "constant" parametrizing the interaction of the quark fields q with a gluon field $\mathcal{A}_\mu$. The latter is associated with the $SU(3)_c$ gauge field $\mathcal{A}^\mu = A_a^\mu T^a$ for $a = 1, 2, \ldots, 8$, T^a being the $SU(3)$ generators. The Yang-Mills field-strength tensor is $\mathcal{G}_{\mu\nu} = i[D_\mu, D_\nu]/g$. Definitely, the QCD Lagrangian is elegant, compact, and contains only few parameters (the bare quark masses and the coupling g). However, a more detailed study within the framework of renormalization shows that the coupling "constant" of strong interaction depends on momentum scale and becomes arbitrarily large in the low-energy regime. This fact forbids the usual approach of perturbation theory and causes QCD to be not analytically solvable. On top of that, only "white", i.e. color singlet states (the hadrons) are the asymptotic states of the theory, which have been measured in detectors. This property goes under the name of confinement.

Content from this work may be used under the terms of the Creative Commons Attribution 3.0 licence. Any further distribution of this work must maintain attribution to the author(s) and the title of the work, journal citation and DOI.
Published under licence by IOP Publishing Ltd

In order to describe hadrons, theoretical physicists had to look for alternatives. One such possibility is given by effective approaches to QCD, such as chiral perturbation theory (e.g. Refs. [4]) or linear Sigma Models (see below). In this process one uses Lagrangians which contain hadrons as degrees of freedom, instead of quarks and gluons as in QCD. To get the basic idea of an effective model, we make a short digression by asking under which conditions a glass of water freezes. The answer is easy: it depends on the temperature of water. But why is this answer so simple? Actually, the exact description of the system requires to solve coupled differential equations for each molecule. This is indeed very complicated, actually impossible also for supercomputers. However, by using the effective description provided by thermodynamics, one can solve (some) problems in a much simpler way. In doing so, we lost the information about the motion of each molecule, but we retained the information that is important for describing the whole system.

Let us then turn back to hadrons with half-integer spin: baryons. A baryon is a very complicated system of quarks and gluons arising from Eq. (1). For a fast moving baryon, as for instance a proton, one may use (generalized) parton distribution functions to analyze its substructure in terms of quarks and gluons, see e.g. Ref. [5]. However, for a baryon in its rest frame, this is a very hard task, possibly even harder than following each molecule of water mentioned above. The use of effective descriptions highly simplify the task. Typically, one uses the concept of 'constituent quarks': an almost massless bare quark is dressed by clouds of gluons and quarks and becomes a quasiparticle with an effective mass of about 300 MeV [3]. A baryon is then described as a bound state of three constituent quarks. Within this framework, also the concept of diquark as the strong correlation of two quarks is important, because often baryons are also regarded as quark-diquark objects.

Certain types of effective models, such as linear Sigma Models, are based on Lagrangians which contain from the very beginning only hadrons (mesons and baryons). Quark and gluon fields do not appear. Yet, these Lagrangians are constructed in such a way that (some of the) symmetries of QCD of Eq. (1) are taken into account at he composite level. In particular, chiral symmetry [3] and its spontaneous and explicit breaking are at the basis of such approaches. Recently, the so-called extended Linear Sigma Model (eLSM) has been developed for both mesons and baryons. It contains from the very beginning scalar, pseudoscalar, vector, and axial-vector mesons. Moreover, it shows chiral symmetry and its breaking but also dilatation symmetry and its anomalous breaking (this is the non-perturbative origin of an energy scale of QCD, the renowned $\Lambda_{QCD} \simeq 200$ MeV). (Additional chiral multiplets such as tensor and pseudotensor mesons or pseudovector and excited vector mesons could be also included in a chirally invariant way.)

In the meson sector, the three-flavour case $N_f = 3$ has been investigated in detail in Ref. [6, 7]. Yet, until recently, the baryonic sector was only studied for $N_f = 2$ in Ref. [8]. Here, following Ref. [1], we enlarge (Sec. 2) the eLSM to $N_f = 3$ in the baryonic sector. This is a non-trivial step which naturally leads to the consideration of four baryonic octets. Next, (Sec. 3) we consider the limiting case in which only baryons with quarks u and d are considered and discuss the mixing patterns and the identification of the chiral partner of the nucleon. Finally, in Sec. 4 we present our conclusions and outlooks.

2. The eLSM and its implications

The mesonic part of the eLSM Lagrangian containing (pseudo)scalar and (axial-)vector mesons is given in Refs. [6, 7]. The inclusion of baryons was performed for $N_f = 2$ in Refs. [8] and investigated at finite density in Ref. [9]. Recently, the development of the baryonic eLSM to $N_f = 3$ was undertaken in Ref. [1]. The basic idea is to construct baryonic fields in a chiral quark-diquark picture. For $N_f = 3$ diquarks transform as antiquarks, thus one may construct baryons in a similar way as mesons. This assumption results in matrices with elements of different flavor content, which are related to the octet-baryonic fields:

$$\underbrace{\begin{pmatrix} u \\ d \\ s \end{pmatrix}}_{\text{quark}} \underbrace{([d,s]\,,[s,u]\,,[u,d])}_{\text{diquark}} \hat{=} \begin{pmatrix} uds & uus & uud \\ dds & uds & udd \\ dss & uss & uds \end{pmatrix} \sim \begin{pmatrix} \frac{\Lambda}{\sqrt{6}}+\frac{\Sigma^0}{\sqrt{2}} & \Sigma^+ & p \\ \Sigma^- & \frac{\Lambda}{\sqrt{6}}-\frac{\Sigma^0}{\sqrt{2}} & n \\ \Xi^- & \Xi^0 & -\frac{2\Lambda}{\sqrt{6}} \end{pmatrix}. \tag{2}$$

When enlarging the present discussion by taking into account the chirality of quarks and diquarks (see Ref. [10]) and requiring chirally invariant mass terms, we naturally obtain four baryonic multiplets, two of which transform in a standard way under chiral transformations and two in a so-called "mirror" way. These four multiplets are represented by four matrices analogous to Eq. (2). Two of these matrices labeled N_1 and N_2 behave under chiral transformations as

$$\begin{aligned} N_{1R} &\to U_R N_{1R} U_R^\dagger\,, & N_{1L} &\to U_L N_{1L} U_R^\dagger, \\ N_{2R} &\to U_R N_{2R} U_L^\dagger\,, & N_{2L} &\to U_L N_{2L} U_L^\dagger\,, \end{aligned} \tag{3}$$

where U_L and U_R are 3×3 representation matrices of $U(3)_L$ and $U(3)_R$. The remaining two matrices M_1 and M_2 show a chiral transformation from the left that is 'mirror-like' compared to the aforementioned:

$$\begin{aligned} M_{1R} &\to U_L M_{1R} U_R^\dagger\,, & M_{1L} &\to U_R M_{1L} U_R^\dagger \\ M_{2R} &\to U_L M_{2R} U_L^\dagger\,, & M_{2L} &\to U_R M_{2L} U_L^\dagger\,. \end{aligned} \tag{4}$$

These transformations comply a so-called "mirror assignment" [11] which allows one to introduce chirally invariant baryon mass terms in the Lagrangian. The Lagrangian describing these baryonic fields and their interactions with mesonic degrees of freedom is invariant under chiral symmetry $U(3)_R \times U(3)_L$ as well as parity and charge-conjugation transformations [1]. It reads

$$\begin{aligned} \mathcal{L}_{N_f=3,bar} = \; & \mathrm{Tr}\left\{\bar{N}_{1L} i\gamma_\mu D^\mu_{2L} N_{1L} + \bar{N}_{1R} i\gamma_\mu D^\mu_{1R} N_{1R} + \bar{N}_{2L} i\gamma_\mu D^\mu_{1L} N_{2L} + \bar{N}_{2R} i\gamma_\mu D^\mu_{2R} N_{2R}\right\} \\ & + \mathrm{Tr}\left\{\bar{M}_{1L} i\gamma_\mu D^\mu_{4R} M_{1L} + \bar{M}_{1R} i\gamma_\mu D^\mu_{3L} M_{1R} + \bar{M}_{2L} i\gamma_\mu D^\mu_{3R} M_{2L} + \bar{M}_{2R} i\gamma_\mu D^\mu_{4L} M_{2R}\right\} \\ & - g_N \,\mathrm{Tr}\left\{\bar{N}_{1L}\Phi N_{1R} + \bar{N}_{1R}\Phi^\dagger N_{1L} + \bar{N}_{2L}\Phi N_{2R} + \bar{N}_{2R}\Phi^\dagger N_{2L}\right\} \\ & - g_M \,\mathrm{Tr}\left\{\bar{M}_{1L}\Phi^\dagger M_{1R} + \bar{M}_{1R}\Phi M_{1L} + \bar{M}_{2L}\Phi^\dagger M_{2R} + \bar{M}_{2R}\Phi M_{2L}\right\} \\ & - m_{0,1}\,\mathrm{Tr}\left\{\bar{N}_{1L} M_{1R} + \bar{M}_{1R} N_{1L} + \bar{N}_{2R} M_{2L} + \bar{M}_{2L} N_{2R}\right\} \\ & - m_{0,2}\,\mathrm{Tr}\left\{\bar{N}_{1R} M_{1L} + \bar{M}_{1L} N_{1R} + \bar{N}_{2L} M_{2R} + \bar{M}_{2R} N_{2L}\right\} \\ & - \kappa_1 \,\mathrm{Tr}\left\{\bar{N}_{1R}\Phi^\dagger N_{2L}\Phi + \bar{N}_{2L}\Phi N_{1R}\Phi^\dagger\right\} - \kappa_1' \,\mathrm{Tr}\left\{\bar{N}_{1L}\Phi N_{2R}\Phi + \bar{N}_{2R}\Phi^\dagger N_{1L}\Phi^\dagger\right\} \\ & - \kappa_2 \,\mathrm{Tr}\left\{\bar{M}_{1R}\Phi M_{2L}\Phi + \bar{M}_{2L}\Phi^\dagger M_{1R}\Phi^\dagger\right\} - \kappa_2' \,\mathrm{Tr}\left\{\bar{M}_{1L}\Phi^\dagger M_{2R}\Phi + \bar{M}_{2R}\Phi M_{1L}\Phi^\dagger\right\}\,. \end{aligned} \tag{5}$$

The traces are invariant under cyclic permutation, which ensures their symmetry under the chiral transformations (3) and (4). The covariant derivatives are given by $D^\mu_{kR} = \partial^\mu - ic_k R^\mu$ and $D^\mu_{kL} = \partial^\mu - ic_k L^\mu$ for $k = 1,\ldots,4$, where the left- and right-handed matrices L^μ and R^μ represent (axial-)vector mesonic degrees of freedom. (Pseudo)scalar mesonic fields are incorporated via the Φ matrix. The mesonic matrices transform under chiral transformations as $R^\mu \to U_R R^\mu U_R^\dagger$, $L^\mu \to U_L L^\mu U_L^\dagger$, and $\Phi \to U_L \Phi U_R^\dagger$. The mass parameters $m_{0,1}$ and $m_{0,2}$ are particularly important, since they allow to shed light on the origin of the nucleonic masses. They emerge from (dilatation-invariant) interactions upon the condensation of glueball and/or four-quark states, see e.g. Ref. [9].

The baryonic fields in Eq. (5) are not parity eigenstates, therefore we construct the fields of definite parity,

$$B_N = \frac{N_1 - N_2}{\sqrt{2}}, \qquad B_{N*} = \frac{N_1 + N_2}{\sqrt{2}}, \qquad B_M = \frac{M_1 - M_2}{\sqrt{2}}, \qquad B_{M*} = \frac{M_1 + M_2}{\sqrt{2}}. \tag{6}$$

where now B_N and B_M have positive parity and B_{N*} and B_{M*} have negative parity. In the limit of zero mixing, B_N describes the ground-state baryonic fields of Eq. (2), i.e., $\{N(939), \Lambda(1116), \Sigma(1193), \Xi(1338)\}$, B_M the positive-parity fields $\{N(1440), \Lambda(1600), \Sigma(1660), \Xi(1690)\}$, B_{N*} can be assigned to the negative-parity fields $\{N(1535), \Lambda(1670), \Sigma(1620), \Xi(?)\}$ and, finally, B_{M*} to $\{N(1650), \Lambda(1800), \Sigma(1750), \Xi(?)\}$. In general the fields describing physical particles emerge as a mixture of B_N, B_{N*}, B_M, and B_{M*}. The detailed study of this mixing will be performed below for the two-flavor case.

3. Results

In order to determine the twelve parameters of the model, we reduce Eq. (5) to $N_f = 2$. This leaves us with four isodoublets (instead of the baryonic 3×3 matrices), which mix to produce the experimentally observed nucleon $N(939)$, $N(1440)$, $N(1535)$, and $N(1650)$. For the fit, we use thirteen quantities: masses and decay widths of the resonances, the axial coupling constant of the nucleon from [2], as well as the lattice results for the remaining axial coupling constants [12]. Using a standard χ^2-square procedure we found that three acceptable and almost equally deep minima exist, see Ref. [1]. The first two minima lead to small absolute values of $m_{0,1}$ and $m_{0,2}$, while the third one features absolute values of these constants comparable with the nucleon's mass (in agreement with the recent study of Ref. [13]).

Quite remarkably, for all three minima the decay $N(1535) \to N\eta$ cannot be described (it is a factor 10 too small [1]). Thus, further studies are needed to understand $N(1535)$. It may contain a sizable admixture of $s\bar{s}$, see Ref. [14], or the problem might be connected to the role of chiral anomaly in the baryonic sector [15]. The assignment of chiral partners can also be investigated by computing the masses as a function of the chiral condensate φ_N, because masses of chiral partners become degenerate for $\varphi_N \to 0$. For all minima, the result shows that the masses of the $N(939)$ and $N(1535)$ as well as $N(1440)$ and $N(1650)$ merge as $\varphi_N \to 0$, therefore these form two pairs of chiral partners.

Finally, as an illustration, we present here one of the resulting mixing matrices (Minimum 1 in [1]):

$$\begin{pmatrix} N(939) \\ \gamma^5 N(1535) \\ N(1440) \\ \gamma^5 N(1650) \end{pmatrix} = \begin{pmatrix} \mathbf{-0.996} & -0.025 & -0.046 & -0.074 \\ 0.075 & \mathbf{-0.492} & 0.039 & \mathbf{-0.867} \\ -0.050 & -0.057 & \mathbf{0.995} & 0.073 \\ 0.010 & \mathbf{0.869} & 0.086 & \mathbf{-0.488} \end{pmatrix} \begin{pmatrix} B_N \\ \gamma^5 B_{N*} \\ B_M \\ \gamma^5 B_{M*} \end{pmatrix}. \tag{7}$$

In fact, although the mixing has been determined for $N_f = 2$ only, our calculations show that it is a good first approximation for all the members of the octet.

4. Summary and outlook

The eLSM is an effective model of QCD whose building blocks are hadrons. Starting from its Lagrangian, one can perform calculations that are not yet possible within QCD. In this work we have generalized the eLSM to the three-flavor case. Requiring chirally invariant mass terms one is lead to use the so-called "mirror assignment", and naturally obtains four baryonic multiplets. In order to determine the parameters, we have performed a reduction to the $N_f = 2$ case and a fit to experimentally known quantities. Three minima produce results that are in good agreement with experiment (expect for the decay width $N(1535) \to N\eta$). Furthermore, we concluded that the pairs $N(939)$, $N(1535)$ and $N(1440)$, $N(1650)$ form chiral partners. The most important open problem is to decide which of the minima is preferable. To this end, we will investigate the complete $N_f = 3$ case by performing an overall fit to measure physical quantities. As a consequence, interesting information for both vacuum physics, such as scattering processes involving strange hadrons [16], and at nonzero density, such as the role of hyperons in compact stars [17], will be obtained.

Acknowledgments

L.O. acknowledges support by HGS-HIRe/HQM. M.Z. was supported by the Hungarian OTKA Fund No. K109462 and HIC for FAIR.

References

[1] Olbrich L, Zétényi M, Giacosa F and Rischke D H 2016 *Phys. Rev.* D **93** 034021
[2] Olive K A *et al* [Particle Data Group Collaboration] 2014 *Chin. Phys.* C **38** 090001
[3] Thomas A W and Weise W, *The Structure of the Nucleon*, Berlin, Germany: Wiley-VCH (2001) 389 p
[4] Scherer S 2003 *Adv. Nucl. Phys.* **27** 277
Fearing H W, Scherer S 1996 *Phys. Rev.* D **53** 315
[5] Belitsky A V and Radyushkin A V 2005 *Phys. Rept.* **418** 1
[6] Parganlija D, Kovacs P, Wolf G, Giacosa F and Rischke D H 2013 *Phys. Rev.* D **87** 014011
[7] Parganlija D, Giacosa F and Rischke D H 2010 *Phys. Rev.* D **82** 054024; Janowski S, Parganlija D, Giacosa F and Rischke D H 2011; *Phys. Rev.* D **84** 054007
Janowski S, Giacosa F and Rischke D H 2014 *Phys. Rev.* D **90** 11, 114005
Eshraim W I, Giacosa F and Rischke D H 2015 *Eur. Phys.* J. A **51** 9, 112
Heinz A, Giacosa F and Rischke D H, *Nucl. Phys.* A **933** (2015), Pages 34-42
[8] Gallas S, Giacosa F and Rischke D H, *Phys. Rev.* D **82** (2010) 014004;
Gallas S and Giacosa F 2014 *Int. J. Mod. Phys.* A **29** no.17, 1450098
[9] Gallas S, Giacosa F and Pagliara G 2011 *Nucl. Phys.* A **872** 13;
[10] Giacosa F 2007 *Phys. Rev.* D **75** 054007
[11] DeTar C E and Kunihiro T 1989 *Phys. Rev.* D **39** 2805
Jido D, Oka M, and Hosaka A 2001 *Prog. Theor. Phys.* **106** 873;
Jido D, Nemoto Y, Oka M and Hosaka 2000 *Nucl. Phys.* A **671** 471
Zschiesche D *et al* 2007 *Phys. Rev.* C **75** 055202
Sasaki C and Mishustin I 2010 *Phys. Rev.* C **82** 035204
Glozman L Y 2007 *Phys. Rept.* **444** 1
Cohen T D and Glozman L Y 2002 *Int. J. Mod. Phys.* A **17** 1327
Glozman L Y 2002 *Phys. Lett.* B **539** 257
[12] Takahashi T T and Kunihiro T 2008 *Phys. Rev.* D **78** 011503
Takahashi T T and Kunihiro T 2007 *eConf* **C070910** 297 [*Mod. Phys. Lett.* A **23** (2008) 2340]
[13] Nishihara H and Harada M 2015 *Phys. Rev.* D **92** 5, 054022
[14] An C S and Zou B S 2009 *Sci. Sin.* G **52** 1452
Liu B C and Zou B S 2006 *Phys. Rev. Lett.* **96**, 042002
Cao X, Xie J J, Zou B S and Xu H S 2009 *Phys. Rev.* C **80** 025203
[15] Eshraim W I, Janowski S, Peters A, Neuschwander K and Giacosa F 2012 *Acta Phys. Polon. Supp.* **5** 1101
Eshraim W I, Janowski S, Giacosa F and Rischke D H 2013 *Phys. Rev.* D **87** 5, 054036.
[16] Rozek T 2006 *Phys. Lett.* B**643** 251
Valdau Yu 2010 *Phys. Rev.* C **81** 045208
Balestra F *et al* 2001 *Phys. Rev.* C **63** 024004
[17] Weissenborn S, Chatterjee D and Schaffner-Bielich J 2012 *Nucl. Phys.* A **881** 62
Drago A, Lavagno A, Pagliara G and Pigato D 2014 *Phys. Rev.* C **90** 6, 065809
[18] Deinet W, Teilab K, Giacosa F and Rischke D H, arXiv:1603.04312 [nucl-th]

Journal of Physics: Conference Series **742** (2016) 012019 doi:10.1088/1742-6596/742/1/012019

From dripline to dripline: Nuclear astrophysics in the laboratory

Zach Meisel

Department of Physics and Joint Institute for Nuclear Astrophysics, University of Notre Dame, Notre Dame, 46556 Indiana, USA

E-mail: zmeisel@nd.edu

Abstract. For the better part of a century the field of nuclear astrophysics has aimed to answer fundamental questions about nature, such as the origin of the elements and the behavior of high-density, low-temperature matter. Sustained and concerted efforts in nuclear experiment have been key to achieving progress in these areas and will continue to be so. Here I will briefly review recent accomplishments and open questions in experimental nuclear astrophysics.

1. Nearly 100 years of nuclear astrophysics

With Eddington's brilliant conjecture in 1920 that the Sun's energy reservoir "can scarcely be other than sub-atomic energy", based on ground-breaking experimental work by Aston and Rutherford, the field of nuclear astrophysics was born [1]. Work accomplished in the following decades demonstrated that nuclear physics was not only key to describing how stars lived and died, but also how they made the elements we see (and are made of) today [2, 3]. To date, great strides have been made in a worldwide effort (See Fig. 1.) toward answering fundamental questions about our universe, chief among them: *Where were the elements made?*, *How does nuclear energy generation impact stars and stellar explosions?*, and *How does matter behave at high density and low temperature?*. This progress has relied on the study of atomic nuclei over nearly the entire nuclear landscape, the region bounded by the so-called proton and neutron driplines, where protons or neutrons 'drip' out of the nucleus due to the extreme mismatch in their respective numbers.

The dramatic enhancements of experimental capabilities offered by next generation radioactive ion beam facilities such as the Facility for Rare Isotope Beams (FRIB) [4] and the NuSTAR experiments at the Facility for Antiproton and Ion Research (FAIR) [5], coupled with advances in observational and computational capabilities (e.g. Refs. [6, 7]), are certain to deepen our understanding of nature and likely yield more than a few surprises. The following sections will briefly touch on recent accomplishments in experimental nuclear astrophysics and outstanding questions across the nuclear landscape. Due to space limitations, many exciting works and research topics have been omitted. For more comprehensive reviews, see Refs. [8–10].

2. Progress on the proton-rich side

Various astrophysical environments feature nuclear reactions on the neutron-deficient side of the valley of β-stability that are responsible for nuclear energy generation and element formation. These proton-rich conditions, caused by an injection of large amounts of hydrogen from a stellar

Content from this work may be used under the terms of the Creative Commons Attribution 3.0 licence. Any further distribution of this work must maintain attribution to the author(s) and the title of the work, journal citation and DOI.
Published under licence by IOP Publishing Ltd

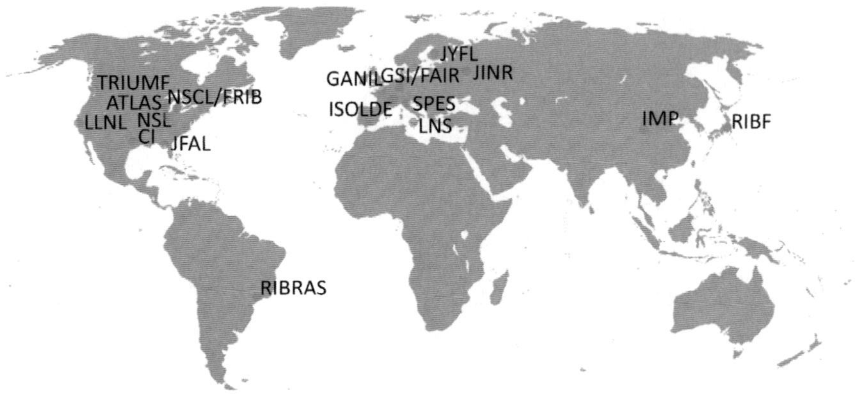

Figure 1. Global census of radioactive ion beam facilities.

envelope or transmutation of neutrons into protons via neutrino capture, enable sequences of proton-capture reactions to proceed at high-temperature. These reactions drive the rapid proton-capture (rp)-process that largely powers type-I x-ray bursts [11] and classical novae [12] and are a main player in the reaction network leading to nucleosynthesis via the neutrino-p (νp)-process in core-collapse supernovae [13].

Classical novae are thermonuclear explosions on the surfaces of white dwarf stars that recur due to the reaccumulation of hydrogen fuel from a binary companion star [12]. These explosions, aside from generating astronomical displays occasionally visible to the naked eye, contribute to the creation of light elements in the universe. Even though these objects have been a focus of intense study for decades, including numerous observational, theoretical, and experimental efforts, their exact contribution to the cosmic abundances is still unknown. Major advances have been made recently by using the power of pure, exotic radioactive ion beams to investigate the origins of presolar dust grains that may have been produced by novae (e.g. Ref. [14]). In fact, studies such as the aforementioned work make novae one of the few astrophysical phenomena for which most nuclear reaction rates are based on experimental data.

Type-I x-ray bursts are similar but much more powerful explosions recurring on the surfaces of hydrogen and helium accreting neutron stars, where the larger surface gravity of the neutron star relative to the white dwarf ultimately leads to the enhanced energy release [11]. These observables are one of the main tools used for understanding neutron stars, unique astronomical laboratories that provide insight into the behavior of dense matter at low temperature. Recent work has focused on understanding reactions that trigger the rp-process (e.g. Ref. [15]) and the locations of the nuclear landscape where the rp-process reaction sequence is significantly stalled, termed waiting-point nuclides (e.g. Ref. [16]). Though the strengths of all waiting-points are soon to be well constrained, theoretical work has shown that a host of other nuclear uncertainties remain that must ultimately be removed or reduced by future experiments [17].

3. New and old directions for neutron-rich nuclides

On the opposite side of the nuclear chart, an array of nuclear reaction sequences operate in astronomical environments to various ends, such as the forging of new elements and the alteration of dense objects' thermal and compositional structures. The oldest and most well known of these is the rapid neutron-capture (r)-process that is responsible for creating roughly half of the elements heavier than iron in as-yet undetermined astrophysical sites [18]. However, other reaction sequences such as the α-process [19] and i-process [20] have recently joined the club of potential mechanisms for nucleosynthesis on the neutron-rich side of stability. Equally exciting are the new developments that have shown individual nuclear properties are critical to understanding the various observables from neutron stars that provide unique windows into the behavior of high-density, low-temperature matter [21].

The r-process, the rapid neutron-capture sequence likely operating in core-collapse supernovae and/or neutron star mergers, has long been out of reach of even the most advanced radioactive ion beam facilities. To date, much of the relevant experimental work has focused on constraining key nuclear quantities nearer to stability so that these results can guide theoretical estimates for nuclides on the r-process path [18]. However, recent innovative techniques have allowed some of the first experimental constraints to be made for nuclear reactions on the path itself (e.g. Ref. [22]). This and other approaches will be particularly powerful when coupled with the extended reach toward the neutron dripline anticipated for FRIB (See Fig. 2.) and FAIR and promise to dramatically advance our understanding of this long-studied problem.

In the past decade other avenues for nucleosynthesis have been identified that likely operate nearer to stability on the neutron-rich side. The α-process, which may operate via a sequence of α-capture, neutron-emission reactions [23], and i-process, a reaction sequence similar in spirit to the r-process but at lower neutron densities, have only just begun to be investigated. Numerous experimental studies are anticipated in the near future that will drastically expand our knowledge of these processes and provide critical tests of their viability as mechanisms of element formation.

Lately it has been shown that neutron-rich nuclides are just as important as their proton-rich cousins with regards to their impact on our understanding of dense matter. When the proton-rich ashes of the rp-process are compressed on the neutron star surface by subsequent accretion from a binary companion, electrons are forced into nuclei, converting protons into neutrons and fundamentally altering the neutron star thermal and compositional outer structure [21]. It has been shown that it is critical to determine the properties of individual neutron-rich nuclides in order to accurately describe the accreted neutron star ocean and crust [24–26]. Efforts in the near future are planned to focus on nuclides which have been identified to have the greatest potential impact on astronomical observables [27].

4. Nuclear astrophysics near stability

Although most nuclides in and near the valley of β-stability have been accessible in the laboratory for some time, several outstanding questions in nuclear astrophysics require their further study. These involve quiescent and explosive stellar environments and require experimental approaches using both indirect and direct techniques to study the most important reactions [28]. Open questions include the extent to which photodisintegration reactions impact the cosmic elemental abundances [29], the exact abundance yield from slow neutron capture in stellar envelopes [30], and the role of electron captures in high-density astrophysical environments [31]. Progress in these areas has been driven by several nuclear physics labs around the world, especially the many stable-ion beam facilities which are far too numerous to include in Fig. 1.

Precisely describing the nuclear reaction sequences of stars has remained a challenge since Eddington first approached the subject [1]. Partially due to triumphs of astrophysical modeling and observations, such as asteroseismology and measurements of neutrinos from our sun, high precision experimental studies are needed to advance our understanding of quiescent nuclear burning in stars [28]. Specialized equipment such as recoil mass separators and underground laboratories (e.g. Refs.[32] and [33], respectively) have played and will continue to play a major role in this effort. When direct measurements via these and other methods are not possible, as is the case for some branch-point nuclides in the slow neutron-capture (s)-process reaction network, indirect techniques will be required to experimentally constrain important reactions [30].

In spite of their association with the most exotic nuclides, models of stellar explosions require a thorough understanding of nearly the entire nuclear landscape, including nuclides along and near stability. A case-in-point is the photodisintegration-driven *p*-process operating in supernovae, which is currently the favored creation mechanism of the so-called *p*-nuclides whose origins cannot be explained by the *s* and *r* processes [29]. Sustained efforts have reduced the nuclear physics uncertainties of this process, where the focus has generally been on constraining the Wolfenstein-Hauser-Feshbach reaction theory that provides essential input to astrophysics models in the absence of experimental data (e.g. Refs. [34, 35]). Additional measurements on and near stability have focused on reducing the uncertainties in nuclear weak rates that limit the ability to describe the mechanisms through which supernovae operate (e.g. Ref. [36]). Here theory calculations have provided important guidance, identifying the most essential nuclear data and filling in the large gaps left by insufficient experimental information [31].

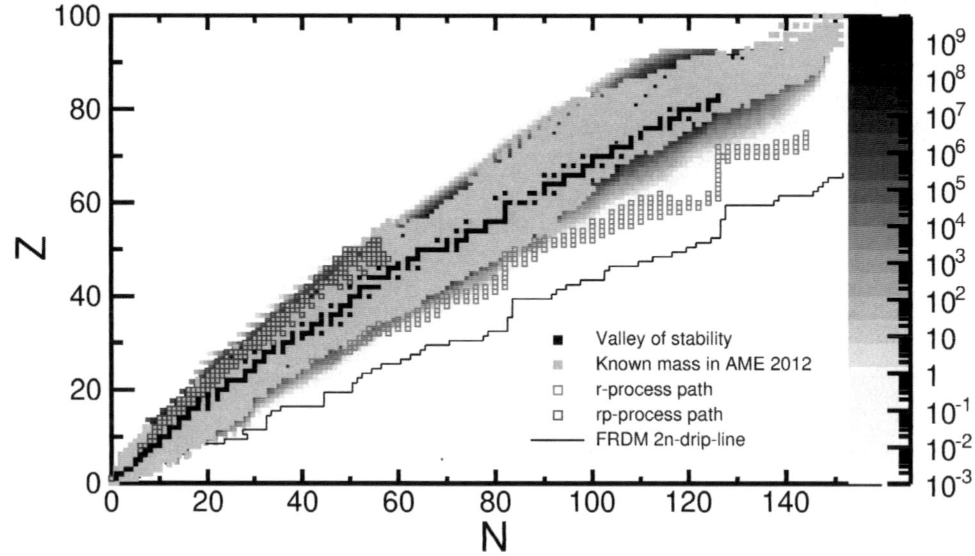

Figure 2. Predicted FRIB production rates in particles per second [37]. See Ref. [38] for a similar prediction for FAIR.

5. FRIB, FAIR, and the future

Roughly 100 years after its inception, nuclear astrophysics research continues to enhance our understanding of nature. At present the field is poised to build upon our current body of knowledge by leaps and bounds, in no small part due to upcoming developments such as new recoil separators [32, 39], underground laboratories [33], and storage rings dedicated to nuclear physics studies [38]. Frontier nuclear physics facilities such as FRIB and the NuSTAR experiments at FAIR will play a central role in this advancement by providing unprecedented access to ever more exotic nuclides (See Fig. 2.). Meanwhile, stable beam facilities will continue to play a complementary role in answering astrophysical questions both new and old. In the near future, together with advances in observation and theory, experimental nuclear astrophysics studies from dripline to dripline promise to offer profound insight into how our universe operates.

Acknowledgements

This work was supported by the National Science Foundation Grants No. 1419765 and 1430152.

References

[1] Eddington A 1920 *The Observatory* **43** 341
[2] Burbidge M E, Burbidge G R, Fowler W A and Hoyle F 1957 *Rev. Mod. Phys.* **29** 547
[3] Cameron A G W 1957 *Publ. Astron. Soc. Pac.* **69** 201
[4] http://www.frib.msu.edu
[5] http://www.fair-center.eu/for-users/experiments/nustar.html
[6] http://www.nustar.caltech.edu
[7] http://www.unedf.org
[8] Jose J and Iliadis C 2011 *Rep. Prog. Phys.* **74** 096901
[9] Wiescher M, Käppeler F and Langanke K 2012 *Annu. Rev. Astron. Astr.* **50** 165
[10] Schatz H and Wiescher M (eds) 2016 *White Paper on Nuclear Astrophysics and Low Energy Nuclear Physics: Part 1, Nuclear Astrophysics* arXiv:1603.02213
[11] Parikh A, José J, Sala G and Iliadis C 2013 *Prog. Part. Nucl. Phys.* **69** 225
[12] Bode M F and Evans A (eds) 2012 *Classical Novae* 2nd ed (Cambridge Univ. Press)
[13] Fröhlich C *et al.* 2006 *Phys. Rev. Lett.* **96** 142502
[14] Bennett M B *et al.* 2016 *Phys. Rev. Lett.* **116** 102502
[15] Parpottas Y *et al.* 2005 *Phys. Rev. C* **72** 025802
[16] Del Santo M, Meisel Z *et al.* 2014 *Phys. Lett. B* **738** 453
[17] Cyburt R H, Amthor A M, Heger A, Johnson E, Meisel Z, Schatz H and Smith K 2015 *Dependence of x-ray burst models on nuclear reaction rates.* Submitted.
[18] Mumpower M *et al.* 2016 *Prog. Part. Nucl. Phys.* **86** 86
[19] Arcones A and Montes F 2011 *Astrophys. J.* **731** 5
[20] Jones S *et al.* 2016 *Mon. Not. R. Astron. Soc.* **455** 3848
[21] Schatz H *et al.* 2014 *Nature* **505** 62
[22] Spyrou A, Liddick S N, Larsen A C, Guttormsen M *et al.* 2014 *Phys. Rev. Lett.* **113** 232502
[23] Periera J, Montes F, Arcones A and Meisel Z 2014 *Proc. of Science* **PoS (NIC XIII)** 15
[24] Estradé A, Matoš M, Schatz H *et al.* 2011 *Phys. Rev. Lett.* **107** 172503
[25] Meisel Z, George S *et al.* 2015 *Phys. Rev. Lett.* **115** 162501
[26] Meisel Z, George S *et al.* 2016 *Phys. Rev. C* **93** 035805
[27] Deibel A, Meisel Z, Schatz H, Brown E F and Cumming A 2016 *Urca cooling pairs in the neutron star ocean and their effect on superbursts.* Submitted., arXiv:1603.01281
[28] Brune C R and Davids B 2015 *Annu. Rev. Nucl. Part. S.* **65** 87
[29] Rauscher T *et al.* 2013 *Rep. Prog. Phys.* **76** 066201
[30] Reifarth R, Lederer C and Käppeler F 2014 *J. Phys. G* **41** 053101
[31] Sullivan C *et al.* 2016 *Astrophys. J.* **816** 44
[32] Couder M *et al.* 2008 *Nucl. Instrum. Meth. A* **587** 35
[33] Robertson D *et al.* 2016 *Eur. Phys. J. Web Conf.* **109** 09002
[34] Quinn S J, Spyrou A, Simon A *et al.* 2015 *Phys. Rev. C* **92** 045805
[35] Yalçin C *et al.* 2015 *Phys. Rev. C* **91** 034610
[36] Noji S, Zegers R G T, Austin S M *et al.* 2014 *Phys. Rev. Lett.* **112** 252502
[37] Bollen G, Hausmann M, Sherrill B M and Tarasov O B 2011 *FRIB Estimated Rates v1.06*
[38] Franzke B, Geissel H and Münzenberg G 2008 *Mass Spectrom. Rev.* **27** 428
[39] Meisel Z, Gilardy G, Moran M, Schmitt J, Seymour C, Stech E and Couder M 2016 *Recoil energy acceptance of the St. George recoil separator.* In preparation.

Journal of Physics: Conference Series **742** (2016) 012020 doi:10.1088/1742-6596/742/1/012020

Measurements of baryon form factors at BESIII

Cui Li for the BESIII Collaboration
Uppsala University, 75105 Uppsala, Sweden
E-mail: cui.li@physics.uu.se

Abstract. The momentum transfer dependence of the electromagnetic form factors is an important probe of the structure of hadrons at different scales. Using data samples collected with the BESIII detector at the BEPCII collider, we study the process of $e^+e^- \to p\bar{p}$ at 12 c.m. energies from 2232.4 to 3671.0 MeV. The Born cross section at these energy points are measured as well as the corresponding effective electromagnetic form factors. Furthermore, the ratio of electric to magnetic form factors, $|G_E/G_M|$ and $|G_M|$ are measured at the c.m. energies where the data samples are the largest. We also report preliminary results of $e^+e^- \to \Lambda\bar{\Lambda}$, which is analysed with the same method. Moreover, future prospects of the measurement of baryon electromagnetic form factors from a unique high luminosity data scan by BESIII, are given.

1. Introduction

The structure of light baryons is very important, but very difficult, to understand. Electromagnetic form factors (EMFFs) are key ingredients to describe the internal structure. The electromagnetic structure of a hadron with spin S is given by $2S+1$ form factors. Nucleon form factors can be measured by the $e^-N \to e^-N$ (space-like EMFFs), or by the $e^+e^- \to N\bar{N}$ and the reversed process $N\bar{N} \to e^+e^-$ (time-like EMFFs). The form factors are analytic functions of the momentum transfer squared, q^2, between the incident and the final states. Form factors are real in the space-like region and complex in the time-like region. This means that there is a non-zero relative phase between the electric and magnetic form factors that manifests itself in polarization of the outgoing baryons [1]. In the present context of QCD and the quark-gluon structure of hadrons, it is particularly interesting to measure form factors of hyperons which are expected to reveal the effects of SU(3) breaking. However, hyperon form factors can only be measured in the time-like region, since their short life-time make hyperon targets unfeasible. Hyperons have advantage compared to nucleons though: the weak, parity violating decay of the hyperons, that causes the decay particles to be emitted in the direction of the spin of the hyperon. This makes the polarisation of the hyperon experimentally accessible. As a consequence, the time-like form factors of hyperons can be fully determined.

Baryons with spin 1/2 have two electromagnetic form factors. In the experiments, the so-called Sachs form factors $G_E(q^2)$ and $G_M(q^2)$ are widely used. The Sachs form factors correspond to the Fourier transformations of the charge and magnetic spatial distributions in the Breit frame. The Born differential cross section for the process $e^+e^- \to \gamma* \to B\bar{B}$ is given by [2]

$$\frac{d\sigma_{Born}(q^2)}{d\Omega} = \frac{\alpha^2\beta C}{4q^2}[|G_M(q^2)|^2(1+cos^2\theta_B) + \frac{1}{\tau}|G_E(q^2)|^2(sin^2\theta_B)], \tag{1}$$

Content from this work may be used under the terms of the Creative Commons Attribution 3.0 licence. Any further distribution of this work must maintain attribution to the author(s) and the title of the work, journal citation and DOI.
Published under licence by IOP Publishing Ltd

Journal of Physics: Conference Series **742** (2016) 012020 doi:10.1088/1742-6596/742/1/012020

where $\alpha \approx \frac{1}{137}$ is the fine structure constant, θ_B is the polar angle of the baryon, $\beta = \sqrt{1 - \frac{1}{\tau}}$ is the velocity, $\tau = \frac{4m_B^2}{q^2}$, and m_B is the mass of baryon. The Coulomb factor, $C = \frac{\pi\alpha}{\beta}\frac{1}{1-exp(-\pi\alpha/\beta)}$, accounts for the electromagnetic $B\bar{B}$ interactions of point-like baryons [3] and is equal to 1 for neutral baryon pairs. This gives a total cross section of

$$\sigma_{Born}(q^2) = \frac{4\pi\alpha^2\beta C}{3q^2}[|G_M(q^2)|^2 + \frac{1}{2\tau}|G_E(q^2)|^2]. \tag{2}$$

The effective form factors are given by

$$|G(q^2)|^2 = \frac{2\tau|G_M(q^2)|^2 + |G_E(q^2)|^2}{2\tau + 1} = \frac{\sigma_{Born}(q^2)}{(1 + \frac{1}{2\tau})(\frac{4\pi\alpha^2\beta C}{3q^2})} \tag{3}$$

Above threshold, the ratio $|G_E(q^2)/G_M(q^2)|$ can be extracted from the angular distribution of baryons.

The initial state radiation (ISR) channel $e^+e^- \to B\bar{B}\gamma_{ISR}$ also could be used to study baryon form factors. The cross section for the ISR process is two orders of magnitude smaller than for direct production [4]. However, depending on the energy of ISR photon, the hadronic mass in the final state is reduced and the hadronic cross section can be extracted for all masses below the actual c.m. energy of the collider up to the production threshold of the hadronic state. The ISR method is therefore suitable for experiments where a lot of data are collected at one single energy, *e.g.* the $\psi(3770)$ mass.

In this report, we present recent measurements of $e^+e^- \to p\bar{p}$ [5], $\Lambda\bar{\Lambda}$ based on data samples collected with the Beijing Spectrometer III (BESIII) at the Beijing Electron Positron Collider II (BEPCII) [6].

2. The BESIII Experiment and data sets

BEPCII is a double ring e^+e^- collider operating at $2.0 - 4.6$GeV c.m. energies with a design luminosity of 1×10^{33} cm^{-2}s^{-1} at c.m. energy of 3.773 GeV. BESIII is a 4π detector located at the BEPCII, and has accumulated world-leading samples of J/ψ, $\psi(2S)$, $\psi(3770)$ events for study of light hadron and charmonium spectroscopy. Furthermore, BESIII has collected the largest sample of scan data for study of *R value* and hadronic time-like form factors, and in the high mass charmonium states regions for study of X, Y and Z particles [7].

3. Nucleon Form Factors

Very recently, BESIII has published the measurement of the process of $e^+e^- \to p\bar{p}$ at 12 c.m. energies ranging from 2.2324 to 3.617 GeV [5]. These data were collected in 2011 and 2012 and the total integrated luminosity is 157 pb^{-1}. Experimentally, the Born cross section of $e^+e^- \to p\bar{p}$ is calculated by

$$\sigma_{Born} = \frac{N_{obs} - N_{bkg}}{L \cdot \epsilon \cdot (1 + \delta)}, \tag{4}$$

where N_{obs} is the observed number of candidate events, N_{bkg} is the expected number of background events, L is the integrated luminosity, ϵ is the detection efficiency, and $(1 + \delta)$ is the radiative correction factor. The ConExc generator [8] is used both for the efficiency and the radiative correction factor evaluation. The accuracy in the cross section measurements is between 6.0% and 18.9% up to $\sqrt{s} < 3.08$ GeV. Results of Born cross section and the effective form factors extracted according to Eq. 3 are shown in Fig. 1 in comparisons with previous experimental results [9, 10, 11, 12, 13, 14]. Compared to previous data from BaBar [9], the precision of the Born cross section is improved by 30% for data sets fulfilling $\sqrt{s} < 3.08$ GeV.

Journal of Physics: Conference Series **742** (2016) 012020 doi:10.1088/1742-6596/742/1/012020

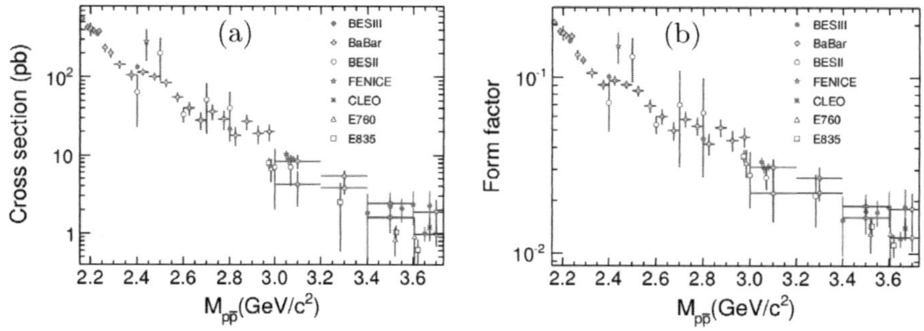

Figure 1. Comparison of the Born cross section (a) and the effective form factors $|G|$ (b) between this measurement and previous experiments, shown on a logarithmic scale for invariant $p\bar{p}$ masses from 2.20 to 3.70 GeV/c^2.

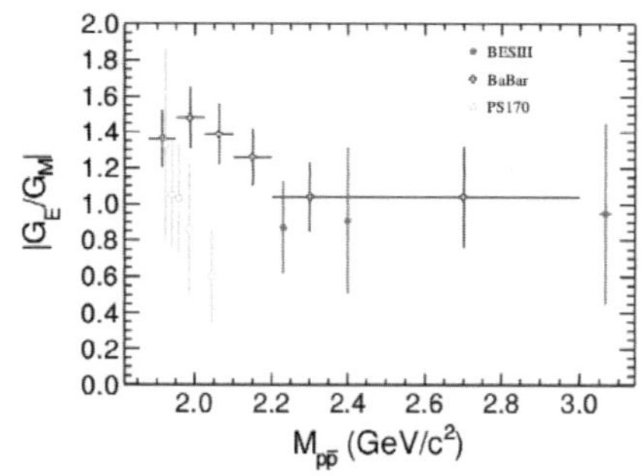

Figure 2. The measured ratio of electric to magnetic form factors $|G_E/G_M|$.

The precision of the corresponding effective form factors is also improved. A fit to the angular distribution of the proton is performed according to Eq. 1 and the ratio of the form factors and $|G_M|$ are determined at the data points which have the largest statistics. The corresponding results are shown in Fig. 2. A different method, named the method of moments (MM), is applied to extract the ratio $|G_E/G_M|$ and $|G_M|$. A comparison of the results obtained by the two methods is shown in Table 1. The results are well consistent and the statistical uncertainty is found to be comparable between the two different methods due to the same number of events.

At present, the precision in the ratio of the proton form factors is dominated by statistical uncertainty. In a new, unique energy scan performed in 2015, an unprecedented data sample was collected. This means that statistical uncertainties below 10% can be achieved. As a complement to the scan data, samples collected at different charmonium and XYZ states above 3.773 GeV (7.4 fb^{-1}), can be used, applying the ISR technique. The data sample collected by BESIII (7.4 fb^{-1}) corresponds to a higher effective ISR luminosity than for instance the BaBar sample of 500 fb^{-1} for the study of the $e^+e^- \to p\bar{p}\gamma_{ISR}$. This is thanks to the fact that BESIII can also perform untagged analysis of these events where the ISR photon emitted at very low polar angles. The lowest-lying data point is at q =2.0 GeV. Comparable precision to BaBar in the form factors measurement is therefore expected from the study of $e^+e^- \to p\bar{p}\gamma_{ISR}$ in BESIII. The analysis of the $e^+e^- \to p\bar{p}\gamma_{ISR}$ is in progress. Experimental results on the annihilation cross section into $n\bar{n}$ are very scarce [15, 16]. Like the case of proton, two different measurement are possible in BESIII: $e^+e^- \to n\bar{n}$ and $e^+e^- \to n\bar{n}\gamma_{ISR}$. Thus, BESIII has a unique chance to shed light on the neutron form factors.

Table 1. Results of $|G_E/G_M|$ and $|G_M|$ by fitting and by the method of moments.

$\sqrt{s}(MeV)$	$\|G_E/G_M\|$	$\|G_M\|(\times 10^{-2})$
	Fitting	
2232.4	0.87±0.24±0.05	18.42±5.09±0.98
2400.0	0.91±0.38±0.12	11.30±4.73±1.53
(3050.0, 3080.0)	0.95±0.45±0.21	3.61±1.71±0.82
	Method of moments	
2232.4	0.83±0.24	18.60±5.38
2400.0	0.85±0.37	11.52±5.01
(3050.0, 3080.0)	0.88±0.46	3.34±1.72

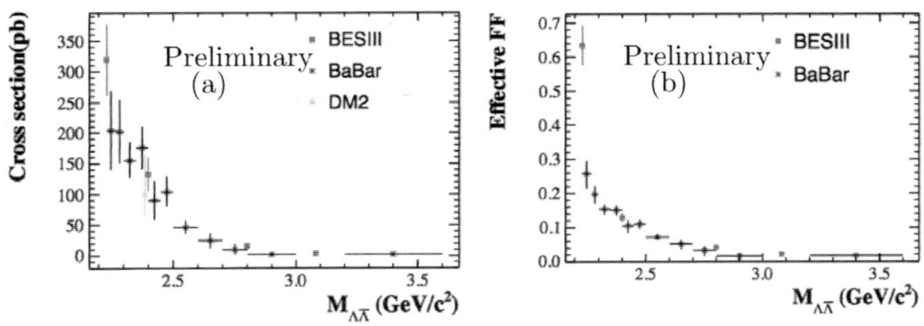

Figure 3. Comparison of the Born cross section (a) and the effective form factors $|G|$ (b) between this measurement and previous experiments for invariant $\Lambda\bar{\Lambda}$ masses from 2.0 to 3.60 GeV/c^2.

4. Hyperon Form Factor

BESIII has preliminary results on the measurement of the channel $e^+e^- \to \Lambda\bar{\Lambda}$. The analysis is based on 40.5 pb^{-1} collected in 4 different scan points during 2011 and 2012. The lowest energy point is at 2.2324 GeV, which is just 1.0 MeV above the $\Lambda\bar{\Lambda}$ threshold. The measurement of cross section at threshold by reconstructing $\Lambda/\bar{\Lambda}$ from the charged $\Lambda \to p\pi^-/\bar{\Lambda} \to \bar{p}\pi^+$ and the neutral $\bar{\Lambda} \to \bar{n}\pi^0$ give consistent results. The combined result is 319.5 ± 57.6 pb, which is much larger than the phase space expectations. The non-vanishing cross section confirms the measurement close to the threshold by BaBar with ISR technique [17]. Measurements of the other three energy points are performed using the charged decay mode. The preliminary results on the measurement of the Born cross section are shown in Fig. 3 together with previous measurements [**?**, 17].

With the higher luminosity scan data of BESIII from 2015, it is possible to fully determine the Λ form factors. A nonzero relative phase between electric $|G_E|$ and magnetic $|G_M|$ leads to the polarization of Λ in the process of $e^+e^- \to \Lambda\bar{\Lambda}$ even though the initial states e^+e^- un-polarized [1]. The polarization is highly depend on polar angle of Λ:

$$P_n = -\frac{sin2\theta_\Lambda/\sqrt{\tau}}{(1+cos^2\theta_\Lambda)+\frac{R^2}{\tau}(1-cos^2\theta_\Lambda)}Rsin\Delta\phi \quad (5)$$

where $\Delta\phi$ is the relative phase and R is the ratio $|G_E/G_M|$ of Λ. Thanks to the self-analyzing weak decay, $\Lambda \to p\pi^-$, the proton from the Λ decay has a tendency to be emitted along the polarization axis of the Λ. The differential cross section of the decay proton angle is given by

$$\frac{d\sigma}{dcos\theta_p} = \frac{1}{2}(1 + \alpha_\Lambda P_n cos\theta_p) \qquad (6)$$

where θ_p is the angle between the decay proton in the Λ frame and the polarization axis of the Λ, α_Λ is the asymmetry parameter. The polarization could be extracted by

$$P_n = \frac{3}{\alpha_\Lambda} < cos\theta_p > \qquad (7)$$

Since CP conservation implies that $\alpha_\Lambda = -\alpha_{\bar{\Lambda}}$, the Λ sample and $\bar{\Lambda}$ sample could be added together which means that the statistic can be increased by a factor of 2. The polarization should be measured as a function of θ_Λ since the polarization depends on this angle, which is clear from Eq. 5.

The expected statistical uncertainties of the analysis of $e^+e^- \to \Lambda\bar{\Lambda}$ with the BESIII scan data in 2015 range between 6% and 17%. The corresponding accuracies for the ratio of the Λ form factors range between 14% and 29%. Similar measurements might also be possible in other hyperons like $e^+e^- \to \Lambda\bar{\Sigma}^0, \Sigma^0\bar{\Sigma}^0, \Sigma^+\bar{\Sigma}^-, \Sigma^-\bar{\Sigma}^+, \Xi^0\bar{\Xi}^0$.

5. Summary

BESIII is an excellent laboratory for the measurement of hadron form factors, since both ISR and scan method can be performed. We have results of proton form factors with scan data which was collected in the year of 2011 and 2012 published recently. There are also preliminary results on Λ form factors based on scan data from 2011 and 2012. In 2015, BESIII performed a unique high luminosity scan (~555 pb^{-1}) in the energy region from 2.0 GeV to 3.08 GeV, mainly in order to measure baryon form factors with higher precision. Measurements of baryon form factors by the ISR technique based on data above 3.77 GeV are on-going.

Reference

[1] A. Z. Dubnickova, S. Dubnicka and M. P. Rekalo, Nuovo Cim. A **109** 241 (1996).
[2] A. Zichichi, S. M. Berman, N. Cabibbo and R. Gatto, Nuovo Cimento **24** 170 (1962).
[3] C. Tzara, Nucl. Phys. B **18** 216-252 (1970).
[4] G. Bonneau and F. Martin, Nucl. Phys. B **27** 381 (1971).
[5] M. Ablikim *et al.* (BESIII Collaboration), Phys. Rev. D **91**, 112004 (2015).
[6] M. Ablikim *et al.* (BESIII Collaboration), Nucl. Instrum. Meth. A **614**, 345 (2010).
[7] K. Chao and Y. Wang, Physics at BESIII, Modn. Phys. A Supplement, V **24**, (2009), Scientific World.
[8] R. G. Ping, Chin. Phys. C **38**, 083001 (2014).
[9] J. P. Lees *et al.* (BaBar Collaboration), Phys. Rev. D **87**, 092005 (2013); J. P. Lees *et al.* (BaBar Collaboration), Phys. Rev. D **88**, 072009 (2013).
[10] T. A. Armstrong *et al.* (E760 Collaboration), Phys. Rev. Lett. **70**, 1212 (1993).
[11] M. Ambrogiani *et al.* (E835 Collaboration), Phys. Rev. D **60**, 032002 (1999); M. Andreotti *et al.*, Phys. Lett. B **559**, 20 (2003).
[12] G. Bardin *et al.* (PS170 Collaboration), Nucl. Phys. B **411**, 3 (1994).
[13] M. Ablikim *et al.* (BES Collaboration), Phys. Lett. B **630**, 14 (2005).
[14] T. K. Pedlar *et al.* (CLEO Collaboration) Phys. Rev. Lett. **95**, 261803 (2005).
[15] A. Antonelli *et al.* (FENICE Collaboration), Nucl. Phys. B **517**, 3 (1998).
[16] N. Achasov *et al.* (SND Collaboration), Phys. Rev. D **90**, 112007 (2014).
[17] B. Aubert *et al.* (BaBar Collaboration), Phys. Rev. D **76**, 092006 (2007).

Journal of Physics: Conference Series **742** (2016) 012021 doi:10.1088/1742-6596/742/1/012021

Phenomenological study of exclusive binary light particle production from antiproton-proton annihilation at FAIR/PANDA

WANG Ying

Univ Paris-Sud, CNRS/IN2P3, Institut de Physique Nucléaire, UMR 8608, 91405 Orsay, France

E-mail: `wangying@ipno.in2p3.fr`

Abstract. Exclusive binary annihilation reactions induced by antiprotons of momentum from 1.5 to 15 GeV/c can be extensively investigated at FAIR/PANDA [1]. We are especially interested in the channel of charged pion pairs. Whereas this very probable channel constitutes the major background for other processes of interest in the PANDA experiment, it carries unique physical information on the quark content of proton, allowing to test different models (quark counting rules, statistical models,..). To study the binary reactions of light meson formation, we are developing an effective Lagrangian model based on Feynman diagrams which takes into account the virtuality of the exchanged particles. Regge factors [2] and form factors are introduced with parameters which may be adjusted on the existing data. We present preliminary results of our formalism for different reactions of light meson production leading to reliable predictions of cross sections, energy and angular dependencies in the PANDA kinematical range.

1. Introduction

Large experimental and theoretical efforts have been going on since decades in order to understand and classify high energy processes driven by strong interaction. We revisit here hadronic reactions at incident energies above 1 GeV, and focus in particular on two body processes. Antiprotons are a very peculiar probe, due to the fact that scattering and annihilation reactions may occur in the same process, with definite kinematical characteristics. We discuss the annihilation reaction of antiproton-proton into two charged pions and the crossed channels of pion-proton elastic scattering. These reactions have been studied in the past at FermiLab and at LEAR at lower energies.

Charged pion production data are scarce, and do not fill continuously in a large angular or energy range [1, 2, 3]. According to the foreseen performances of the PANDA experiment at FAIR, a large amount of data related to light meson pair production from $\bar{p}p$ annihilation is expected in the next future. The best possible knowledge of light meson production is also requested prior to the experiment, as pions constitute an important background for many other channels making a timely development of a reliable model.

We develop here an effective Lagrangian model (ELM), with meson and baryon exchanges in s, t, and u channels, applicable in the energy range $2 \leq \sqrt{s} \leq 15$ GeV, that is the accessible domain for the PANDA experiment at FAIR. It is known that first order Born diagrams give cross sections much larger than measured, as Feynman diagrams assume on-shell point-like particles.

Content from this work may be used under the terms of the Creative Commons Attribution 3.0 licence. Any further distribution of this work must maintain attribution to the author(s) and the title of the work, journal citation and DOI.
Published under licence by IOP Publishing Ltd

Form factors are added in order to take into account the composite nature of the interacting particles at vertices. Their form is, however, somehow arbitrary, and parameters for masses of the exchanged particles, coupling constants or cutoff are adjusted to reproduce the data. Therefore these models should be considered in an effective way to take into account microscopic degrees of freedom and quark exchange diagrams. A "Reggeization" of the trajectories is added to reproduce the very forward and very backward scattering angles.

To get maximum profit from the available data, we consider also existing elastic scattering $\pi^{\pm}p$ data, and apply crossing symmetry in order to compare the predictions based on the annihilation channel, at least in a limited kinematical range.

2. Formalism and Comparison with data

The annihilation reactions are best described in the centre of mass (CMS) frame, whereas the kinematics of elastic scattering is simpler in Laboratory (Lab) frame. We consider the reaction:

$$\bar{p}(p_1) + p(p_2) \to \pi^-(k_1) + \pi^+(k_2) \tag{1}$$

The notation of four momenta is indicated in the parenthesis. The following notations are used: $q_t = (-p_1 + k_1)$, $q_t^2 = t$, $q_u = (-p_1 + k_2)$, $q_u^2 = u$ and $q_s = (p_1 + p_2)$, $q_s^2 = s$. where s, t and u are the Mandelstam variables, $s + t + u = 2M_p^2 + 2m_\pi^2$, M_p is the proton mass, m_π is the pion mass.

The general expression for the differential cross section in CMS of reaction (1) is:

$$\frac{d\sigma}{d\Omega} = \frac{1}{2^8\pi^2}\frac{1}{s}\frac{\beta_\pi}{\beta_p}\overline{|\mathcal{M}|^2}, \quad \frac{d\sigma}{d\cos\theta} = 2E^2\beta_p\beta_\pi\frac{d\sigma}{dt} \tag{2}$$

where $\beta_p(\beta_\pi)$ is the velocity of the the proton(pion), and E is the energy in CMS. $d\Omega = 2\pi\ d\cos\theta$ due to the azimuthal symmetry of binary reactions.

Crossing symmetry relates annihilation and scattering cross sections. Crossing symmetry states that the amplitudes of the crossed processes are the same, i.e., the matrix element $\mathcal{M}(s,t)$ for the scattering (s) process $\pi(-k_1) + p(p_2) \to \pi(k_2) + p(-p_1)$ and the annihilation (a) (1), is the same, at corresponding s and t values. In order to find the correspondence, kinematical replacements should be done, as $s \leftrightarrow t$. The cross sections are related by:

$$\sigma^a = \frac{1}{2}\frac{|\vec{k}_s|^2}{|p_a|^2}\sigma^s. \tag{3}$$

where p_a is the CM momentum for $\bar{p}p$ annihilation and $|k_s|$ is the CM momentum for π^-p scattering, evaluated at the same s value:

$$|\vec{k}_s|^2 = \frac{1}{4s}\left[m_\pi^4 - 2m_\pi^2(M_p^2 + s) + (M_p^2 - s)^2\right]. \tag{4}$$

If the scattering cross section is measured at a value $s^s = s_1$ different from $s^a = s$, at small t values one can rescale the cross section, using the empirical dependence: $\sigma^s \simeq const \cdot s^{-2}$. An example of cross section for annihilation and scattering processes at similar incident momenta are reported in Fig. 1.

In order to calculate $\mathcal{M}$, one needs to specify a model for the reaction. In this work we consider the formalism of effective meson Lagrangians. The following contributions to the cross section for reaction (1) are calculated: baryon exchange: t-channel nucleon (neutron) exchange, t-channel Δ^0 exchange, u-channel Δ^{++} exchange, s-channel ρ-meson exchange. The total amplitude is written as a coherent sum of all the amplitudes:

$$\mathcal{M} = \mathcal{M}_N + \mathcal{M}_{\Delta^0} + \mathcal{M}_{\Delta^{++}} + \mathcal{M}_\rho. \tag{5}$$

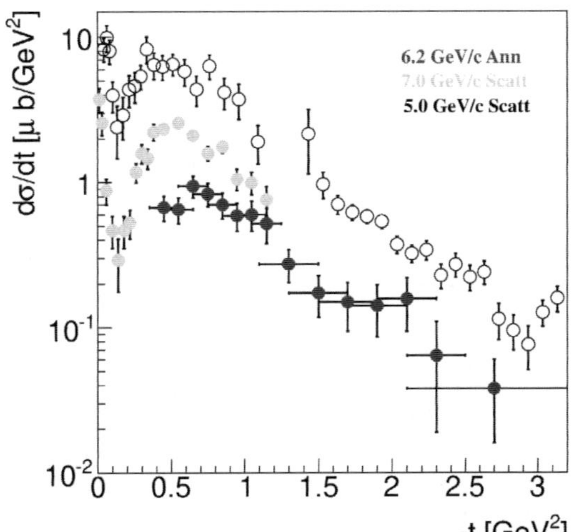

Figure 1. Data for $\bar{p}+p \to \pi^- + \pi^+$, from Ref. [2] (blue circles), for $\pi^+ + p \to \pi^+ + p$ from Ref. [4] (green solid circles), and from Ref. [5] (black empty circles) - for π^+ emission at small t-values (or large u-values). The $\pi^+ p$ data have been scaled by the crossing constant factor 0.589 according to formula (3).

In case of charged pions, the dominant contribution in forward direction is N exchange, whereas Δ^{++} mostly contribute to backward scattering. We neglect the difference of masses between different charge states of particles, like nucleons, pion and Δ. Central scattering is driven by s-channel exchange of vector mesons, with the same quantum numbers as the photon. We limit our considerations to ρ-meson exchange.

The expressions for the amplitudes and their interferences follow the Feynman rules. The coupling constants are fixed from the known decays of the particles, or we use the values from effective potentials like in ref. [6]. Masses and widths are taken from [7]. The effects of strong interaction in the initial state coming from the exchange of vector and (pseudo)scalar mesons between proton-antiproton are essential and effectively lead to the Regge form of the amplitude. The t and u diagrams are modified by adding a general Regge factor R_x (where $x = t, u$) with the following form:

$$R(x) = \left(\frac{s}{s_0}\right)^{2[\alpha(x)-1]}, \quad \alpha(x) = \frac{1}{2} + r\frac{\alpha_s}{\pi}\frac{x - M^2}{M^2}; \tag{6}$$

where $s_0 \simeq 1$ GeV^2 can be considered a fitting parameter [8] and $r\alpha_s/\pi \simeq 0.7$ is fixed by the slope of the Regge trajectory. And M here is the mass of exchange particle. In the present model the values have been set at $s_0 = 1.4$ GeV^2 and $r\alpha_s/\pi = 0.7$ for the nucleon.

A form factor of the form: $F = 1/(x - p^2_{N,\Delta})^2$, was introduced in the $Np\bar{p}$ and $Np\bar{\Delta}$ vertices, with $p_N = 0.8$ GeV and $p_\Delta = 5$ GeV.

The ρNN vertex includes the proton structure in the vector current form with two form factors (FF) $F^\rho_{1,2}$:

$$\Gamma_\mu(q_s) = F_1^\rho(q_s^2)\gamma_\mu(q_s) + \frac{i}{2M_N}F_2^\rho(q_s^2)\sigma_{\mu\nu}q_s^\nu, \tag{7}$$

where $\sigma_{\mu\nu} = \frac{i}{2}[\gamma_\mu\gamma_\nu - \gamma_\nu\gamma_\mu]$ is the antisymmetric tensor. We fix the form, the constants and

Journal of Physics: Conference Series **742** (2016) 012021 doi:10.1088/1742-6596/742/1/012021

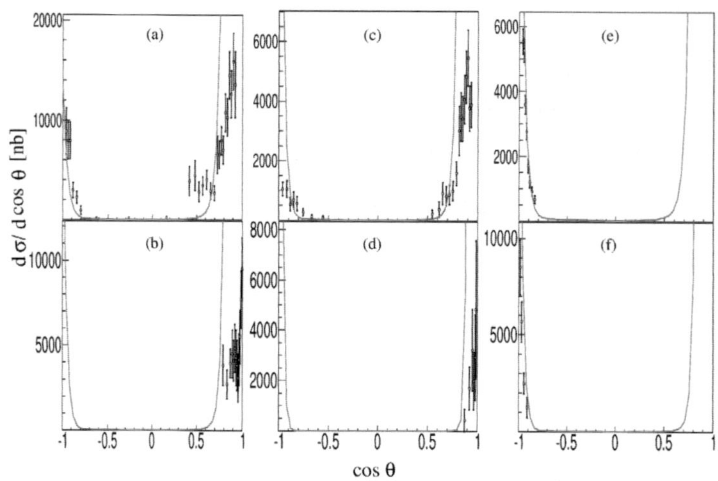

Figure 2. Angular dependence for the reaction $\bar{p}+p \to \pi^- + \pi^+$, for different value of the total CMS energy $\sqrt{s}$: (a) $\sqrt{s} = 3.362$ GeV from Ref. [3], (b) $\sqrt{s} = 3.627$ GeV from Ref. [11], (c) $\sqrt{s} = 3.680$ GeV from Ref. [2], (d) $\sqrt{s} = 4.559$ GeV from Ref. [12]. The corresponding data from the elastic reactions $\pi^{\pm}p \to \pi^{\pm}p$ are also reported: (e) $\sqrt{s} = 3.463$ GeV from Ref. [13], (f) $\sqrt{s} = 3.747$ GeV from Ref. [4]. Black points are data taken from each reference, and the red lines are from the formula (2).

the parameters of $F^{\rho}_{1,2}(s)$ according to [6, 9, 10] as:

$$F^{\rho}_1(s) = g_{\rho NN}\frac{\Lambda^4}{\Lambda^4 + (s - M^2_{\rho})^2},\ F^{\rho}_2(s) = \kappa_{\rho}F^{\rho}_1(s), \tag{8}$$

with normalization $F^{\rho}_1(M^2_{\rho}) = g_{\rho NN}$, where the constant $g_{\rho NN}$ corresponds to the coupling of the vector meson ρ with the nucleon $(g^2_{\rho NN}/(4\pi) = 0.55)$, $\kappa_{\rho} = 3.7$ is the anomalous magnetic moment of the proton with respect to the coupling with ρ, and $\Lambda = 0.911$ is an empirical cut-off.

The angular dependence for the reaction $\bar{p} + p \to \pi^- + \pi^+$ is shown in Fig. 2 (a-d) with satisfactory agreement. The results for the crossed channels $\pi^{\pm}$ elastic scattering are also reported in Fig. 2 (e-f), where data for the differential cross section span a small very forward or very backward angular region, bringing an additional test of the model.

The angular distribution for $\sqrt{s}$= 3.680 GeV is shown in Fig. 3. The total result (black, solid line) gives a good description of the data (red open circles) from Ref. [2] for charged pion production. All components and their interferences are illustrated. The main contribution at central angles is given by ρ s-channel exchange, whereas n exchange dominates forward angles (t channel) followed by Δ^0 exchange. Δ^{++} represents the largest contribution for backward angles (u channel). The interferences are also shown. Their contribution affects the shape of the angular distribution, some of them being negative in part of the angular region.

3. Conclusions

A model, based on effective meson Lagrangian, has been built in order to reproduce the existing data for two pion production in proton-antiproton annihilation at moderate and large energies. Form factors and Regge factors are implemented and parameters adjusted to the existing data for charged pion pair production. Coupling constants are fixed from the known properties of the corresponding decay channels. The agreement with a large set of data is satisfactory for the angular dependence as well as the energy dependence of the cross section.

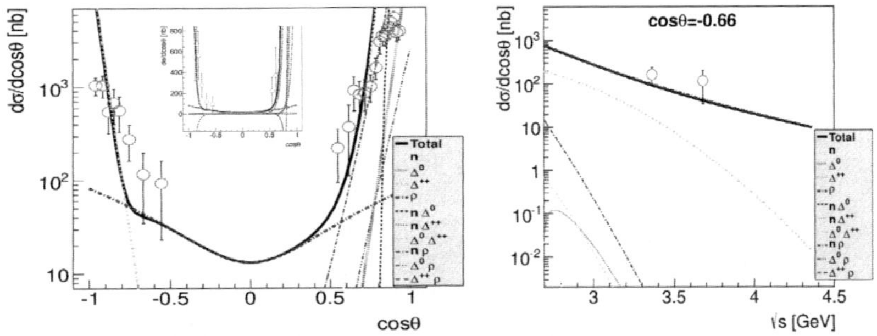

Figure 3. (color online) $\cos\theta$-dependence for the reaction $\bar{p}+p \to \pi^- + \pi^+$ for $\sqrt{s}$= 3.680 GeV (black, solid line) where the different components are illustrated (see text): n-exchange (yellow thick short dash line) and Δ_0 (read thick dotted line) Δ^{++} (green thick dash-dotted line), ρ channel (blue thick long dash line) dominates at large angles, The interferences are $n\Delta^0$ (thin black short-dash line) , $n\Delta^{++}$ (thin red dotted line), $\Delta^0\Delta^{++}$(green thin short dash-dotted line), $n\rho$(blue thin long-dashed dotted line) , $\Delta^0\rho$ (blue thin dash-dotted line), $\Delta^{++}\rho$ (blue thin long dash line). Data are taken from Ref. [2].

A comparison with data from elastic $\pi^{\pm}p \to \pi^{\pm}p$, using crossing symmetry prescriptions shows a good agreement within the uncertainty, which verifies that crossing symmetry works well at backward angles, where one diagram is dominant.

This model can be extended to other binary channels, with appropriate changes of constants. The implementation to Monte Carlo simulations for predictions and optimization future experiments is also foreseen.

4. Acknowledgments
Thanks are due to E. Tomasi-Gustafsson, D. Marchand, Y. Bystritskiy and A. Dorokhov, for supervision and useful discussion. The author is supported by the China Scholarship Council.

5. References
[1] E. Eisenhandler, W. Gibson, C. Hojvat, P. Kalmus, L. L. C. Kwong, et al., Nucl.Phys. B96 (1975) 109.
[2] T. Buran, A. Eide, P. Helgaker, P. Lehmann, A. Lundby, et al., Nucl.Phys. B116 (1976) 51.
[3] A. Eide, P. Lehmann, A. Lundby, C. Baglin, P. Briandet, et al., Nucl.Phys. B60 (1973) 173–220.
[4] W. Baker, K. Berkelman, P. Carlson, G. Fisher, P. Fleury, D. Hartill, R. Kalbach, A. Lundby, S. Mukhin, R. Nierhaus, K. Pretzl, J. Woulds, Nuclear Physics B 25 (2) (1971) 385 – 410.
[5] L. M. Kurdadze, A. P. Onuchin, S. I. Serednyakov, V. A. Sidorov, S. I. Eidelman, Phys. Lett. B 42 (1972) 515–518.
[6] R. Machleidt, Phys. Rev. C 63 (2001) 024001.
[7] K. Olive, et al., Chin.Phys. C38 (2014) 090001.
[8] A. B. Kaidalov, arXiv:hep-ph/0103011.
[9] E. Kuraev, Y. Bystritskiy, V. Bytev, E. Tomasi-Gustafsson, arXiv:1012.5720.
[10] C. Fernandez-Ramirez, E. Moya de Guerra, J. M. Udias,Annals Phys. 321 (2006) 1408–1456.
[11] N. Stein, R. Edelstein, D. Green, H. Halpern, E. Makuchowski, et al., Phys.Rev.Lett. 39 (1977) 378–381.
[12] A. Berglund, T. Buran, P. Carlson, C. Damerell, I. Endo, et al., Nucl.Phys. B137 (1978) 276–282.
[13] D. P. Owen, F. C. Peterson, J. Orear, A. L. Read, D. G. Ryan, D. H. White, A. Ashmore, C. J. S. Damerell, W. R. Frisken, R. Rubinstein, Phys. Rev. 181 (1969) 1794–1807.

The fixed-target experiment at STAR

Kathryn C Meehan for the STAR Collaboration
UC Davis, One Shields Avenue, Davis, CA 95616, USA
E-mail: kcmeehan@ucdavis.edu

Abstract. The RHIC Beam Energy Scan (BES-I) program, which covers center-of-mass energies 7.7 GeV to 39 GeV, was proposed to look for the turn-off of signatures of the quark gluon plasma, search for a possible QCD critical point, and study the nature of the phase transition between hadronic and partonic matter. RHIC BES-I has shown that the partonic interactions are dominant at center-of-mass energies above 20 GeV. Several observables, including v_1 of protons and Lambdas, v_2 of all identified hadrons, and net-proton higher moments, show interesting behavior below 20 GeV and could suggest a transition to a hadron interaction dominated regime. Data from energies lower than 7 GeV could help determine whether these behaviors are indicative of phase transitions or criticality. The goal of the STAR Fixed-Target Program is to extend the collision energy range in BES II to lower energies than what is feasible at RHIC with colliding beams. First physics results from test runs of Au + Au fixed-target collisions will be presented. The results will be compared with published results from similar collision energies at the AGS. The implications for the fixed-target program after the completion of the inner TPC (iTPC) and endcap TOF (eTOF) detector upgrades will also be discussed.

1. Introduction to STAR's Fixed-Target (FXT) Program

The goals of the Beam Energy Scan I (BES-I) Program at STAR include observing the disappearance of QGP signatures, finding evidence of the first-order phase transition, and finding the possible critical point [1]. Results from BES-I suggest a softening of the equation of state [2] and hint at critical fluctuations [3]. By accessing energies below $\sqrt{s_{NN}}$=7.7 GeV, where we expect no QGP formation, STAR could help clarify these hints. However, the luminosity at RHIC is too low for this energy range, making it impractical to operate in collider mode. The fixed-target program will allow this energy range to be accessed.

2. FXT Test Run Results

In 2014 a 4% gold foil target was inserted into the beam pipe, as shown in Fig. 1. During the Au + Au $\sqrt{s_{NN}}$=14.5 GeV BES-I run, data from fixed-target collisions between the beam halo nuclei and the target were taken concurrently. This corresponds to a fixed-target center-of-mass energy of 3.9 GeV per nucleon. The target was positioned at the edge of the TPC in order to use the full TPC acceptance, as shown in Fig. 2. Figure 3 shows the halo nuclei illuminating the target. A Coulomb potential analysis [4] was performed on the top 10% most central events (5000 events) and the result is compared to the result [5] from E895 at a similar energy in Fig. 4. These results agree within uncertainties. These results as well as the results from experiments that studied collisions at other energies and with other species are shown in Fig. 5. The STAR FXT result is consistent with other data from Au + Au and Pb + Pb collisions indicating that the Z of the halo projectile is consistent with that of gold.

Content from this work may be used under the terms of the Creative Commons Attribution 3.0 licence. Any further distribution of this work must maintain attribution to the author(s) and the title of the work, journal citation and DOI.
Published under licence by IOP Publishing Ltd

Journal of Physics: Conference Series **742** (2016) 012022 doi:10.1088/1742-6596/742/1/012022

Figure 1. Photo of the gold target and its support structure.

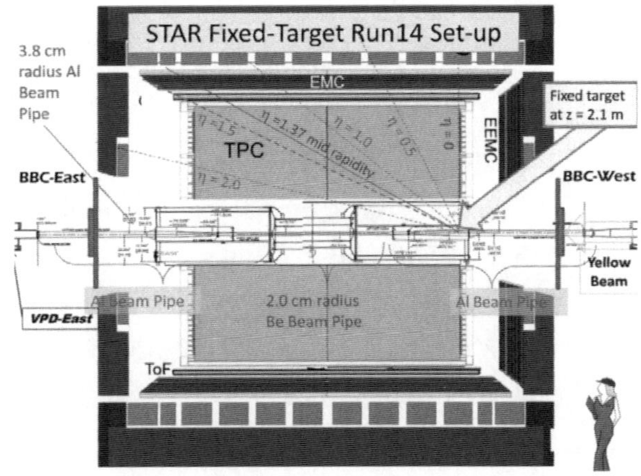

Figure 2. Diagram of the run 14 test run setup. The target is located 210 cm from the interaction region just outside of the TPC.

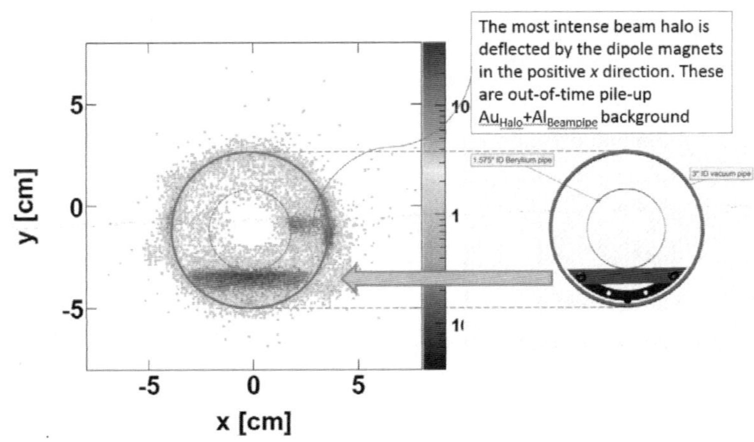

Figure 3. Vertex distribution after removing the beam spot. The target is clearly illuminated.

In 2015 a dedicated fixed-target test run was conducted; only one beam was used and it was lowered to graze the edge of the target. The resulting data set was much cleaner as shown in Fig. 6. Physics results from this run await the estimation of detector efficiencies. The cascade invariant mass, measured without efficiency measurements yet, is shown in Fig. 7. It is expected that the final result will be as statistically significant as the AGS result. Additionally, a test run with an aluminum beam on the gold target has been conducted.

3. Future Plans & Detector Upgrades

Fixed-target physics runs are expected to be conducted during Beam Energy Scan-II (BES-II). Running 1-2 days at each energy would collect sufficient statistics to extend BES-II to lower energies. From the test runs discussed above, up to 50 million events per day are expected. Additionally, three detector upgrades, the event plane detector (EPD), the inner Time Projection Chamber (iTPC) and the endcap Time-of-Flight (eTOF), will be used during BES-II. The event plane detector will improve event plane resolution and can also be used for triggering in fixed-

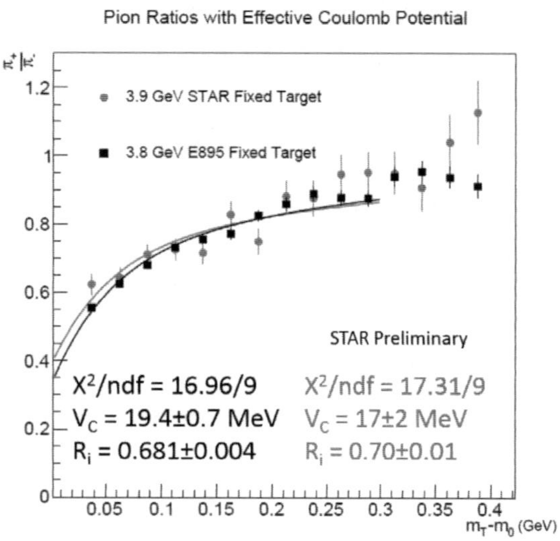

Figure 4. Fit to the pion ratio result for STAR's Au + Au $\sqrt{s_{NN}}$=3.9 GeV fixed-target test run is shown in red. It is compared to the result [5] from Au + Au $\sqrt{s_{NN}}$=3.8 GeV collisions from E895, shown in black.

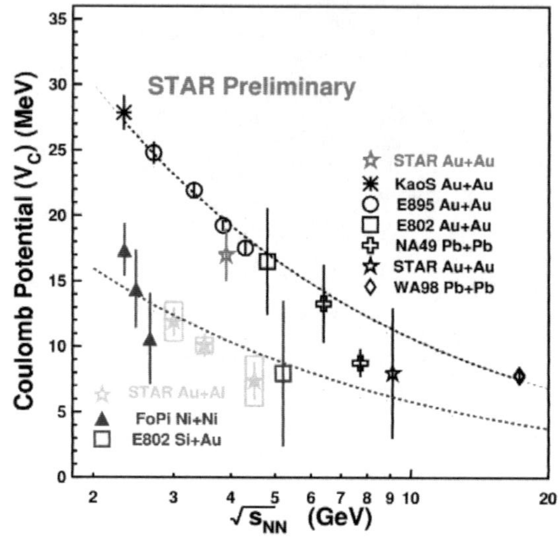

Figure 5. Coulomb potential measurements for several different experiments at both different center-of-mass energies as well as different collision species [4]. The STAR FXT result is represented by the red star, whereas the STAR collider result is represented by the black star. The black dotted line is an empirical fit to the Au + Au and Pb + Pb data, while the blue dotted line is the same empirical fit to the Au + Au and Pb + Pb data scaled by Z/r where Z is the atomic number and r is the size of the overlap region determined from Glauber models.

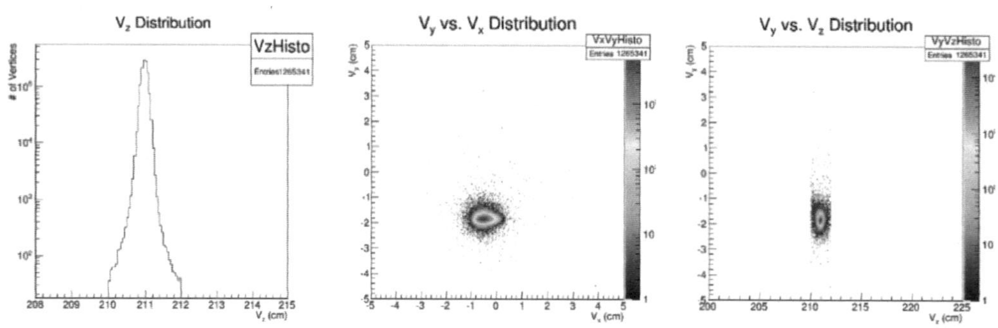

Figure 6. Vertex distribution plots for Au + Au fixed-target $\sqrt{s_{NN}}$=4.5 GeV

target events. It will have full coverage to beam rapidity. The prototype is currently installed and taking data. The iTPC is an upgrade to the inner sectors of the TPC. Since electronics are now smaller, the readout electronics can be packed more densely and more pad rows can be added to the inner sector. The number of pad rows will be increased from 13 to 40. This will provide both better momentum resolution and better dE/dx resolution for particle identification. This

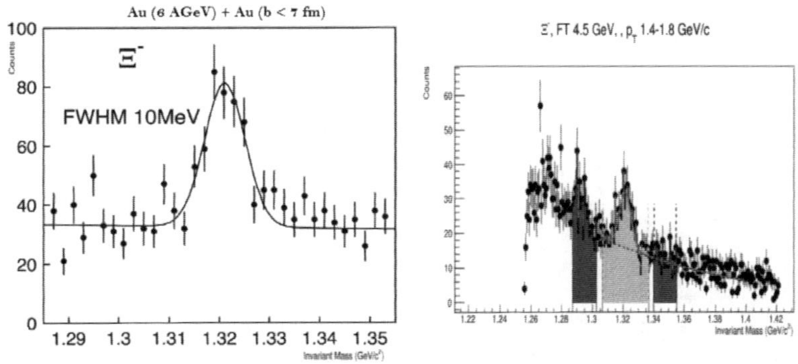

Figure 7. Comparison of cascade invariant mass measurement from the $\sqrt{s_{NN}}$=4.5 GeV fixed-target test run to the AGS measurement [6]. The STAR measurement has not taken into account efficiency yet.

Figure 8. Diagram of the STAR detector with the proposed future upgrades.

increased coverage will extend the acceptance from $|\eta| \leq 1.0$ to $|\eta| \leq 1.5$. The eTOF upgrade complements the iTPC upgrade by extending the particle identification capabilities of STAR. The Compressed Baryonic Matter (CBM) experiment at the Facility for Antiproton and Ion Research (FAIR) has entered a cooperative agreement with STAR. CBM is loaning 108 MRPCs to be used in STAR for BES-II. This allows STAR to have an endcap TOF for this program while giving CBM a unique opportunity to do a large-scale integration test of their TOF system.

The fixed-target acceptance for mid-rapidity pions and protons both with the current TPC and TOF systems and with the upgrades are shown in Fig. 9. The FXT program will extend BES-II down to 3.0 GeV, but, as the acceptance plots show, the iTPC and eTOF upgrades are required to have acceptance at mid-rapidity at the higher fixed-target energies. These detector upgrades will enable the FXT program to extend its energy range up to 7.7 GeV. Since there are also collider data at this energy, this will allow comparisons between fixed-target and collider analyses.

4. Conclusions

Several fixed-target test runs have been conducted. Coulomb potentials were measured and are consistent with results from previous experiments. Detector upgrades will extend the FXT program up to $\sqrt{s_{NN}}$=7.7 GeV which will allow for comparison with collider mode analyses at the same energy. The FXT program will allow us to extend BES-II down to $\sqrt{s_{NN}}$=3.0 GeV and access the higher μ_B region of the QCD phase diagram and the accompanying physics, shown in Fig. 10.

Journal of Physics: Conference Series **742** (2016) 012022 doi:10.1088/1742-6596/742/1/012022

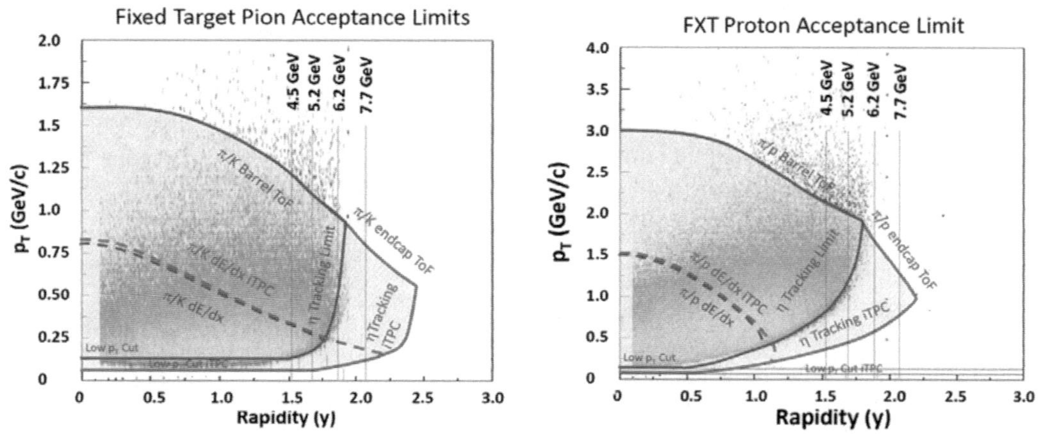

Figure 9. Pion and proton acceptance at mid-rapidity for different fixed-target energies. The current acceptance is within the blue curves and the acceptance of the proposed upgrades is within the red lines.

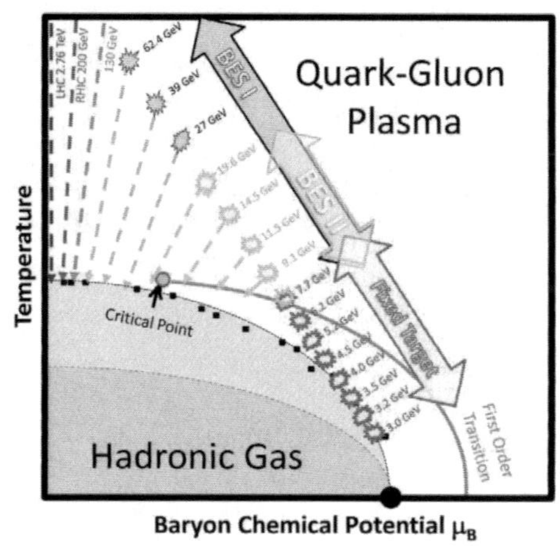

Figure 10. QCD phase diagram showing the ranges of the BES programs and the FXT program. The critical point and phase transition lines are currently conjecture

Acknowledgments
This material is based upon work supported by the National Science Foundation under Grant No. 1404281. Any opinions, findings, and conclusions or recommendations expressed in this material are those of the authors and do not necessarily represent the views of the National Science Foundation.

References
[1] 2014 STAR Internal Note SN-0598
[2] Adamczyk L *et al.* 2014 *Phys. Rev. Lett.* **112** 162301
[3] Luo X 2015 *J. Phys.: Conf. Ser. 599* 012023 (*Preprint* `arXiv:1501.03010`)
[4] Cebra D *et al.* 2014 (*Preprint* `nucl-ex/14081369`)
[5] Klay J L *et al.* 2003 *Phys. Rev. C* **68** 054905
[6] Chung P *et al.* 2003 *Phys. Rev. Lett.* **91** 202301

Journal of Physics: Conference Series **742** (2016) 012023 doi:10.1088/1742-6596/742/1/012023

Performance study of the anisotropic flow and reaction plane reconstruction in the CBM experiment

V Mikhaylov[1,2,3], A Kugler[1], V Kushpil[1], I Selyuzhenkov[4], P Tlustý[1] (for the CBM Collaboration)

[1] Nuclear Physics Institute, Czech Academy of Sciences, Řež, Czech Republic
[2] Czech Technical University in Prague, Faculty of Nuclear Sciences and Physical Engineering, Prague, Czech Republic
[3] National Research Tomsk Polytechnic University, Department of Electronics and Automation of Nuclear Plants, Tomsk, Russia (on leave from)
[4] GSI Helmholtzzentrum für Schwerionenforschung GmbH, Darmstadt, Germany

E-mail: mikhaylov@ujf.cas.cz

Abstract. The Projectile Spectator Detector (PSD) is a subsystem of the CBM experiment at the future FAIR facility designed to determine centrality and reaction plane orientation in the heavy-ion collisions. It will be done by measurement of the energy distribution of the heavy nucleons and nuclei fragments emitted close to the beam rapidity in forward direction. For the anticipated beam energies of FAIR SIS100 and SIS300 accelerators, different event generators (iQMD, UrQMD, DCM-QGSM, LA-QGSM and HSD) were used for the study of directed and elliptic proton flow in Au+Au collisions. Produced particles were transported with the GEANT4 Monte-Carlo using the CBM detector geometry. Performance of the reaction plane determination is shown for different PSD setups to demonstrate effects of the detector granularity and magnetic field. Simulation results are compared with the FOPI, AGS E877, E895 and STAR experimental data.

1. Introduction

Experimental estimate of global event characteristics in nucleus-nucleus collisions such as the centrality of the collision and the reaction plane orientation are challenging tasks for any high-energy heavy-ion experiment. One of the examples is the future Compressed Baryonic Matter (CBM) experiment at the Facility for Antiproton and Ion Research (FAIR) in Darmstadt, Germany designed to provide a heavy-ion beams of intensity up to 10^9 ions/sec and beam energy range $E_b = 1 - 30$ AGeV with SIS18/SIS100/SIS300 accelerator rings [1]. The CBM experiment (Fig. 1) will work in a high luminosity environment and will allow for a precision systematic measurement of production yields, phase-space distributions, correlations, collective flow and fluctuation observables for various particle species produced in nuclear collisions [2].

In order to understand the degree of thermalisation, the features of equation of state, and the in-medium properties of strange particles in collisions of different systems at beam energies in the SIS100-SIS300 energy range, a multi-differential flow measurements are needed for a number of hadron species [4]. Moreover, the accuracy of the collision centrality and the reaction plane

Content from this work may be used under the terms of the Creative Commons Attribution 3.0 licence. Any further distribution of this work must maintain attribution to the author(s) and the title of the work, journal citation and DOI.
Published under licence by IOP Publishing Ltd

Journal of Physics: Conference Series **742** (2016) 012023 doi:10.1088/1742-6596/742/1/012023

determination depends on the multiplicity and the energy distribution of fragments, and on the strength of the directed flow that they carry.

The Projectile Spectator Detector (PSD) of the CBM experiment is a compensating lead-scintillator calorimeter, which will measure the energy distribution of the projectile nuclei fragments and nucleons (spectators) produced close to the beam rapidity [3]. The main design requirements of the PSD are forward rapidity coverage, fine azimuthal granularity and sufficient energy resolution to allow for precise collision centrality determination and reaction plane reconstruction. The 44 module design of the PSD shown in Fig. 2 covers large transverse area around the beam spot position such that most of the projectile spectator fragments deposit their energy in the PSD.

Directed and elliptic flows of protons are simulated for the collision events generated by iQMD, UrQMD, DCM-QGSM, LA-QGSM and HSD event generators in the energy range E_b from 1 to 30 AGeV. The DCM-QGSM event generator appear to reproduce available data from the FOPI at GSI, E877 and E895 at AGS, and STAR at RHIC. The reaction plane reconstruction performance in terms of the reaction plane correction factor is studied using the DCM-QGSM generator. Produced particles and fragments were transported through realistic CBM detector geometry with help of GEANT4 Monte-Carlo simulation. The bias of reaction plane determination introduced by detector effects, such as PSD azimuthal granularity and particle tracks deflection due to a particle deflection in the magnetic field is accessed. Other event generators are used for a cross-check.

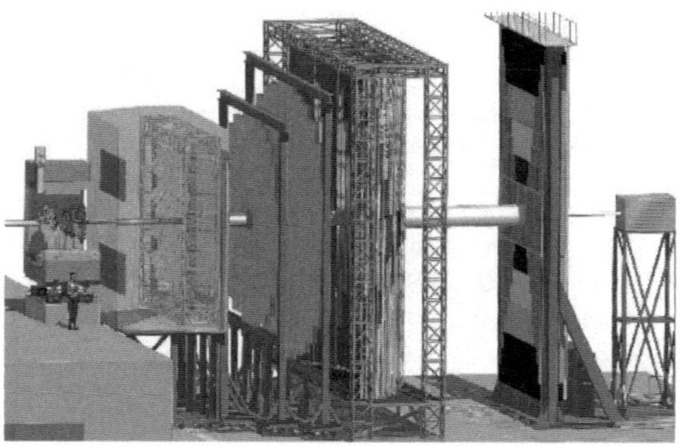

Figure 1. The CBM experiment layout.

Figure 2. The PSD detector layout.

2. Modelling the directed and elliptic flow at FAIR energies

Directed and elliptic flow in Au+Au collisions are deduced exploiting for five different heavy-ion collision event generators, namely iQMD [5], UrQMD [6], HSD [7], DCM-QGSM and LA-QGSM [8]. Simulations are performed for semi-central Au+Au collisions in the projectile beam energy range $E_b = 1 - 30$ AGeV.

The slope of proton directed flow (dv_1/dy) and elliptic flow (v_2) at midrapidity as function of beam energy E_b simulated with different generators is compared in Fig. 3 with the FOPI [9], E895 [10], E877 [11] and STAR [12] data.

The impact parameter range is chosen to be $b = 3.3-6.0$ fm for $E_b = 1.2$ AGeV, $b = 5-7$ fm for $E_b = 2-8$ AGeV, and $b = 4.5-9.2$ fm for $E_b = 30$ AGeV. The lower p_T cut is 0.3 GeV/c for $E_b = 1.2$ AGeV, 0.1 GeV/c for $E_b = 2$, 4 AGeV, 0.2 GeV/c for $E_b = 6$, 30 AGeV and 0.4 GeV/c for $E_b = 8$ AGeV. The p_T and impact parameter (centrality) cuts are chosen to match those used in the experimental data analysis. The flow signals vary strongly with event generators. At the lowest energy $E_b = 1.2$ AGeV, which corresponds to that measured by FOPI and recently by HADES at SIS18, the iQMD agrees well with the data. For the SIS100/SIS300 energy range $E_b = 2 - 30$ AGeV which corresponds to collisions anticipated for CBM at FAIR, the DCM-QGSM model is the best in describing the E895, E877 and STAR data.

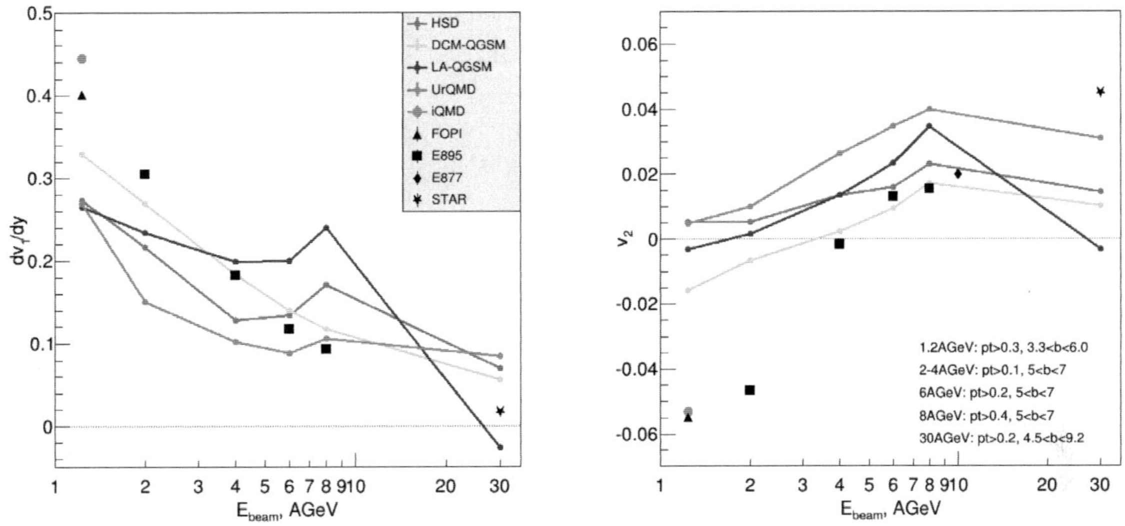

Figure 3. Directed (left) and elliptic (right) proton flow at midrapidity vs beam energy calculated with different models and compared with data.

3. Reaction plane reconstruction performance with CBM

The collective flow develops along the reaction plane, which is spanned by the beam direction and the impact parameter of the collision. The vector plane can be estimated by nucleons and fragments, which do not participate in the collision (spectators). The reaction plane is determined from the position and energy deposited by the spectators in the PSD modules. The PSD have good energy resolution and segmentation to measure the transverse distribution energy with high accuracy.

The finite number of fragments and the fluctuation of the particle multiplicity from one collision to another result in a difference between the event plane deduced from the spectators and the reaction plane orientation. The event plane resolution correction factor R_{EP} for directed flow corrects for the finite event plane angle resolution relatively to the reaction plane and is defined as

$$R_{\mathrm{EP}} = \langle cos(\Psi_{\mathrm{EP}} - \Psi_{\mathrm{RP}}) \rangle. \quad (1)$$

In case of the simulation, the event plane angle Ψ_{EP} is deduced from PSD data, while the true reaction plane angle Ψ_{RP} is taken directly from the event generator input. R_{EP} value ranges between zero (very poor resolution) and unity (ideal resolution).

The PSD performance for the reaction plane reconstruction is studied within CBMROOT environment using GEANT4 Monte-Carlo simulations and various detector setups and event generators. Figure 4 shows the concept of the simulation setup without the detector geometry and corresponding GEANT transport procedure, where the MC-tracks from generators are chosen to fit to the PSD acceptance by corresponding θ-angle cut (refereed as "PSD-accept." in Fig. 6 (left panel)). Figure 5 shows the simulated geometry of the CBM detector including

Journal of Physics: Conference Series **742** (2016) 012023 doi:10.1088/1742-6596/742/1/012023

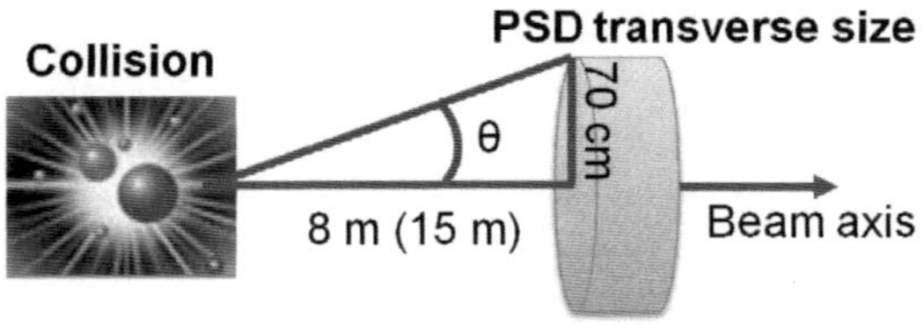

Figure 4. The concept of simple simulation setup without CBM PSD geometry.

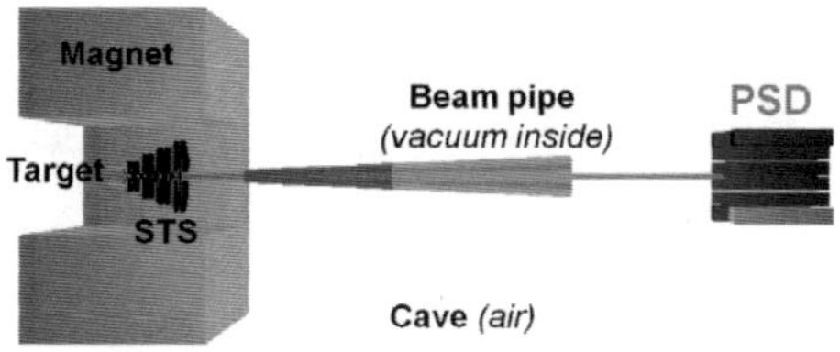

Figure 5. The realistic simulated CBM PSD geometry.

Au target, dipole magnet, silicon tracking stations (STS) detector, the beam pipe and the PSD. In Fig. 6 (left panel) this setup is referred as "PSD-geom.,B=0" and "PSD-geom.,B>0" with magnetic field turned-off/on respectively.

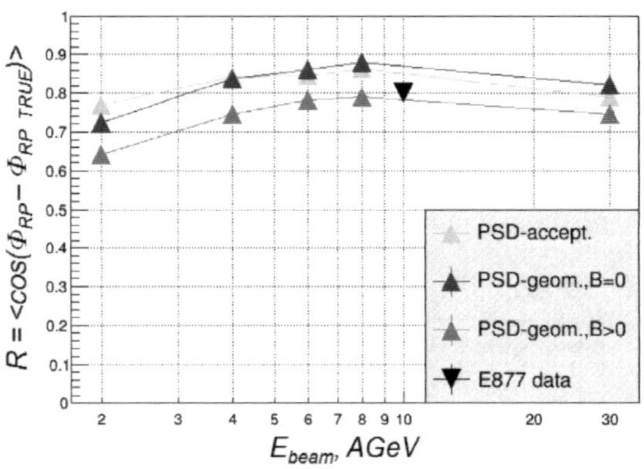

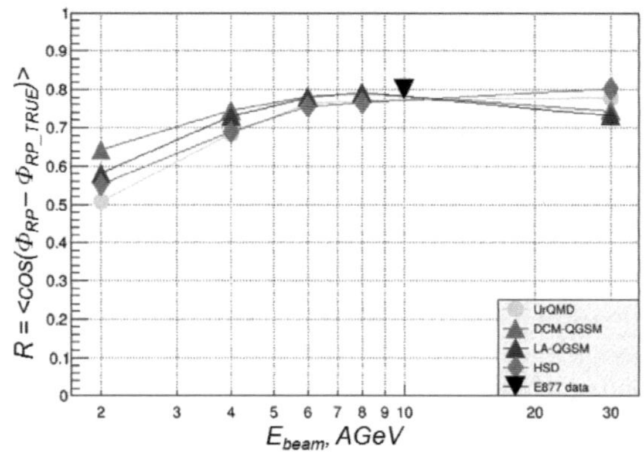

Figure 6. Reaction plane resolution correction factor vs beam energy for various biasing effects (left) and various event generators (right).

Figure 6 (left panel) shows the reaction plane resolution correction factor of the PSD as a function of the beam energy simulated with the DCM-QGSM event generator for three different detector setups. The comparison of the reaction plane resolution correction factor achieved with "PSD-accept." and "PSD-geom.,B=0" setups shows that the azimuthal segmentation of PSD does not introduce any significant bias into the measurement. The splitting of spectators into neutral and charged fragments due to magnetic field decreases resolution correction factor over the whole beam energy range by about $5-10$ %.

Figure 6 (right panel) shows the reaction plane resolution correction factor of the PSD as a function of the beam energy simulated with various event generators for the realistic detector setup with turned-on magnetic field. Despite the strong variation of the collective flow magnitude in models, reaction plane resolution correction factor does not differ much between various event generators. The correction factor around $E_b = 10$ AGeV (and thus the PSD reaction plane resolution) is comparable to that of the target (TCal) and participant (PCal) calorimeters used for the collective flow measurements in Au+Au collisions by the E877 experiment at AGS, which had a maximum of about $R_{EP} = 0.8$ in semi-central collisions (see Fig. 5 in [11]).

4. Conclusion

Performance of the PSD reaction plane reconstruction for the CBM experiment is studied using various detector setups and heavy-ion event generators and GEANT Monte-Carlo response of the detector components. The simulation is performed for semi-central Au+Au collisions in the beam energy range $E_b = 1 - 30$ AGeV using iQMD, UrQMD, DCM-QGSM, LA-QGSM and HSD heavy-ion event generators. The slope of the proton directed flow at mid-rapidity simulated with various event generators at various energies differs significantly from the experimental data. The DCM-QGSM generator seems to be the best in describing experimental data on proton flow in SIS100/SIS300 energy range. The PSD reaction plane resolution correction factor for semi-central collisions is similar for different generators and geometry setups and is comparable with E877 data.

Acknowledgments

We would like to thank for the help in simulating a large ensemble of heavy-ion collisions and helpful discussions: Yvonne Leifels (iQMD), Volker Friese (UrQMD), Marina Golubeva (DCM-QGSM), Konstantin Gudima (LA-QGSM), Elena Bratkovskaya (HSD). This work has been supported by the European Community FP7–Capacities, contract HadronPhysics3 n°0283286, GACR 13-06759S of the Czech Science Foundation, grant LM201549 of the Ministry of Education of the Czech Republic and partially by TPU P.L.U.S. program.

References

[1] FAIR Baseline Technical Report 2006, www.fair-center.eu/fileadmin/fair/publications_FAIR/FAIR_BTR_6.pdf (2006).
[2] The CBM experiment at FAIR facility, http://www.fair-center.eu/public/experiment-program/nuclear-matter-physics/cbm.html.
[3] F. Guber et.al. [CBM Collaboration], Technical Design Report for the CBM Projectile Spectator Detector (PSD), https://repository.gsi.de/record/109059.
[4] B. Friman, C. Hohne, J. Knoll, S. Leupold, J. Randrup, R. Rapp and P. Senger, Lect. Notes Phys. **814**, pp. 980 (2011).
[5] Ch. Hartnack and J. Aichelin, Investigation of the radial expansion of nuclear matter in collisions of heavy nuclei at SIS energies, Rapport Interne SUBATECH 96-08.
[6] M. Bleicher *et al.*, J. Phys. G **25**, 1859 (1999) [hep-ph/9909407].
[7] The HSD model, fias.uni-frankfurt.de/ brat/HSD/index1.html.
[8] A. V. Dementyev and N. M. Sobolevsky, SHIELD, a Monte Carlo Hadron Transport Code, http://www.inr.troitsk.ru/shield/introd-eng.html.
[9] W. Reisdorf *et al.* [FOPI Collaboration], Nucl. Phys. A **876**, 1 (2012) [arXiv:1112.3180 [nucl-ex]].
[10] H. Liu *et al.* [E895 Collaboration], Phys. Rev. Lett. **84**, 5488 (2000) [nucl-ex/0005005].
[11] E877 Collaboration, J. Barrette et al., Energy and charged particle flow in a 10.8 AGeV/c Au + Au collisions, Phys.Rev. C55, 1420 (1997), arXiv:nucl-ex/9610006, doi:10.1103/PhysRevC.55.1420,10.1103/PhysRevC.56.2336.
[12] L. Adamczyk *et al.* [STAR Collaboration], Phys. Rev. Lett. **112**, no. 16, 162301 (2014) [arXiv:1401.3043 [nucl-ex]].

Journal of Physics: Conference Series **742** (2016) 012024 doi:10.1088/1742-6596/742/1/012024

Modeling early time dynamics of relativistic heavy ion collisions

Lucia Oliva[1,2], Marco Ruggieri[1,3], Armando Puglisi[1,2], Salvatore Plumari[1,2], Francesco Scardina[1,2], Gabriele Coci[1,2] and Vincenzo Greco[1,2]

[1] Department of Physics and Astronomy, University of Catania, Via S. Sofia 64, 95123 Catania, Italy

[2] INFN-Laboratori Nazionali del Sud, Via S. Sofia 62, 95123 Catania, Italy

[3] College of Physics, University of Chinese Academy of Sciences, Yuquanlu 19A, Beijing 100049, China

E-mail: lucia.oliva@lns.infn.it

Abstract. We studied isotropization and thermalization of the quark-gluon plasma produced by decaying color-electric flux tubes created at the very early stages of relativistic heavy ion collisions. We coupled the dynamical evolution of the initial field, which decays to a plasma by the Schwinger mechanism, to the dynamics of the many particles system produced by the decay. The evolution of such a system is described by relativistic transport theory at fixed values of the viscosity over entropy density ratio. Within a single self-consistent calculation scheme we computed quantities which serve as indicators of the equilibration of the plasma for a 1+1 dimensional expanding geometry. We find that the initial color-electric field decays within 1 fm/c and particles production occurs in less than 1 fm/c; however, in the case of large viscosity oscillations of the field appear along the entire time evolution of the system, affecting also the behaviour of the ratio between longitudinal and transverse pressure. In case of small viscosity we find that the isotropization time is about 0.8 fm/c and the thermalization time is about 1 fm/c, in agreement with the common lore of hydrodynamic approaches.

1. Introduction

Ultra-relativistic heavy ion collisions represent the only way to have experimental access to the high temperature and small barion density region of the QCD phase diagram and study the formation and the evolution of the quark-gluon plasma. One of the most compelling issues in this framework is the understanding of the early stages of such collision processes which are strongly out of equilibrium. In the general picture of relativistic heavy ion collisions, just after the two colliding nuclei had passed one through each other, a peculiar configuration of longitudinal color-electric and color-magnetic fields is produced. This state of classical color flux tubes, known as Glasma, should become an isotropic and thermalized quark-gluon plasma in a very short time, a small fraction of fm/c. It is very interesting to understand which are the mechanisms responsible of such particle formation and equilibration. In our approach a longitudinal color-electric field decays to a particle plasma through the Schwinger effect; the evolution equation of the initial field is coupled to the relativistic transport equation which describes the evolution of the many particle system and leads to an isotropic and thermalized state by means of collisions. Thus, we

Content from this work may be used under the terms of the Creative Commons Attribution 3.0 licence. Any further distribution of this work must maintain attribution to the author(s) and the title of the work, journal citation and DOI.
Published under licence by IOP Publishing Ltd

model early time dynamics of relativistic heavy ion collisions treating in a self-consistent way color-electric field and particles produced by its decay.

2. Abelian flux tube model and Schwinger mechanism

In our simulation of the very early stage of relativistic heavy ion collisions we implement the Abelian flux tube model [1, 2, 3], focusing on a single flux tube of a given transverse area. Although this geometry is a strong simplification of the realistic initial condition relevant for heavy ion collision experiments, in which one should consider an ensemble of several tubes more similar to the Glasma configuration, our model allows to better understand the physics which underlies the formation of the quark-gluon plasma and to compute quantities which serve as indicators of thermalization and isotropization of the system and of timescales of such processes. Moreover, we neglect for simplicity initial longitudinal color-magnetic fields which should be taken into account in a more appropriate description of the Glasma state. Thus, our initial condition is a purely longitudinal color-electric field E which then decays into particle quanta by the Schwinger mechanism, which consists in a vacuum instability towards the creation of particle pairs by a strong electric field and is related to the existence of an imaginary part in the quantum effective action of a pure electric field [4, 5, 6].

The decay probability per unit of spacetime and invariant momentum space produced by the decay of the electric field through the Schwinger effect is

$$\frac{dN_{jc}}{d\Gamma} \equiv p_0 \frac{dN_{jc}}{d^4x d^2p_T dp_z} = \frac{\mathcal{E}_{jc}}{4\pi^3} \left| \ln\left(1 \pm e^{-\pi p_T^2/\mathcal{E}_{jc}}\right) \right| \delta(p_z) p_0, \tag{1}$$

assuming massless quanta and the plus (minus) sign corresponding to the creation of a boson (fermion-antifermion) pair. In this equation p_T and p_z refer to each of the two particles created from the vacuum; $\mathcal{E}$ is the effective force which acts on the tunneling pair, depends on color and flavor and is given by $\mathcal{E}_{jc} = (g|Q_{jc}E| - \sigma_j)\,\theta\,(g|Q_{jc}E| - \sigma_j)$, where σ_j corresponds to the string tension depending on the flavor considered and $p_0 = \sqrt{p_T^2 + p_z^2}$ is the single particle kinetic energy. The Q_{jc} are color-flavor charges which in the case of quarks are $Q_{j1} = 1/2,\ Q_{j2} = -1/2,\ Q_{j3} = 0$, for $j = 1, N_f$; for antiquarks, corresponding to negative values of j, the color-flavor charges are just minus the corresponding charges for quarks; finally for gluons (which in our notation correspond to $j = 0$) the charges are $Q_{01} = 1,\ Q_{02} = 1/2,\ Q_{03} = -1/2,\ Q_{04} = -Q_{01},\ Q_{05} = -Q_{02},\ Q_{06} = -Q_{03}$. We have only six gluons, corresponding to the non-diagonal color generators; the two gluon fields corresponding to the diagonal color generators have vanishing coupling with the background field, hence they cannot be produced by the Schwinger mechanism.

Since the dynamics is assumed to be Abelian, classical field dynamics is governed by Maxwell equations, in which the back-reaction of particle production and propagation on the color field is taken into account by means of polarization and conductive currents. We limit our simulation to a one dimensional expansion along the direction of the initial color-electric field which, being longitudinal at the beginning, remains longitudinal along all the time evolution, since transverse currents are not produced.

3. Relativistic transport theory and field equations

In order to study the evolution of the system self-consistently we couple the dynamical evolution of the color field to the dynamics of the many-particle system produced by the decay. The latter is described by the relativistic Boltzmann transport equation which, in presence of a gauge field $F^{\mu\nu}$, is given by

$$\left(p^\mu \partial_\mu + g Q_{jc} F^{\mu\nu} p_\nu \partial_\mu^p\right) f_{jc}(x,p) = \frac{dN_{jc}}{d\Gamma} + C_{jc}\,[f], \tag{2}$$

Journal of Physics: Conference Series **742** (2016) 012024 doi:10.1088/1742-6596/742/1/012024

where $f_{jc}(x,p)$ is the distribution function for flavor j and color c and $F^{\mu\nu}$ is the color-electromagnetic tensor. On the right hand side $dN/d\Gamma$ is the source term, which describes the creation of quarks, antiquarks and gluons due to the decay of the color-electric field and is given by Eq. 1, and $C[f]$ represents the collision integral, which accounts for change of f due to collision processes and is responsible for a finite viscosity ($\eta/s \neq 0$). To solve numerically the Boltzmann equation we divide the space into a tridimensional lattice and use the standard test particle method to sample $f(x,p)$; then the collision integral is solved by means of a stochastic algorithm [7, 8, 9, 10, 11, 12, 13, 14, 15]. Within one single theoretical approach one can follow the entire dynamical evolution of the system produced in relativistic heavy ion collisions. Usually input of a transport approach are cross sections of a fixed set of microscopic processes, but in our model we start from a fixed value of the viscosity over entropy density ratio $\eta/s \neq 0$ and compute cross sections according to the Chapman-Enskog equation [12], simulating a fluid with the desired shear viscosity. In this way we make a more direct link between transport theory, based on a description in terms of parton distribution functions, and hydrodynamical formulations, in which the dynamics is governed by macroscopic quantities.

Following previous work on this subject, we assume that the dynamics of the color field is Abelian and invariant for boosts along the longitudinal directions [3]; hence the color-electric field E satisfies the classical Maxwell equations which assume a boost-invariant formulation:

$$\frac{dE}{d\tau} = -j_M - j_D, \tag{3}$$

where τ is the proper time and the right hand side corresponds to the opposite of the total current computed in the local rest frame of the fluid, given by two contributions: the conductive current which depends on the motion of the charges and the polarization current which is due to the dipoles created by the Schwinger effect. Color charges and currents depend on parton distribution functions as well as the kinetic equations, hence allowing to solve field and particles equations in a self-consistent way and to take into account the back-reaction of particles on the field. Refer to [1, 2] for more details.

4. Results for a 1+1D expansion

In this section we show the effect of a boost-invariant longitudinal expansion on pair production due to the decay of a color-electric flux tube.

In the left panel of Fig. 1 we plot the color-electric field strength as a function of time for three different values of η/s. The electric field is averaged in the central rapidity region $|y| < 0.5$. The initial value of the color-electric field used in the simulations is $E_0 = 2.2$ GeV^2, in agreement with the CERN Large Hadron Collider case of Ref. [2]. We find that the color-electric field shows a quick decay for small values of viscosity, due to the fact that in this case the coupling among particles is large, meaning collisions are very efficient in randomizing particle momenta in each cell, hence damping conductive currents that might sustain the field. For larger values of η/s the electric field experiences stronger fluctuations during all its time evolution. In the inset of the left panel of Fig. 1 we plot the number of particles produced per unit of transverse area and rapidity versus time. We find that regardless of the value of the viscosity, the particles are produced at very early times, approximately within 0.5 fm/c, with only few more particles produced at later times in the case $\eta/s = 10/4\pi$. We have checked that changing the initial value E_0 of the electric field does not modify the production time in a considerable way unless E_0 is very small, namely $E_0 \ll 1$ GeV^2. Moreover, the value of η/s affects the conversion of the initial classical field to particles only within a few percent. This indicates the Schwinger effect provides an efficient mechanism to produce a plasma from a classical field.

In the right panel of Fig. 1 we plot the ratio P_L/P_T, where P_L and P_T correspond to the longitudinal and transverse pressures respectively. These quantities are computed cell by cell in

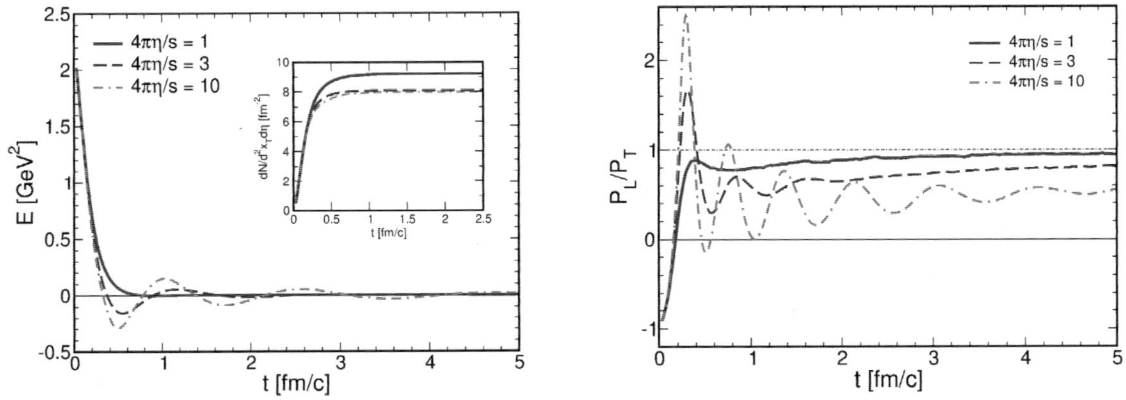

Figure 1. *Left panel.* Color-electric field strength averaged in the central rapidity region $|y| < 0.5$ (main panel) and particle number produced per unit of transverse area and rapidity (inset panel) versus time; the different curves correspond to different values of η/s. *Right panel.* Longitudinal over transverse pressure ratio as a function of time for several values of η/s. Figures are adapted from [1].

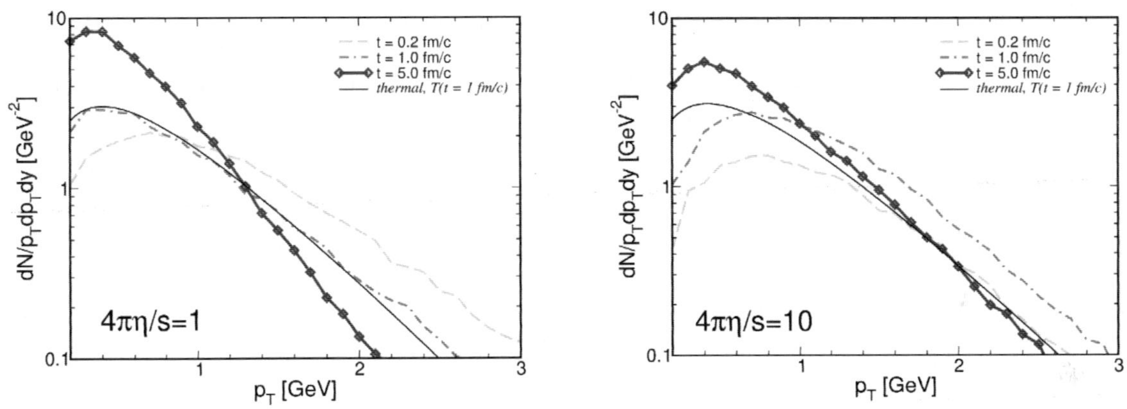

Figure 2. Particle spectra at midrapidity $|y| < 0.5$ for $\eta/s = 1/4\pi$ (left panel) and $\eta/s = 10/4\pi$ (right panel). For each values of η/s the spectrum at three different times is shown. The black thin solid lines correspond to the thermal spectrum at $t = 1$ fm/c. Figures are adapted from [1].

the local rest frame of the fluid and then averaged in the midrapidity region $|y| < 0.5$. The initial longitudinal pressure is negative and $P_L/P_T = -1$ because at initial time the system is made of pure longitudinal color-electric field. As soon as particles are produced, they give a positive contribution to the longitudinal pressure and the field magnitude decreases leading to a positive pressure. For all the values of viscosity considered, we find that the total longitudinal pressure becomes positive in about 0.2 fm/c. Moreover, in the case $\eta/s = 1/4\pi$ the strong interactions among the particles remove the initial pressure anisotropy quite efficiently and quickly, then the ratio tends to increase towards 1, which corresponds to an isotropic system. This would justify the use of viscous hydrodynamics with an initial time of 0.6 fm/c in which the pressure ratio is about 0.7. However, the larger the viscosity of the fluid the larger are the oscillations of P_L/P_T; for example, in the case $\eta/s = 10/4\pi$ the pressure ratio experiences several oscillations which follow the alternation of maxima of $|E|$ (corresponding to minima of P_L because the field gives a negative contribution to P_L) and zeros of E (corresponding to maxima of P_L). Moreover, at large times P_L/P_T is quite smaller than 1.

In Fig. 2 we plot the gluon spectra at midrapidity $|y| < 0.5$ for two different values of viscosity: left panel corresponds to $\eta/s = 1/4\pi$, right panel to $\eta/s = 10/4\pi$. For each value of the viscosity the spectrum at three different times is shown. The thin solid black line corresponds to a thermal spectrum, namely

$$\frac{dN}{p_T dp_T dy} \propto p_T e^{-\beta p_T}, \tag{4}$$

which describes a thermalized system in three spatial dimensions at the temperature $T = 1/\beta$ corresponding to the same energy density of the produced particles spectrum at $t = 1$ fm/c. We find that for $\eta/s = 1/4\pi$ the system efficiently thermalizes in $t = 1$ fm/c, as evident by comparing the thermal spectrum (black thin line) with simulation data (dot-dashed thin red line). For $\eta/s = 10/4\pi$ the spectrum of produced particles at $t = 1$ fm/c is in disagreement with the corresponding thermal spectrum, meaning the system is not completely thermalized in three dimensions. Moreover, the very mild change in the slope of the spectrum we measure from $t = 1$ fm/c to $t = 5$ fm/c shows that the system does not cool down efficiently in this case; this is expected because the larger the viscosity the larger is the energy dissipated into heat, hence the system cools down more slowly than in the case of small viscosity.

5. Conclusions

In this work we have reported some of our results on simulations of the evolution of color-electric flux tubes which decay to a quark-gluon plasma by means of the Schwinger mechanism, coupling relativistic transport theory to Maxwell equations for a self-consistent solution. We have focused on quantities like decay time of the color-electric field, isotropization and thermalization rates of the fluid produced by the decay of the field in the tube, which serve as indicators of the equilibration of the particle plasma. These simulations are important in the context of ultra-relativistic nuclear collisions, where flux tubes of strong color fields are expected to be produced in the very early stage. We have studied the case of flux-tube decay in a box with a longitudinal expansion. The approach developed here, based on a stochastic solution of the relativistic Boltzmann equation, has the advantage to be easily extended to more realistic simulations, making it possible to study the impact of the early dynamics on observables like elliptic flow, photon and dilepton production.

Acknowledgments

V. G., S. P., M. R. and F. S. acknowledge the ERC-StG funding under a QGPDyn grant.

References

[1] Ruggieri M, Puglisi A, Oliva L, Plumari S, Scardina F and Greco V 2015 *Phys. Rev.* C **92** 064904
[2] Ryblewski R and Florkowski W 2013 *Phys. Rev.* D **88** 034028
[3] Florkowski W 2010 *Phenomenology of Ultra-Relativistic Heavy-Ion Collisions* (Singapore: World Scientific) p 416
[4] Heisenberg W and Euler H 1936 *Z. Phys.* **98** 714-32
[5] Schwinger J S 1951 *Phys. Rev.* **82** 664-79
[6] Dunne G V in 2005 *From Fields to Strings: Circumnavigating Theoretical Physics* vol 1 ed Shifman M, Vainshtein A and Wheater J (Singapore: World Scientific) pp 445522
[7] Xu Z, Greiner C and Stocker H 2008 *Phys. Rev. Lett.* **101** 082302
[8] Xu Z and Greiner C 2009 *Phys. Rev.* C **79** 014904
[9] Bratkovskaya E L, Cassing W, Konchakovski V P and Linnyk O 2011 *Nucl. Phys.* A **856** 162-82
[10] Ferini G, Colonna M, Di Toro M and Greco V 2009 *Phys. Lett.* B **670** 325-9
[11] Plumari S and Greco V 2012 *AIP Conf. Proc.* **1422** 56-61
[12] Plumari S, Puglisi A, Scardina F and Greco V 2012 *Phys. Rev.* C **86** 054902
[13] Plumari S, Puglisi A, Colonna M, Scardina F and Greco V 2013 *J. Phys. Conf. Ser.* **420** 012029
[14] Ruggieri M, Scardina F, Plumari S and Greco V 2013 *Phys. Lett.* B **727** 177-81
[15] Ruggieri M, Scardina F, Plumari S and Greco V 2014 *Phys. Rev.* C **89** 054914

Journal of Physics: Conference Series **742** (2016) 012025 doi:10.1088/1742-6596/742/1/012025

Poincaré invariance in low-energy effective field theories for QCD

Sungmin Hwang

Physik-Department, Technische Universität München, James-Franck-Str. 1, D-85747 Garching and Max-Planck-Institut für Physik, Föhringer Ring 6, D-80805 München, Germany

E-mail: sungmin.hwang@tum.de

Abstract. We present the calculations on deriving relations between the Wilson coefficients in non-relativistic quantum chromodynamics and potential non-relativistic quantum chromodynamics by exploiting the Poincaré invariance. Implications of the constraints are briefly discussed in the context of the effective string theory.

1. Introduction

Quantum Chromodynamics (QCD) has theoretically and experimentally been verified as the solid theory for explaining colour interactions at subatomic scales, during the last decades. However, it contains both perturbative and non-perturbative aspects, manifested by the asymptotic freedom [1, 2, 3] and the confinement [4], respectively. While perturbative QCD (pQCD) [5] is the proper framework for describing interactions at high energies, it is not able to give any feasible description of its dynamics around the confinement phase. Thus, it is helpful to separate the perturbative and non-perturbative regimes.

Separately treating those two different aspects of physics calls upon a framework of the effective field theories (EFTs), which requires to observe a hierarchy of scales in the physical systems of one's interest. In case of heavy mesons, such as the bottomonium or charmonium states, hierarchy is given such that the rest mass of single quark/antiquark M is much greater than the momentum transfer Mv between them (v being the relative velocity) and the confinement scale Λ_{QCD}: $M \gg Mv, \Lambda_{QCD}$. As the heavy scale M is integrated out, dynamics of the heavy quark fields enter into the non-relativistic regime, so that the resulting effective theories are either *non-relativistic QCD* (NRQCD) [6, 7] for heavy quark-antiquark systems or the *heavy quark effective theory* (HQET) [8, 9] for heavy quark and light quark states, whose Lagrangians are organised with respect to $1/M$ expansion, $\mathcal{L} = \sum_n c_n \mathcal{O}_n / M^{d_n - 4}$, where operators $\mathcal{O}_n$ consist of effective degrees of freedom (DOFs) with its mass dimension d_n, and c_n are undetermined scalar functions, namely the *Wilson* (or matching) coefficients. These coefficients contain the information of its underlying (or UV) theory, whose exact values are to be determined via matching to the UV theory at the corresponding energy scale.

NRQCD is the useful EFT for analysing productions (or annihilations) of the heavy quark-antiquark states, but it shows some difficulties in explaining the mass splitting of the bound states. In order to resolve these issues, one can construct a valid EFT for such cases, which is the *potential* NRQCD (pNRQCD)[10]. When the hierarchy of scale is established as such, $M \gg Mv \gg Mv^2$, (Mv^2 being the relative kinetic energy between quark and antiquark),

Content from this work may be used under the terms of the Creative Commons Attribution 3.0 licence. Any further distribution of this work must maintain attribution to the author(s) and the title of the work, journal citation and DOI.
Published under licence by IOP Publishing Ltd

pNRQCD is constructed by integrating out the relative momentum Mv from NRQCD, and its Lagrangian is organised by the multipole expansion in the relative distance r upon the $1/M$ expansion from NRQCD. Just like in NRQCD/HQET, the pNRQCD Lagrangian contains a series of the effective operators along with the Wilson coefficients, and these coefficients are to be determined via matching to NRQCD, the underlying theory of pNRQCD. It is apparent that calculations can easily become more complicated and lengthier if one tries to perform the matching directly from QCD to NRQCD and then to pNRQCD, due to the proliferation of the effective operators; it is, there, worthwhile to look for some ways to reduce the amount of the matching calculations, such as by deriving constraints between the coefficients.

There have been two separate ways of finding constraints between Wilson coefficients, both of which exploit symmetry of the EFTs. The first symmetry is the *reparametrisation invariance* [11], and the second is the *Poincaré invariance* [12, 13], which we explore in detail here. Since QCD is a Poincaré invariant theory, it is natural to observe that its low-energy counterpart, NRQCD in particular, is also to observe Poincaré invariance [1]. When the Poincaré symmetry is realised in NRQCD, pNRQCD is also to preserve the same symmetry, thereby constraints between the Wilson coefficients are established.

Concerning the derivations of the constraints, this report is organised as follows: in Sec. 2, general principles of Poincaré transformation of non-relativistic fields are discussed, which we apply to NRQCD and derive the constraints between the coefficients. In Sec. 3, we present similar arguments on the pNRQCD, but certain complications arise due not only to the proliferations of the effective operators, but also to the transformation behaviour of the fields; these issues are addressed in the same section. Implications of the constraints in pNRQCD are briefly discussed in Sec. 4, with the framework of the *effective string theory* [14], which is the EFT for the flux tube model in the heavy quark-antiquark bound system.

2. Poincaré invariance in NRQCD

In relativistic field theories, a transformation of a field ϕ under the Poincaré group is given: $\phi_a(x) \to \mathcal{M}_{ab}(\Lambda)\phi_b(\Lambda^{-1}x)$, with $\mathcal{M}_{ab}(\Lambda)$ being the finite dimensional representation of the Poincaré group and Λ being the transformation matrix which acts on the coordinate $x'^{\mu} = \Lambda^{\mu}_{\nu}x^{\nu}$. There are ten independent bases for the transformation matrix, three of which are responsible for the spatial rotations, three are for the spatial boosts, and four of them are the spacetime translations. Since the representation of the Poincaré group in the case of the spatial boosts is no longer linear at the non-relativistic regime, let us pay our attention to the boost transformation of a free relativistic field, which can be expressed with $\phi_a(x) \to (e^{\mp\boldsymbol{\eta}\cdot\boldsymbol{\Sigma}})_{ab}\phi_b(\mathcal{B}^{-1}x)$, for η is the parameter for the boost, Σ is the spin representation, and $\mathcal{B}$ is the boost transformation matrix that acts on spacetime coordinates. In case of a massive field with mass M, one can rewrite the boost transformation in a slightly different manner when a rest frame of the massive particle, $v = (1, 0, 0, 0)$, is chosen as its reference [13]

$$\phi_a(x) \to \exp\left[\mp\boldsymbol{\eta}\cdot\frac{\boldsymbol{\Sigma}\times\boldsymbol{\partial}}{M+\sqrt{M^2-\boldsymbol{\partial}^2}}\right]_{ab}\phi_b(\mathcal{B}^{-1}x). \tag{1}$$

The transformation has now changed from global to local (appearance partial derivatives), and one can take a non-relativistic expansion of this expression. For an interacting case, the boost transformation of the field with regard to symmetries under charge, parity and time reversal,

[1] Although the fields in the EFT Lagrangian are non-relativistic (it will be clarified in the next section), the Lagrangian itself can observe Poincaré invariance order by order in the expansion parameter.

Journal of Physics: Conference Series **742** (2016) 012025 doi:10.1088/1742-6596/742/1/012025

(C, P, T in short), as well as spatial rotations, are given by (up to a cubic order in $1/M$) [15]

$$\phi_a(x) \to \Big\{1 + iM\boldsymbol{\eta}\cdot\boldsymbol{x} - \frac{ik_1\boldsymbol{\eta}\cdot\boldsymbol{D}}{2M} + \frac{k_2(\boldsymbol{\Sigma}\times\boldsymbol{\eta})\cdot\boldsymbol{D}}{2M} - \frac{ik_3\boldsymbol{\eta}\cdot\boldsymbol{D}\boldsymbol{D}^2}{4M^3} + \frac{k_4(\boldsymbol{\Sigma}\times\boldsymbol{\eta})\cdot\boldsymbol{D}}{2M}\frac{\boldsymbol{D}^2}{4M^2} + \mathcal{O}(g, 1/M^4)\Big\}\phi_a(\mathcal{B}^{-1}x). \tag{2}$$

for k's are undetermined coefficients; it is clear that $\boldsymbol{D}$ is the gauge covariant derivatives from the interacting theory. Let us now apply this expression to NRQCD whose bilinear parts of the Lagrangian in the heavy quark and antiquark fields are given by [6, 7]

$$\mathcal{L}_{\text{NRQCD}} \ni \psi^\dagger\Big\{iD_0 + c_1\frac{\boldsymbol{D}^2}{2M} + c_2\frac{\boldsymbol{D}^4}{8M^3} + c_F g\frac{\boldsymbol{\sigma}\cdot\boldsymbol{B}}{2M} + c_D g\frac{[\boldsymbol{D}\cdot, \boldsymbol{E}]}{8M^2} + ic_s g\frac{\boldsymbol{\sigma}\cdot[\boldsymbol{D}\times, \boldsymbol{E}]}{8M^2}\Big\}\psi + \chi^\dagger\Big\{iD_0 - c_1\frac{\boldsymbol{D}^2}{2M} - c_2\frac{\boldsymbol{D}^4}{8M^3} - c_F g\frac{\boldsymbol{\sigma}\cdot\boldsymbol{B}}{2M} + c_D g\frac{[\boldsymbol{D}\cdot, \boldsymbol{E}]}{8M^2} + ic_s g\frac{\boldsymbol{\sigma}\cdot[\boldsymbol{D}\times, \boldsymbol{E}]}{8M^2}\Big\}\chi \tag{3}$$

where c's are the Wilson coefficients, ψ annihilates the heavy quark, and χ creates the heavy antiquark. From the Eq. (2), the quark and antiquark fields transform under the boost as (up to $1/M$ order)

$$\psi(x) \to \Big\{1 + iM\boldsymbol{\eta}\cdot\boldsymbol{x} - \frac{ik_1}{2M}\boldsymbol{\eta}\cdot\boldsymbol{D} + \frac{k_2}{4M}\boldsymbol{\eta}\cdot(\boldsymbol{D}\times\boldsymbol{\sigma}) + \mathcal{O}(g, M^{-2})\Big\}\psi(\mathcal{B}^{-1}x) \tag{4}$$

$$\chi(x) \to \Big\{1 - iM\boldsymbol{\eta}\cdot\boldsymbol{x} + \frac{ik_1}{2M}\boldsymbol{\eta}\cdot\boldsymbol{D} - \frac{k_2}{4M}\boldsymbol{\eta}\cdot(\boldsymbol{D}\times\boldsymbol{\sigma}) + \mathcal{O}(g, M^{-2})\Big\}\chi(\mathcal{B}^{-1}x) \tag{5}$$

and applying Eq. (4) and Eq. (5) to Eq. (3), we obtain the original form of the bilinear sectors of the Lagrangian and additional terms up to total derivatives (also up to the order of $1/M$)

$$\delta\mathcal{L}_{2\psi} = \psi^\dagger\Big[i(1-c_1)\boldsymbol{\eta}\cdot\boldsymbol{D} - \frac{1}{2M}(k_1 - c_1)\{D_0, \boldsymbol{\eta}\cdot\boldsymbol{D}\} + \frac{1}{4M}(1 - 2c_F + c_s)\boldsymbol{\eta}\cdot(g\boldsymbol{E}\times\boldsymbol{\sigma})\Big]\psi. \tag{6}$$

This expression is to vanish according to the Poincaré invariance, which implies $k_1 = c_1 = 1, c_s = 2c_F - 1$. Note that the coefficient k_2 from the boost transformations, Eq. (4) and Eq. (5), is still to be determined. It can be fixed by using the commutation relations between boost generators which is to be equivalent to the rotation generator up to $1/M$ order; k_2 equals to the unity. These results confirm the values from the literature [12], but the amount of calculations involved here is smaller and simpler than the hitherto known methods, which means that one can apply it for the higher loop calculations with much ease. Similar procedures are applied to the four-quark sector of NRQCD, and the results are in accord with known values [16]; several new features are also to be shown. Full results on the four-quark sector will be discussed in the upcoming paper.

3. Poincaré invariance in pNRQCD

pNRQCD is obtained by integrating out the momentum transfer $Mv \sim 1/r$ in NRQCD [10], from which arises a slight change in its coordinate system: spatial dependence of the colour singlet and octet fields is on the centre-of-mass $\boldsymbol{R} = (\boldsymbol{x}_1 + \boldsymbol{x}_2)/2$ as well as on the relative coordinates $\boldsymbol{r} = \boldsymbol{x}_1 - \boldsymbol{x}_2$, where $\boldsymbol{x}_1$ and $\boldsymbol{x}_2$ are the spatial coordinates of the heavy quark and heavy antiquark, respectively. Also the ultrasoft gluons have only spatial dependence on the relative coordinates; in short, the non-relativistic DOFs in pNRQCD are the singlet $S(t, \boldsymbol{R}, \boldsymbol{r})$ and the octet fields $O^a(t, \boldsymbol{R}, \boldsymbol{r})$, with a being the colour index, and we denote the gluons with $A^a_\mu(t, \boldsymbol{R})$. Power counting schemes are followed $\nabla_r, 1/r \sim Mv$, $\partial_0, \nabla_R, A_\mu \sim Mv^2$, $\boldsymbol{E}, \boldsymbol{B} \sim M^2v^4$, so the bilinear parts of the singlet field (up to $1/M^2$) are organised by [12]

$$\begin{aligned}S^\dagger \mathcal{K}_{SS} S =& S^\dagger \Big(i\partial_0 + \frac{1}{2M}\left\{c_s^{(1,-2)}, \boldsymbol{\nabla}_r^2\right\} + \frac{c_s^{(1,0)}}{4M}\boldsymbol{\nabla}_R^2 - V_S^{(0)} - \frac{V_S^{(1)}}{M} + \frac{V_{p^2Sa}}{8M^2}\boldsymbol{\nabla}_R^2 + \frac{1}{2M^2}\left\{\boldsymbol{\nabla}_r^2, V_{p^2Sb}\right\}\\ &+ \frac{V_{L^2Sa}}{4M^2r^2}(\boldsymbol{r}\times\boldsymbol{\nabla}_R)^2 + \frac{V_{L^2Sb}}{4M^2r^2}(\boldsymbol{r}\times\boldsymbol{\nabla}_r)^2 - \frac{V_{S_{12}S}}{M^2r^2}\Big(3(\boldsymbol{r}\cdot\boldsymbol{\sigma}^{(1)})(\boldsymbol{r}\cdot\boldsymbol{\sigma}^{(2)}) - \boldsymbol{r}^2\boldsymbol{\sigma}^{(1)}\cdot\boldsymbol{\sigma}^{(2)}\Big)\\ &- \frac{V_{S^2S}}{4M^2}\boldsymbol{\sigma}^{(1)}\cdot\boldsymbol{\sigma}^{(2)} + \frac{iV_{LSSa}}{4M^2}(\boldsymbol{r}\times\boldsymbol{\nabla}_R)\cdot(\boldsymbol{\sigma}^{(1)}-\boldsymbol{\sigma}^{(2)}) + \frac{V_{LSSb}}{4M^2}(\boldsymbol{r}\times\boldsymbol{\nabla}_r)\cdot(\boldsymbol{\sigma}^{(1)}+\boldsymbol{\sigma}^{(2)})\Big)S.\end{aligned} \tag{7}$$

Note that c's and V's are the undetermined Wilson coefficients; superscripts on c's denote the orders of $1/M$ and r, respectively, and V's are the potential terms. Also, the superscript to the Pauli matrices $\boldsymbol{\sigma}^{(1/2)}$ denotes the spin for the heavy quark/antiquark, respectively. Our task is to construct the most general form of the spatial boost which acts on the singlet field with the symmetries under C, P, T transformations in mind, up to the order of $1/M$ and r, and it is shown as [15]

$$\begin{aligned}S'(t,\boldsymbol{R},\boldsymbol{r}) =& \Bigg(1 - 2iM\boldsymbol{\eta}\cdot\boldsymbol{R} + \frac{ik_D^{(1,0)}}{4M}\boldsymbol{\eta}\cdot\boldsymbol{\nabla}_R + \frac{i}{4M}\left\{k_{a'}^{(1,0)}\boldsymbol{\eta}\cdot\boldsymbol{r}, \boldsymbol{\nabla}_R\cdot\boldsymbol{\nabla}_r\right\}\\ &+ \frac{i}{4M}\left\{k_{a''}^{(1,0)}\boldsymbol{r}\cdot\boldsymbol{\nabla}_R, \boldsymbol{\eta}\cdot\boldsymbol{\nabla}_r\right\} + \frac{i}{4M}\left\{k_{a'''}^{(1,0)}\boldsymbol{r}\cdot, \boldsymbol{\nabla}_r(\boldsymbol{\eta}\cdot\boldsymbol{\nabla}_R)\right\} + \frac{i}{4M}\left\{\frac{k_b^{(1,0)}}{r^2}(\boldsymbol{\eta}\cdot\boldsymbol{r})(\boldsymbol{r}\cdot\boldsymbol{\nabla}_R)\boldsymbol{r}\cdot, \boldsymbol{\nabla}_r\right\}\\ &- \frac{k_c^{(1,0)}}{8M}\boldsymbol{\eta}\cdot\boldsymbol{\nabla}_R\times\left(\boldsymbol{\sigma}^{(1)}+\boldsymbol{\sigma}^{(2)}\right) + \frac{k_{d'}^{(1,0)}}{8Mr^2}(\boldsymbol{\eta}\cdot\boldsymbol{r}\times\boldsymbol{\nabla}_R)\left(\boldsymbol{r}\cdot\left(\boldsymbol{\sigma}^{(1)}+\boldsymbol{\sigma}^{(2)}\right)\right)\\ &- \frac{k_{d''}^{(1,0)}}{8Mr^2}\left(\boldsymbol{\eta}\cdot\boldsymbol{r}\times\left(\boldsymbol{\sigma}^{(1)}+\boldsymbol{\sigma}^{(2)}\right)\right)(\boldsymbol{r}\cdot\boldsymbol{\nabla}_R) - \frac{1}{8M}\left\{k_a^{(1,-1)}, \boldsymbol{\eta}\cdot\boldsymbol{\nabla}_r\times\left(\boldsymbol{\sigma}^{(1)}-\boldsymbol{\sigma}^{(2)}\right)\right\}\\ &+ \frac{1}{8M}\left\{\frac{k_{b'}^{(1,-1)}}{r^2}\left(\boldsymbol{r}\cdot\left(\boldsymbol{\sigma}^{(1)}-\boldsymbol{\sigma}^{(2)}\right)\right)(\boldsymbol{\eta}\times\boldsymbol{r})\cdot, \boldsymbol{\nabla}_r\right\}\\ &- \frac{1}{8M}\left\{\frac{k_{b''}^{(1,-1)}}{r^2}\left(\boldsymbol{\eta}\cdot\boldsymbol{r}\times\left(\boldsymbol{\sigma}^{(1)}-\boldsymbol{\sigma}^{(2)}\right)\right)\boldsymbol{r}\cdot, \boldsymbol{\nabla}_r\right\}\Bigg) S(t',\boldsymbol{R}',\boldsymbol{r}').\end{aligned} \tag{8}$$

At first glance, more problems have arisen due to the emergence of the unknown coefficients k in the boost transformation. Eq. (8) can, however, be simplified by the means of the *field redefinitions*, unitary transformations in particular: $S = \mathcal{U}_S\tilde{S}$ where $\tilde{S}$ is the newly defined singlet field and $\mathcal{U}_S$ is the unitary operator for it. A natural choice for $\mathcal{U}_S$ is an exponential function with its exponents being anti-hermitian operators. Order of $1/M$ expansion of the exponents is quadratic from the boost transformation of the new singlet field: $\mathcal{U}_S = \exp\left[-\frac{1}{4M^2}\left\{q_{a'}^{(1,0)}\boldsymbol{r}\cdot\boldsymbol{\nabla}_R, \boldsymbol{\nabla}_r\cdot\boldsymbol{\nabla}_R\right\} - \frac{1}{4M^2}\left\{q_{a''}^{(1,0)}\boldsymbol{r}\cdot\boldsymbol{\nabla}_R, \boldsymbol{\nabla}_r\cdot\boldsymbol{\nabla}_R\right\} + \cdots\right]$, where the qs are parameters we can freely choose. When we apply the unitary transformation to the singlet field and choose qs to eliminate as many terms in the boost as possible, there only remain four boost coefficients, $k_D^{(1,0)}, k_{a'}^{(1,0)}, k_c^{(1,0)}, k_{d''}^{(1,0)}$. As shown in the case of NRQCD, there exist additional constraints one can impose on the boost generator. First of which is the commutation relations between the boost generators acting onto the singlet field, $[1 - i\boldsymbol{\xi}\cdot\boldsymbol{k}, 1 - i\boldsymbol{\eta}\cdot\boldsymbol{k}]S = i(\boldsymbol{\xi}\times\boldsymbol{\eta})\cdot\boldsymbol{j}S$, and it fixes three coefficients: $k_{a'}^{(1,0)} = k_c^{(1,0)} = 1, k_{d''}^{(1,0)} = 0$. The last remaining one $k_D^{(1,0)}$ is fixed in the end along with deriving the constraints between the Wilson coefficients. The boost transformation of the singlet bilinear part of the pNRQCD Lagrangian is obtained by the simplified expression of the boost generator, thereby the original structure of the Lagrangian is kept intact up to total derivatives (organised up to the order of $1/M$)

$$\begin{aligned}\delta\mathcal{L}_{2S} = S^\dagger\Big(& i\Big(1 - c_S^{(1,0)}\Big)\boldsymbol{\eta}\cdot\boldsymbol{\nabla}_R - \frac{1}{2M}\Big(k_D^{(1,0)} - c_S^{(1,0)}\Big)\boldsymbol{\eta}\cdot\boldsymbol{\nabla}_R\partial_0 - \frac{i}{M}\Big(V_{p^2Sa} + V_{L^2Sa} + \frac{1}{2}V_S^{(0)}\Big)\boldsymbol{\eta}\cdot\boldsymbol{\nabla}_R\\ &+ \frac{i}{Mr^2}\Big(V_{L^2Sa} + \frac{r}{2}\partial_r V_S^{(0)}\Big)(\boldsymbol{\eta}\cdot\boldsymbol{r})(\boldsymbol{r}\cdot\boldsymbol{\nabla}_R) + \frac{1}{2M}\Big(V_{LSSa} + \frac{1}{2r}\partial_r V_S^{(0)}\Big)\boldsymbol{\eta}\cdot(\boldsymbol{\sigma}^{(1)}-\boldsymbol{\sigma}^{(2)})\times\boldsymbol{r}\Big)S\end{aligned} \tag{9}$$

which implies that all of the expressions inside the bracket in Eq. (9) are to vanish. Similar procedures are applied to the octet bilinear as well as to the singlet-octet sector, and they also yield relations between the coefficients. The significance of this procedure is that the form of the boost generator we started with is greatly simplified and eventually coincides with the one inspired by Wigner's *little group* formalism [13, 17], due to a particular choice of the unitary transformation.

4. Outlook: effective string theory

Poincaré invariance in low-energy EFTs for QCD sheds light on an another effective description of the heavy quark-antiquark bound system, namely the *effective string theory* (EST) [14]. The EST is valid at the long distance separation, $r\Lambda_{QCD} \gg 1$, and the transversal excitations of the long strings with its two ends fixed are the DOFs. There has been a hypothesis that Wilson loops in NRQCD are related to the EST at long distance as well as at large time scale [18]. On the other hand, Wilson loops in NRQCD at large time are related to the static potential in pNRQCD [19], which implies that one can relate the heavy quark-antiquark potentials to the string field variables. There arise dimensionful parameters in the string description [20], and Poincaré invariance in pNRQCD induces constraints between the parameters in the EST [21]. In [21], constraints are derived, but string variables are only considered up to the leading order, which are subjected to be modified when subleading terms are included; it is clear that $1/r^5$ potentials would change its coefficients, which is a part of our work in progress. Full EFT systematics to the EST concerning the inclusion of the subleading terms is to be discussed in the follow-up paper.

5. Acknowledgments

I thank Matthias Berwein, Nora Brambilla, and Antonio Vairo for the collaboration on this work, and International Max Planck Research School (IMPRS) program from Max-Planck-Institut für Physik, München and Excellence Cluster "Universe" for the financial support.

[1] Gross D. and Wilczek F. 1973 *Phys. Rev. Lett.* **30**, 1343
[2] Gross D. and Wilczek F. 1973 *Phys. Rev.* D **8**, 3633
[3] Politzer H. 1973 *Phys. Rev. Lett.* **30**, 1346
[4] Wilson K. 1974 *Phys. Rev.* D **10**, 2445
[5] Collins J. 2013 *Foundations of perturbative QCD* (Cambridge: Cambridge University Press)
[6] Caswell W. and Lepage G. 1986, *Phys. Lett.* B **167**, 437-442
[7] Bodwin G. Braaten E. and Lepage G. 1995 *Phys. Rev.* D **51**, 1125
[8] Isgur N. and Wise M. 1989 *Phys. Lett.* B **232**, 113
[9] Eichten E. and Hill B. 1990 *Phys. Lett.* B **234**, 511
[10] Brambilla N. Pineda A. Soto J. and Vairo A. 2000 *Nucl. Phys.* B **566**, 275-310
[11] Luke M. and Manohar A. 1992 *Phys. Lett.* B **286**, 348-354
[12] Brambilla N. Gromes D. and Vairo A. 2003 *Phys. Lett.* B **576**, 314-327
[13] Heinonen J. Hill R. and Solon M. 2012 *Phys. Rev.* D **86**, 094020
[14] Kogut J. and Parisi G. 1981 *Phys. Rev. Lett.* **47**, 16
[15] Berwein M. Brambilla N. Hwang S. and Vairo A. 2016 (*Preprint* TUM-EFT 74/15)
[16] Brambilla N. Vairo A. and Mereghetti E. 2009 *Phys. Rev.* D **83**, 079904
[17] Wigner E. 1939 *Annals Math.* **40**, 149
[18] Lüscher M. 1981 *Nucl. Phys.* B **180**, 317
[19] Brambilla N. Pineda A. Soto J. and Vairo A. 2000 *Phys. Rev.* D **63**, 014023
[20] Perez-Nadal G. and Soto J. 2009 *Phys. Rev.* D **79**, 114002
[21] Brambilla N. Groher M. Martinez H. and Vairo A. 2014 *Phys. Rev.* D **90**, 114032

Journal of Physics: Conference Series **742** (2016) 012026 doi:10.1088/1742-6596/742/1/012026

Progress in light hadron spectroscopy at BESIII

W C Yan (on behalf of BESIII Collaboration)

University of Science and Technology of China, Hefei, Anhui, 230026, China.

E-mail: wencheng@mail.ustc.edu.cn

Abstract. Using samples of 1.31×10^9 J/ψ events and 1.06×10^8 $\psi(3686)$ events accumulated at the BESIII detector, the recent progresses in the light meson spectroscopy and baryon spectroscopy are presented.

1. Introduction

Most of the hadronic states could be properly understood in the framework of constituent quark model [1], which is very successful in describing the observed spectrum, especially for the heavy flavor sector. However, the situation is quite different for the light hadronic spectrum. That is due to the large effective coupling within Quantum Chromo Dynamics (QCD) in the low energy regime, and the perturbative theory becomes invalid. Therefore, light hadron spectroscopy from the experimental measurement is a unique way to understand and verify the QCD formalism. In addition, QCD also predicts some new exotic hadrons, e.g. glueballs, multiquark states and hybrids. It has been the aim for many years to search for exotic states: many candidates have been reported, but none of them is unambiguously established yet.

The J/ψ and $\psi(3686)$ decays, especially the radiative decays, can provide wealthy information on the light hadron spectroscopy. Since their masses are below the mass threshold of charmed meson pair, the direct decay into charmed mesons is forbidden, which offers an excellent laboratory to study light hadron spectroscopy.

The BESIII detector [2] is a general-purpose magnetic spectrometer operating at BEPCII [3], which is a double-ring e^+e^- collider with the center-of-mass energy between 2.0 and 4.6 GeV. In 2009, samples of 2.25×10^8 J/ψ events and 1.06×10^8 $\psi(3686)$ events were collected with the BESIII detector. To study the light hadron physics comprehensively, in 2012, more data samples of J/ψ and $\psi(3686)$ with a factor 4 of events number were accumulated, and 1.31×10^9 J/ψ events and 4.48×10^8 $\psi(3686)$ events have been totally collected with the BESIII detector. In this talk, we present the recent progresses in light meson spectroscopy and baryon spectroscopy at BESIII.

2. Light meson spectroscopy

For the light meson spectroscopy, the attention will be focused on the observed $X(18xx)$ states and f_J^* states in J/ψ radiative decays.

2.1. X(18xx) States

A strong $p\bar{p}$ mass threshold enhancement, namely $X(p\bar{p})$, was firstly observed by the BESII experiment in $J/\psi \to \gamma p\bar{p}$ [4], and later confirmed by the CLEO-c [5] and BESIII [6]. Based

Content from this work may be used under the terms of the Creative Commons Attribution 3.0 licence. Any further distribution of this work must maintain attribution to the author(s) and the title of the work, journal citation and DOI.
Published under licence by IOP Publishing Ltd

on the 2.25×10^8 J/ψ events accumulated at the BESIII detector, a Partial Wave Analysis (PWA) of $J/\psi \to \gamma p\bar{p}$ [7] is performed, in which the effect of final state interaction (FSI) is considered by using the Jülich formulation [8]. In the optimal PWA fit, as shown in Fig. 1 (a), the J^{PC} of the $X(p\bar{p})$ is assigned to be 0^{-+}, and the statistical significance of the $X(p\bar{p})$ component is much larger than 30σ. The mass and width are determined to be $M = 1832^{+19}_{-5}\,(\mathrm{stat})^{+18}_{-17}\,(\mathrm{syst}) \pm 19\,(\mathrm{model})$ MeV/c^2, $\Gamma = 13 \pm 39\,(\mathrm{stat})^{+10}_{-13}\,(\mathrm{syst}) \pm 4\,(\mathrm{model})$ MeV/c^2 (a total width of $\Gamma < 76$ MeV/c^2 at the 90% confidence level), respectively.

A new resonance, $X(1835)$, was observed in $J/\psi \to \gamma\pi^+\pi^-\eta'$ [9] at the BESII, which was confirmed via the same decay channel by the BESIII experiment [10]. At present, the $X(1835)$ is further confirmed in a PWA of $J/\psi \to \gamma K^0_S K^0_S \eta$ with the sample of 1.31×10^9 J/ψ events [11]. The PWA is performed to the events satisfying $M(K^0_S K^0_S \eta) < 2.8$ GeV/c^2 and $M(K^0_S K^0_S) < 1.1$ GeV/c^2. The PWA fit, as shown in Fig. 1 (b), requires a contribution from $X(1835) \to K^0_S K^0_S \eta$ with a statistical significance greater than 12.9σ, and the spin parity of the $X(1835)$ is determined to be 0^{-+}. The mass and width of the $X(1835)$ are measured to be $M = 1844 \pm 9\,(\mathrm{stat})^{+16}_{-25}\,(\mathrm{syst})$ MeV/c^2 and $\Gamma = 192^{+20}_{-17}\,(\mathrm{stat})^{+62}_{-43}\,(\mathrm{syst})$ MeV/c^2, respectively. The mass and width of the $X(1835)$ are consistent with the values obtained from the decay $J/\psi \to \gamma\pi^+\pi^-\eta'$.

A strong deviation from the doubly OZI suppressed decay $J/\psi \to \gamma\omega\phi$ is observed near the $\omega\phi$ mass threshold [12] using the sample of 2.25×10^8 J/ψ events. Assuming the enhancement is due to the $X(1810)$ resonance, the PWA with a tensor covariant amplitude determine that the spin parity of the $X(1810)$ is 0^{++}, and the statistical significance is more than 30σ, as shown in Fig. 1 (c). The mass and width of the $X(1810)$ are determined to be $M = 1975 \pm 7\,(\mathrm{stat})^{13_{-5}\,(\mathrm{syst})} \pm 19\,(\mathrm{model})$ MeV/c^2 and $\Gamma = 95 \pm 10\,(\mathrm{stat})^{+21}_{-34}\,(\mathrm{syst}) \pm 75\,(\mathrm{model})$ MeV/c^2, respectively.

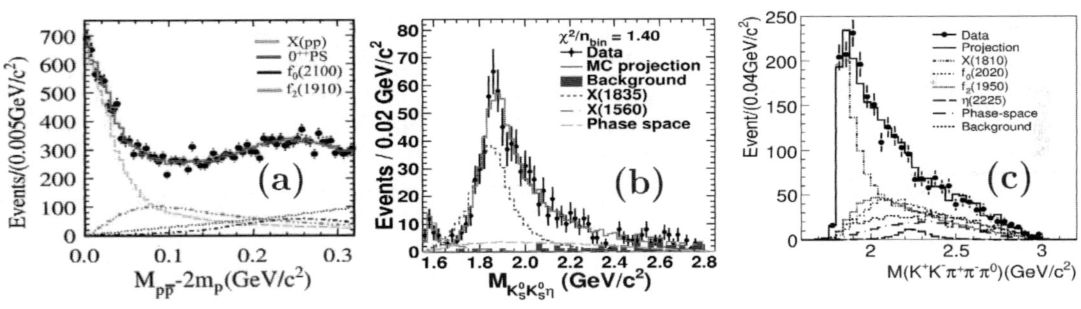

Figure 1. (a)The $X(p\bar{p})$ in the $p\bar{p}$ invariant-mass spectrum in $J/\psi \to \gamma p\bar{p}$, (b)the $X(1835)$ in the $K^0_S K^0_S \eta$ invariant-mass spectrum in $J/\psi \to \gamma K^0_S K^0_S \eta$, (c)the $X(1810)$ in the $K^+K^-\pi^+\pi^-\pi^0$ invariant-mass spectrum in $J/\psi \to \gamma\omega\phi$.

In addition, two new structures around 1.85 GeV/c^2 are also observed in the other J/ψ decays based on the sample of 2.25×10^8 J/ψ events. One is $X(1840)$ in the $3(\pi^+\pi^-)$ mass spectrum of $J/\psi \to \gamma 3(\pi^+\pi^-)$ [13] with a statistical significance of 7.6σ, as shown in Fig. 2 (a). The mass and width are measured to be $M = 1842.2 \pm 4.2\,(\mathrm{stat})^{+7.1}_{-2.6}\,(\mathrm{syst})$ MeV/c^2 and $\Gamma = 83 \pm 14\,(\mathrm{stat}) \pm 11\,(\mathrm{syst})$ MeV/c^2, respectively. The other is $X(1870)$ in the decay of $J/\psi \to \omega\pi^+\pi^-\eta$ [14], which is firstly observed at the BESIII experiment with the statistical significance of 7.2σ. From the fit, as shown in Fig. 2 (b), the mass and width are determined to be $M = 1877.3 \pm 6.3\,(\mathrm{stat})^{+3.4}_{-7.4}\,(\mathrm{syst})$ MeV/c^2 and $\Gamma = 57 \pm 12\,(\mathrm{stat})^{+19}_{-4}\,(\mathrm{syst})$ MeV/c^2, respectively. The spin parities of these two structures are still unknown.

The BESIII results of the masses and widths of the $X(p\bar{p})$, $X(1835)$, $X(1810)$, $X(1840)$ and

Journal of Physics: Conference Series **742** (2016) 012026 doi:10.1088/1742-6596/742/1/012026

$X(1870)$ are compared in Fig. 2 (c). The $X(p\bar{p})$ and $X(1835)$ have the same $J^{PC} = 0^{-+}$, similar central mass but different widths, and there is a speculation that the $X(p\bar{p})$ may be the tail of the $X(1835)$. The $X(1810)$ is not compatible with either the $X(1835)$ or the $X(p\bar{p})$ due to the different masses and spin parities. The mass of the $X(1840)$ is consistent with the $X(p\bar{p})$ and $X(1835)$, but the width is significantly different from either of them, and much smaller than the width of the $X(1835)$. While for the $X(1870)$, whether it is due to the $X(1835)$, the $\eta_2(1870)$, an interference of both, or a new resonance, it still needs further study.

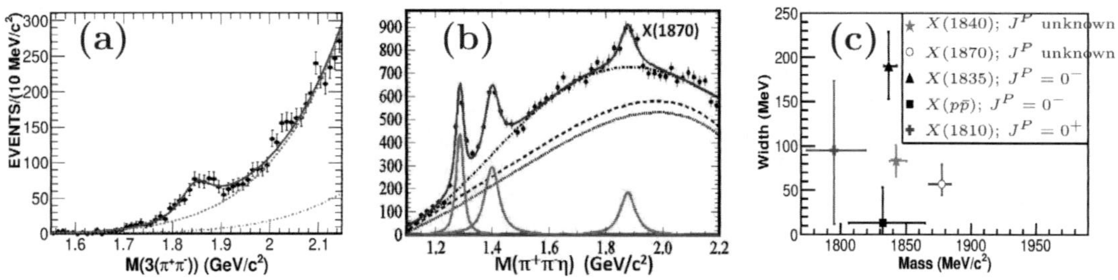

Figure 2. (a)The $X(1840)$ in the $3(\pi^+\pi^-)$ invariant-mass spectrum in $J/\psi \to \gamma 3(\pi^+\pi^-)$, (b)the $X(1870)$ in the $\pi^+\pi^-\eta$ invariant-mass spectrum in $J/\psi \to \omega\pi^+\pi^-\eta$, (c) the comparisons of the observed $X(18xx)$ states at the BESIII.

2.2. f_J^* states

A full PWA on the radiative decay of $J/\psi \to \gamma\eta\eta$ is performed using the relativistic covariant tensor amplitude method based on the sample of 2.25×10^8 J/ψ events [15]. The PWA results, listed in Tab. 1, show that the dominant scalar components are from the $f_0(1500)$, $f_0(1710)$ and $f_0(2100)$, while no evident contributions from $f_0(1370)$ or $f_0(1790)$ are observed. The tensor components, which are dominantly from $f_2'(1525)$, $f_2(1810)$ and $f_2(2340)$, also have a large contribution in $J/\psi \to \gamma\eta\eta$.

Table 1. Summary of the PWA results of $J/\psi \to \gamma\eta\eta$, including the masses and widths for resonances, branching ratios, as well as the significance. The first errors are statistical and the second are systematic.

Resonance	Mass(MeV/c^2)	Width(MeV/c^2)	$\mathcal{B}(J/\psi \to \gamma X \to \gamma\eta\eta)$	Significance
$f_0(1500)$	1468^{+14+23}_{-15-74}	$136^{+41+28}_{-26-100}$	$(1.65^{+0.26+0.51}_{-0.31-1.40}) \times 10^{-5}$	8.2σ
$f_0(1710)$	$1759\pm6^{+14}_{-25}$	$172\pm10^{+32}_{-16}$	$(2.35^{+0.13+1.24}_{-0.11-0.74}) \times 10^{-4}$	25.0σ
$f_0(2100)$	$2081\pm13^{+24}_{-36}$	273^{+27+70}_{-24-23}	$(1.13^{+0.09+0.64}_{-0.10-0.28}) \times 10^{-4}$	13.9σ
$f_2'(1525)$	$1513\pm5^{+4}_{-10}$	75^{+12+16}_{-10-8}	$(3.42^{+0.43+1.37}_{-0.51-1.30}) \times 10^{-5}$	11.0σ
$f_2(1810)$	1822^{+29+66}_{-24-57}	$229^{+52+88}_{-42-155}$	$(5.40^{+0.60+3.42}_{-0.67-2.35}) \times 10^{-5}$	6.4σ
$f_2(2340)$	$2362^{+31+140}_{-30-63}$	$334^{+62+165}_{-54-100}$	$(5.60^{+0.62+2.37}_{-0.65-2.07}) \times 10^{-5}$	7.6σ

Another model-independent PWA is performed on the radiative decay of $J/\psi \to \gamma\pi^0\pi^0$ with the sample of 1.31×10^9 J/ψ events [16]. The goal of this analysis is to provide a description of the scalar and tensor components of the $\pi^0\pi^0$ system while making minimal assumptions about the properties or number of poles in the amplitude. The PWA results indicate that the scalar contributions are mainly from $\sigma(600)$, $f_0(1370)$, $f_0(1500)$, $f_0(1710)$ and $f_0(2020)$. The tensor components are from the dominant contribution of $f_2(1270)$.

Journal of Physics: Conference Series **742** (2016) 012026 doi:10.1088/1742-6596/742/1/012026

3. Baryon spectroscopy

For the baryon spectroscopy, we only report the observed excited N^* states and two hyperons at the BESIII experiment.

3.1. *Excited N^* states*

To search for new excited N^* baryons, a PWA of $\psi(3686) \rightarrow p\bar{p}\pi^0$ is performed based on 1.06×10^8 $\psi(3686)$ events [17]. In this PWA, seven N^* intermediate resonances are observed. Besides the five established resonances, $N(1440)$, $N(1520)$, $N(1535)$, $N(1650)$ and $N(1720)$, two new resonances, $N(2300)$ and $N(2570)$, are also observed with the statistical significance both higher than 11σ. The optimized results of the PWA for the seven N^* states are listed in Tab. 2.

Table 2. The optimized mass, width and significance of the seven significant N^* resonances. ΔS and ΔN_{dof} represent the change of the log likelihood value and the change of the number of free parameters in the fit, respectively. In the second and third columns, the first uncertainty is statistical and the second systematic.

Resonance	Mass(MeV/c^2)	Width(MeV/c^2)	ΔS	ΔN_{dof}	Significance
$N(1440)$	1390^{+11+21}_{-21-30}	$340^{+46+70}_{-40-156}$	72.5	4	11.5σ
$N(1520)$	1510^{+3+11}_{-7-9}	115^{+20+0}_{-15-40}	19.8	6	5.0σ
$N(1535)$	1535^{+9+15}_{-8-22}	120^{+20+0}_{-20-42}	49.4	4	9.3σ
$N(1650)$	1650^{+5+11}_{-5-30}	150^{+21+14}_{-22-50}	82.1	4	12.2σ
$N(1720)$	1700^{+30+32}_{-28-35}	$450^{+109+149}_{-94-44}$	55.6	6	9.6σ
N(2300)	$\mathbf{2300^{+40+109}_{-30-0}}$	$\mathbf{340^{+30+110}_{-30-58}}$	**120.7**	**4**	$\mathbf{15.0}\sigma$
N(2570)	$\mathbf{2570^{+19+34}_{-10-10}}$	$\mathbf{250^{+14+69}_{-24-21}}$	**78.9**	**6**	$\mathbf{11.7}\sigma$

3.2. *Excited hyperons states*

Using a sample of 1.06×10^8 $\psi(3686)$ events collected with the BESIII detector, the processes of $\psi(3686) \rightarrow K^-\Lambda\bar{\Xi}^+ + c.c.$ and $\psi(3686) \rightarrow \gamma K^-\Lambda\bar{\Xi}^+ + c.c.$ are studied for the first time [18]. In the decay $\psi(3686) \rightarrow K^-\Lambda\bar{\Xi}^+ + c.c.$, two new hyperons, $\Xi(1690)^-$ and $\Xi(1820)^-$ shown in Fig. 3, are confirmed in $K^-\Lambda$ mass spectrum with the statistical significance of 4.9σ and 6.9σ, respectively. The masses and widths of the $\Xi(1690)^-$ and $\Xi(1820)^-$ are measured to be $M_{\Xi(1690)^-} = 1687.7 \pm 3.8\,(\text{stat}) \pm 1.0\,(\text{syst})$ MeV/c^2, $\Gamma_{\Xi(1690)^-} = 27.1 \pm 10\,(\text{stat}) \pm 2.7\,(\text{syst})$ MeV/c^2, and $M_{\Xi(1820)^-} = 1826.7 \pm 5.5\,(\text{stat}) \pm 1.6\,(\text{syst})$ MeV/c^2, $\Gamma_{\Xi(1820)^-} = 54.4 \pm 15.7\,(\text{stat}) \pm 4.2\,(\text{syst})$ MeV/c^2, which are consistent with the values from PDG [19] within one standard deviation.

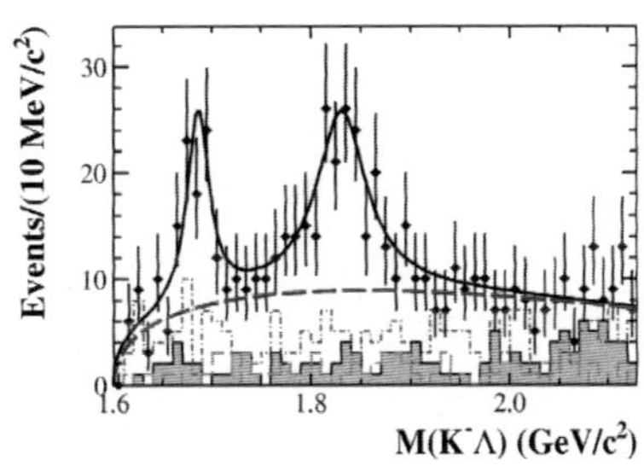

Figure 3. The hyperons $\Xi(1690)^-$ and $\Xi(1820)^-$ in $K^-\Lambda$ invariant mass spectrum.

In the decay $\psi(3686) \to \gamma K^-\Lambda\bar{\Xi}^+ + c.c.$, the measured branching fractions are summarized in Tab. 3. The measurements provide new information on charmonium decays to hyperons and on the resonance parameters of the hyperons, and may help in the understanding of the charmonium decay mechanism.

Table 3. Summary of the branching fractions measurements in the decay $\psi(3686) \to \gamma K^-\Lambda\bar{\Xi}^+ + c.c.$, where the first uncertainty is statistical and the second systematic.

Decay	Branching fraction $\mathcal{B}$
$\psi(3686) \to K^-\Lambda\bar{\Xi}^+$	$(3.86 \pm 0.27 \pm 0.32) \times 10^{-5}$
$\psi(3686) \to \Xi(1690)^-\bar{\Xi}^+,\ \Xi(1690)^- \to K^-\Lambda$	$(5.21 \pm 1.48 \pm 0.57) \times 10^{-6}$
$\psi(3686) \to \Xi(1820)^-\bar{\Xi}^+,\ \Xi(1820)^- \to K^-\Lambda$	$(12.03 \pm 2.94 \pm 1.22) \times 10^{-6}$
$\psi(3686) \to K^-\Sigma^0\bar{\Xi}^+$	$(3.67 \pm 0.33 \pm 0.28) \times 10^{-5}$
$\psi(3686) \to \gamma\chi_{c0},\ \chi_{c0} \to K^-\Lambda\bar{\Xi}^+$	$(1.90 \pm 0.30 \pm 0.16) \times 10^{-5}$
$\psi(3686) \to \gamma\chi_{c1},\ \chi_{c1} \to K^-\Lambda\bar{\Xi}^+$	$(1.32 \pm 0.20 \pm 0.12) \times 10^{-5}$
$\psi(3686) \to \gamma\chi_{c2},\ \chi_{c2} \to K^-\Lambda\bar{\Xi}^+$	$(1.68 \pm 0.26 \pm 0.15) \times 10^{-5}$

4. Summary

Based on the data samples of J/ψ and $\psi(3686)$ accumulated at the BESIII detector, the recent progresses in the light meson spectroscopy and baryon spectroscopy are presented. For the light meson spectroscopy, the observed $X(18xx)$ states near the $p\bar{p}$ mass threshold are reviewed. In addition, the f_J^* states in the PWA of $J/\psi \to \gamma\eta\eta$ and $J/\psi \to \gamma\pi^0\pi^0$ are reported. For the baryon spectroscopy, two new resonances, $N(2100)$, $N(2200)$, are observed for the first time, and two hyperons, $\Xi(1690)^-$ and $\Xi(1820)^-$, are firstly observed in charmonium decay.

The J/ψ and $\psi(3686)$ decays, especially the radiative decays, have been used extensively for the study of the light hadron spectroscopy at the BESIII experiment. With the high statistics data accumulated at the BESIII detector, more interesting results are expected to be coming soon.

References

[1] M. Gell-Mann 1964 *Phys. Lett.* **8**, 214
[2] M. Ablikim *et al.* (BESIII Collaboration) 2010 *Nucl. Instrum. Meth.* A **614**, 345
[3] J. Z. Bai *et al.* (BES Collaboration) 2001 *Nucl. Instrum. Meth.* A **458**, 627
[4] J. Z. Bai *et al.* (BES Collaboration) 2003 *Phys. Rev. Lett.* **91**, 022001
[5] J. P. Alexander *et al.* (CLEO Collaboration) 2010 *Phys. Rev.* D **82**, 092002
[6] M. Ablikim *et al.* (BESIII Collaboration) 2010 *Chin. Phys.* C **34**, 421
[7] M. Ablikim *et al.* (BESIII Collaboration) 2012 *Phys. Rev. Lett.* **108**, 112003
[8] A. Sibirtsev, J. Haidenbauer, S. Krewald, Ulf-G. Meissner, and A. W. Thomas 2005 *Phys. Rev.* D **71**, 054010
[9] M. Ablikim *et al.* (BES Collaboration) 2005 *Phys. Rev. Lett.* **95**, 262001
[10] M. Ablikim *et al.* (BESIII Collaboration) 2011 *Phys. Rev. Lett.* **106**, 072002
[11] M. Ablikim *et al.* (BESIII Collaboration) 2015 *Phys. Rev. Lett.* **115**, 091803
[12] M. Ablikim *et al.* (BESIII Collaboration) 2013 *Phys. Rev.* D **87**, 032008
[13] M. Ablikim *et al.* (BESIII Collaboration) 2013 *Phys. Rev.* D **88**, 091502
[14] M. Ablikim *et al.* (BESIII Collaboration) 2011 *Phys. Rev. Lett.* **107**, 182001
[15] M. Ablikim *et al.* (BESIII Collaboration) 2013 *Phys. Rev.* D **87**, 092009
[16] M. Ablikim *et al.* (BESIII Collaboration) 2015 *Phys. Rev.* D **92**, 052003
[17] M. Ablikim *et al.* (BESIII Collaboration) 2013 *Phys. Rev. Lett.* **110**, 022001
[18] M. Ablikim *et al.* (BESIII Collaboration) 2015 *Phys. Rev.* D **91**, 092006
[19] K. A. Olive *et al.* (Particle Data Group) 2014 *Chin. Phys.* C **38**, 090001

An effective field theory approach to hybrids

M Berwein

Technische Universität München, Physik-Department T30f, James-Franck-Str. 1, 85747 Garching, Germany

E-mail: matthias.berwein@mytum.de

Abstract. Heavy quarkonium hybrids are studied in an effective field theory framework. Coupled and uncoupled Schrödinger equations are obtained for different quantum numbers of the hybrid states. The results are discussed and compared to other approaches.

We summarize here the methods and results of our more detailed paper on hybrids published in [1]. Heavy quarkonia are bound states of a quark Q and an antiquark $\bar{Q}$ of charm or bottom flavor, whose observable J^{PC} quantum numbers for total angular momentum, parity, and charge conjugation are usually interpreted in theoretical descriptions in terms of orbital angular momentum and spin of the $Q\bar{Q}$-system and extended by another quantum number for radial excitations. However, QCD also allows for bound states with gluonic or light quark excitations, which are called hybrids or tetraquarks respectively. Such excitations are believed to account for the exotic states observed in experiment, collectively called X, Y, or Z.

The great advantage in studying systems with heavy quarks is that usually their mass M is larger than any other energy scale in the system. This means that even when the coupling constant g is large and cannot be used as a perturbative parameter, still one can perform an expansion in $1/M$. A systematic way of achieving this is through effective field theories (EFTs), where the expansion is performed at the Lagrangian level. The EFT obtained from QCD by integrating out the heavy quark mass is called nonrelativistic QCD (NRQCD) [2, 3]. The effective degrees of freedom are two Pauli spinor fields ψ and χ, which annihilate a heavy quark and create a heavy antiquark respectively, so-called soft gluons, i.e., with momenta much smaller than M, as well as n_f flavors of light quarks q. The Hamiltonian of NRQCD is also given as an expansion in $1/M$:

$$\mathcal{H}_{\mathrm{NRQCD}} = \mathcal{H}^{(0)} + \mathcal{H}^{(1)} + \mathcal{O}\left(M^{-2}\right) , \tag{1}$$

$$\mathcal{H}^{(0)} = \frac{1}{2}\int d^3x \left(\boldsymbol{E}^a\cdot\boldsymbol{E}^a + \boldsymbol{B}^a\cdot\boldsymbol{B}^a\right) - \sum_{j=1}^{n_f}\int d^3x\, \bar{q}_j i\boldsymbol{D}\cdot\boldsymbol{\gamma} q_j , \tag{2}$$

$$\mathcal{H}^{(1)} = -\frac{1}{2M}\int d^3x\, \psi^\dagger\left(\boldsymbol{D}^2 + c_F g\boldsymbol{B}\cdot\boldsymbol{\sigma}\right)\psi + \frac{1}{2M}\int d^3x\, \chi^\dagger\left(\boldsymbol{D}^2 + c_F g\boldsymbol{B}\cdot\boldsymbol{\sigma}\right)\chi . \tag{3}$$

The physical states are required to satisfy the Gauss law

$$(\boldsymbol{D}\cdot\boldsymbol{E})^a\,|\mathrm{phys}\rangle = g\Big(\psi^\dagger T^a\psi + \chi^\dagger T^a\chi + \sum_{j=1}^{n_f}\bar{q}_j\gamma^0 T^a q_j\Big)|\mathrm{phys}\rangle . \tag{4}$$

Content from this work may be used under the terms of the Creative Commons Attribution 3.0 licence. Any further distribution of this work must maintain attribution to the author(s) and the title of the work, journal citation and DOI.
Published under licence by IOP Publishing Ltd

In the following we will neglect the light quarks, i.e., $n_f = 0$.

The principles of this $1/M$ expansion in physical observables are the same as in perturbation theory of ordinary quantum mechanics. We have a Hamiltonian that is split into a main part $\mathcal{H}^{(0)}$ and a perturbation $\mathcal{H}^{(1)}$. In order to find the spectrum of the full Hamiltonian, we first determine the spectrum of the main part and then calculate corrections from the perturbation. The main part $\mathcal{H}^{(0)}$ corresponds to the theory in the limit of infinite quark mass, also called the static limit. In this limit, the heavy quark and antiquark act only as static color sources, and accordingly they do not appear in the Hamiltonian (although they are still connected to the gluons through the Gauss law). So the positions $\boldsymbol{x}_1$ and $\boldsymbol{x}_2$ for the heavy quark and the antiquark are good quantum numbers in this limit. We can write the spectrum as

$$\mathcal{H}^{(0)}|\underline{n};\boldsymbol{x}_1,\boldsymbol{x}_2\rangle^{(0)} = E_n^{(0)}(r)|\underline{n};\boldsymbol{x}_1,\boldsymbol{x}_2\rangle^{(0)}\,, \tag{5}$$

$$\text{with} \qquad {}^{(0)}\langle\underline{n}';\boldsymbol{x}_1',\boldsymbol{x}_2'|\underline{n};\boldsymbol{x}_1,\boldsymbol{x}_2\rangle^{(0)} = \delta_{n'n}\delta(\boldsymbol{x}_1'-\boldsymbol{x}_1)\delta(\boldsymbol{x}_2'-\boldsymbol{x}_2)\,, \tag{6}$$

where n stands for the set of all other quantum numbers, and the static energies $E_n^{(0)}$ depend only on the relative quark-antiquark distance $r = |\boldsymbol{x}_1 - \boldsymbol{x}_2|$ because of translational and rotational invariance.

In general, the spectrum of $\mathcal{H}^{(0)}$ cannot be obtained analytically, because it would correspond to the solution of QCD without heavy quarks, or Yang-Mills theory for $n_f = 0$. However, at least the static energies can be obtained numerically in the following way: consider some arbitrary state $|X_n\rangle$ in the static theory that has the same quantum numbers as $|\underline{n};\boldsymbol{x}_1,\boldsymbol{x}_2\rangle^{(0)}$; the large time T transition amplitude of this state is given by

$$\langle X_n(T)|X_n(0)\rangle = \langle X_n|e^{-i\mathcal{H}^{(0)}T}|X_n\rangle = \sum_n \int d^3x_1 d^3x_2 \left|\langle X_n|\underline{n};\boldsymbol{x}_1,\boldsymbol{x}_2\rangle^{(0)}\right|^2 e^{-iE_n^{(0)}(r)T}\,. \tag{7}$$

We have inserted the unity operator $\sum_n \int d^3x_1 d^3x_2\, |\underline{n};\boldsymbol{x}_1,\boldsymbol{x}_2\rangle^{(0)(0)}\langle\underline{n};\boldsymbol{x}_1,\boldsymbol{x}_2|$, where the sum contains in principle all quantum numbers n, however, in this equation the sum contains only excited states with the same quantum numbers as $|X_n\rangle$, because the overlap of $|X_n\rangle$ with any other eigenstate is zero. In the large time limit, this correlator is then dominated by the lowest static energy with these quantum numbers, so

$$E_n^{(0)}(r) = \lim_{T\to\infty} \frac{i}{T} \ln\langle X_n(T)|X_n(0)\rangle\,, \tag{8}$$

where some analytic continuation to complex times T is implied. By extending the set of states $|X_n\rangle$, also excited static energies with quantum numbers n can be projected out.

In this way, the lowest static energies have been determined on the lattice in the quenched approximation in [4] and [5], whose data we have used in [1]. The different static energies are classified by their quantum numbers, which represent the transformation properties under the symmetries of the static system. The symmetry group for two static particles of opposite charge is the same as for a cylinder and called $D_{\infty\,h}$, where the parity transformation P has to be extended by charge conjugation C. The elementary transformations of this group are rotations around the quark-antiquark axis, CP transformation, and reflections across a plane that contains the axis. The corresponding quantum numbers are labeled as Λ_η^σ, where Λ is the rotational quantum number and can be interpreted as the absolute value of the projection of the gluonic angular momentum operator on the quark-antiquark axis, the sign η is the eigenvalue under CP transformations, and σ is the sign under reflections. Λ can take integer values $0, 1, 2, \ldots$ for which traditionally capital Greek letters $\Sigma, \Pi, \Delta, \ldots$ are used, for η the labels g for plus and u for minus are used, and the quantum number σ appears only for $\Lambda = 0$, because $\Lambda \geq 1$ representations have two degenerate components with different values of σ, so σ is irrelevant for the static energies. According to the lattice results, the ground state has Σ_g^+ quantum numbers, which corresponds to standard quarkonium, and the lowest gluonic excitations are Π_u and Σ_u^- states.

Complementary information can be obtained in the short quark-antiquark distance limit $r \to 0$. In this limit another EFT called potential NRQCD (pNRQCD) [6, 7] can be obtained by integrating out the scale $1/r$; the resulting expansion in r has the form of a multipole expansion. After integrating out $1/r$, the heavy quark and antiquark fields can no longer be resolved individually, so the effective degrees of freedom are quarkonium fields, which can appear in either a color singlet or octet configuration, called S or O^a respectively, as well as so-called ultrasoft gluons with momenta smaller than $1/r$. The Hamiltonian of pNRQCD will be labeled with a double index, the first referring to the power of $1/M$ and the second to the power of r. The first terms in the multipole expansion of the static Hamiltonian are given by

$$\mathcal{H}^{(0,0)} = \frac{1}{2}\int d^3R \left(\boldsymbol{E}^a \cdot \boldsymbol{E}^a + \boldsymbol{B}^a \cdot \boldsymbol{B}^a\right) + \int d^3R d^3r \left(S^\dagger V_s(r) S + O^{a\,\dagger} V_o(r) O^a\right), \tag{9}$$

$$\mathcal{H}^{(0,1)} = -\int d^3R d^3r \left[\frac{V_A(r)}{\sqrt{6}}\left(O^{a\,\dagger}\boldsymbol{r}\cdot g\boldsymbol{E}^a S + S^\dagger \boldsymbol{r}\cdot g\boldsymbol{E}^a O^a\right) + \frac{V_B(r)}{2} d^{abc} O^{a\,\dagger}\boldsymbol{r}\cdot g\boldsymbol{E}^b O^c\right], \tag{10}$$

where d^{abc} is the symmetric structure constant, $\boldsymbol{R}$ and $\boldsymbol{r}$ are the center-of-mass coordinate and the relative distance respectively, and the quarkonium fields S and O^a depend on both coordinates, while the gluons depend only on $\boldsymbol{R}$.

In this theory it is fairly straightforward to write down suitable $|X_n\rangle$ states to determine the static energies. For the first two gluonic excitations we use

$$|X_{\Pi_u}\rangle = \boldsymbol{r} \times g\boldsymbol{B}^a O^{a\,\dagger}|\text{vac}\rangle \qquad \text{and} \qquad |X_{\Sigma_u^-}\rangle = \boldsymbol{r} \cdot g\boldsymbol{B}^a O^{a\,\dagger}|\text{vac}\rangle\,. \tag{11}$$

We see that the Π_u state has two independent components, while Σ_u^- has only one. The static energies we get from these states are identical at leading order:

$$E_{\Pi_u}^{(0)}(r) = \lim_{T\to\infty}\frac{i}{T}\ln\langle X_{\Pi_u}(T)|X_{\Pi_u}(0)\rangle = V_o(r) + \Lambda_B + \mathcal{O}\left(r^2\right), \tag{12}$$

$$E_{\Sigma_u^-}^{(0)}(r) = \lim_{T\to\infty}\frac{i}{T}\ln\langle X_{\Sigma_u^-}(T)|X_{\Sigma_u^-}(0)\rangle = V_o(r) + \Lambda_B + \mathcal{O}\left(r^2\right), \tag{13}$$

$$\text{with} \qquad \Lambda_B = \lim_{T\to\infty}\frac{i}{T}\ln\langle \text{vac}|B_i^a(T)\phi^{ab}(T,0)B_i^b(0)|\text{vac}\rangle\,, \tag{14}$$

where $\phi(T,0)$ is a straight Wilson line in the adjoint representation from time 0 to T ensuring the gauge invariance of the expression. At $\mathcal{O}\left(r^2\right)$, interactions from $\mathcal{H}^{(0,1)}$ come into play, which introduce a splitting between the two static energies, but the calculation of these corrections requires the nonperturbative evaluation of correlators of four gauge fields, which are currently not available. Already for the so-called magnetic gluelump mass Λ_B one has to rely on nonperturbative determinations, as performed in [5]. So we have determined those quadratic terms instead from a fit to the lattice data of the static energies from [4] and [5].

But what this leading pNRQCD result shows clearly is the approximate degeneracy of the two static energies for small distances, which becomes exact in the $r \to 0$ limit. This requires some care when calculating perturbative corrections to the static result. In naive perturbation theory for a generic theory with Hamiltonian $H^{(0)}$ and spectrum $|n\rangle^{(0)}$, the first two corrections to the energy due to the perturbation $H^{(1)}$ are given by

$$E_n = E_n^{(0)} + {}^{(0)}\langle n|H^{(1)}|n\rangle^{(0)} + \sum_{m\neq n}\frac{\left|{}^{(0)}\langle n|H^{(1)}|m\rangle^{(0)}\right|^2}{E_n^{(0)} - E_m^{(0)}} + \ldots\,. \tag{15}$$

However, if two energies are nearly degenerate, as in our case the short distance static energies of Π_u and Σ_u^-, then the second correction term cannot be considered small compared to the first and the perturbative series does not converge. Instead, one has to find the states $|h\rangle$ in the space spanned by the nearly degenerate states $|n\rangle^{(0)}$ which diagonalize the full Hamiltonian $H^{(0)} + H^{(1)}$ on that space:

$$\langle h'|H^{(0)} + H^{(1)}|h\rangle = E_h^{(1)}\delta_{h'h} \qquad \text{with} \qquad |h\rangle \in \text{span}\left\{|n\rangle^{(0)} \middle| E_n^{(0)} \approx E_{n'}^{(0)}\,\forall n, n'\right\}. \tag{16}$$

The next correction to the energy is then given by

$$E_h = E_h^{(1)} + \sum_{m\notin\{h\}} \frac{\left|\langle h|H^{(1)}|m\rangle^{(0)}\right|^2}{E_h^{(1)} - E_m^{(0)}} + \ldots, \tag{17}$$

and since the sum now contains no nearly degenerate states the denominator does not become large.

In our case there are in fact two sources for degeneracy: on the one hand there is the short distance behavior of the Π_u and Σ_u^- static energies, on the other hand we see that $\mathcal{H}^{(1)}$ contains derivatives, which connect states at an infinitesimal difference in their coordinates, and thus the corresponding static energies are naturally also degenerate. The last point can be readily understood in that $\mathcal{H}^{(1)}$ goes beyond the static limit, and for nonstatic quarks the positions are no longer good quantum numbers. Consequently, the lowest hybrid states at leading order beyond the static limit will be a linear combination of the static Π_u and Σ_u^- states, each equipped with a wave function.

Since the degeneracy between Π_u and Σ_u^- exists only for small distances r, we will restrict the calculation of the first nonstatic corrections to the leading order in r. The corresponding Hamiltonian in pNRQCD is given by

$$\mathcal{H}^{(1,-2)} = -\frac{1}{M}\int d^3R\,d^3r\left(S^\dagger \boldsymbol{\nabla}_r^2 S + O^{a\,\dagger}\boldsymbol{\nabla}_r^2 O^a\right). \tag{18}$$

So at this order $\boldsymbol{R}$ in fact remains a good quantum number and the wave functions have to depend only on $\boldsymbol{r}$. In order to calculate the corrections, we need the actual static states, not just $|X_n\rangle$ states that have a nonvanishing overlap. Even though those are unknown, we can introduce a gluonic operator $\boldsymbol{G}_B$ for them and write

$$|\underline{n}; \boldsymbol{x}_1, \boldsymbol{x}_2\rangle = \hat{\boldsymbol{n}}\cdot\boldsymbol{G}_B^a O^{a\,\dagger}|\text{vac}\rangle + \mathcal{O}(r), \qquad \text{with} \qquad \langle\text{vac}|(G_B)_i^a (G_B)_j^b|\text{vac}\rangle = \frac{\delta_{ij}\delta^{ab}}{8}, \tag{19}$$

where $\hat{\boldsymbol{n}}$ are some unit projection vectors perpendicular or parallel to the quark-antiquark axis for Π_u or Σ_u^- respectively. At leading order, the details of the operator $\boldsymbol{G}_B$ are irrelevant, the only important property is the orthogonality relation given above.

The matrix elements of $\mathcal{H}^{(0)} + \mathcal{H}^{(1)}$ between these hybrid states

$$|\text{Hyb}\rangle = \sum_n \int d^3r\,\hat{\boldsymbol{n}}\cdot\boldsymbol{G}_B^a O^{a\,\dagger}|\text{vac}\rangle\psi_n(\boldsymbol{r}) + \mathcal{O}(r) \tag{20}$$

are then given by

$$\langle\text{Hyb}'|\mathcal{H}^{(0)} + \mathcal{H}^{(1)}|\text{Hyb}\rangle = \sum_{n'n}\int d^3r\,\psi_{n'}'^{\,*}(\boldsymbol{r})\left(E_n^{(0)}(r)\delta_{n'n} - \hat{n}_i'\frac{\nabla_r^2}{M}\hat{n}_i\right)\psi_n(\boldsymbol{r}). \tag{21}$$

The wave functions therefore have to be eigenfunctions of a matrix-valued differential operator, i.e., we have a coupled Schrödinger equation, where the coupling comes from the action of $\boldsymbol{\nabla}_r^2$ on the projection vectors $\hat{\boldsymbol{n}}$ and $\hat{\boldsymbol{n}}'$. Like in ordinary Schrödinger equations, the wave functions can be split into an orbital and a radial part. The orbital wave functions satisfy a modified version of the defining differential equation of the spherical harmonics and are labeled by quantum numbers l and m, which correspond to the eigenvalues of the combined angular momentum operator of the gluons and the relative quark-antiquark motion. The radial Schrödinger equation then is found to decouple into two parts: the first part is still coupled but now has only two components:

$$\left[-\frac{1}{Mr^2}\,\partial_r r^2 \partial_r + \frac{1}{Mr^2}\begin{pmatrix} l(l+1)+2 & 2\sqrt{l(l+1)} \\ 2\sqrt{l(l+1)} & l(l+1) \end{pmatrix} + \begin{pmatrix} E_\Sigma^{(0)} & 0 \\ 0 & E_\Pi^{(0)} \end{pmatrix}\right]\begin{pmatrix} \psi_\Sigma \\ \psi_\Pi \end{pmatrix} = \mathcal{E}\begin{pmatrix} \psi_\Sigma \\ \psi_\Pi \end{pmatrix}, \tag{22}$$

while the second part is an uncoupled Schrödinger equation:

$$\left[-\frac{1}{Mr^2}\,\partial_r\, r^2\, \partial_r + \frac{l(l+1)}{Mr^2} + E_\Pi^{(0)}\right]\psi_\Pi = \mathcal{E}\,\psi_\Pi\,. \tag{23}$$

The two Schrödinger equations correspond to opposite parity solutions; also note that the first equation itself decouples for $l = 0$ and only the solution for ψ_Σ is physical. The energy eigenvalue $\mathcal{E}$ then gives the mass of the hybrid as $m_{\rm Hyb} = 2M + \mathcal{E}$.

In [1] we have compared the results of these Schrödinger equations using the potentials obtained from pNRQCD and the lattice data from [4] and [5] with various other investigations of hybrids. There are several experimental candidates in the mass range we obtain, although at this point it is not possible to make a clear identification. Another calculation of hybrid masses has been done in the framework of the Born-Oppenheimer approximation in [8]. While similar in approach and leading to comparable results, they do not obtain the splitting between opposite parity states that emerges clearly in the EFT treatment. There has also been a direct lattice computation of exotic charmonium states in [9]. The masses of the hybrid candidates identified in this paper have to be averaged over the different spin configurations for a comparison with our results; there seems to be an overall shift, but in particular the mass differences between different J^{PC} multiplets agree well with ours. For a more detailed discussion and graphical representation of our results, the interested reader is referred to [1].

Acknowledgments

We thank Colin Morningstar and Gunnar Bali for giving us access to their lattice data for the static energies. We acknowledge support from the DFG cluster of excellence "Origin and structure of the universe" (www.universe-cluster.de).

References

[1] Berwein M, Brambilla N, Tarrús Castellà J and Vairo A 2015 *Phys. Rev.* D **92** 114019 (*Preprint* 1510.04299)
[2] Caswell W E and Lepage G P 1986 *Phys. Lett.* B **167** 437
[3] Bodwin G T, Braaten E and Lepage G P 1995 *Phys. Rev.* D **51** 1125–1171 [Erratum: 1997 *Phys. Rev.* D **55** 5853] (*Preprint* hep-ph/9407339)
[4] Juge K J, Kuti J and Morningstar C 2003 *Phys. Rev. Lett.* **90** 161601 (*Preprint* hep-lat/0207004)
[5] Bali G S and Pineda A 2004 *Phys. Rev.* D **69** 094001 (*Preprint* hep-ph/0310130)
[6] Pineda A and Soto J 1998 *Nucl. Phys. Proc. Suppl.* **64** 428–432 (*Preprint* hep-ph/9707481)
[7] Brambilla N, Pineda A, Soto J and Vairo A 2000 *Nucl. Phys.* B **566** 275 (*Preprint* hep-ph/9907240)
[8] Braaten E, Langmack C and Smith D H 2014 *Phys. Rev.* D **90** 014044 (*Preprint* 1402.0438)
[9] Liu L, Moir G, Peardon M, Ryan S M, Thomas C E, Vilaseca P, Dudek J J, Edwards R G, Joo B and Richards D G (Hadron Spectrum Collaboration) 2012 *JHEP* **07** 126 (*Preprint* 1204.5425)

Journal of Physics: Conference Series **742** (2016) 012028 doi:10.1088/1742-6596/742/1/012028

Study of Excited Ξ Baryons in $\bar{p}p$-Collisions with the $\bar{P}$ANDA Detector

Jennifer Pütz[1], Albrecht Gillitzer[1], James Ritman[1], Tobias Stockmanns[1] on behalf of the $\bar{P}$ANDA Collaboration

[1] Institut für Kernphysik, Forschungszentrum Jülich, 52425 Jülich, Germany

E-mail: j.puetz@fz-juelich.de

Abstract. Understanding the excitation pattern of baryons is indispensable for the understanding of non-perturbative QCD. Up to now only the nucleon excitation spectrum has been subject to systematic experimental studies, while very little is known on excited states of double or triple strange baryons. In studies of antiproton-proton collisions, the $\overline{P}$ANDA experiment is well-suited for a comprehensive baryon spectroscopy program in the multi-strange and charm sector. In the present study we focus on excited Ξ^- states. For final states containing a Ξ^- $\bar{\Xi}^+$ pair, cross sections of the order of μb are expected, corresponding to production rates of $\sim 10^6$/d at a luminosity $L = 10^{31}\,\mathrm{cm}^{-2}\,\mathrm{s}^{-1}$. Here we present the reconstruction of the reaction $\bar{p}p \rightarrow \Xi(1820)^- \, \bar{\Xi}^+$ with $\Xi(1820)^- \rightarrow \Lambda\, K^-$ and its charged conjugate channel with the $\overline{P}$ANDA detector.

1. Motivation

In QCD the elementary particles of the strong interaction are quarks and gluons, whereas the minimum quark content of hadrons observable in nature is either three quarks (baryons) or a quark-antiquark pair (mesons). How these composite systems emerge is much less understood. To understand the structure of baryons and mesons it is necessary to study the excitation pattern of these particles. In the spectroscopy of baryons one of the main goals is to answer the following questions: Which degrees of freedom are relevant for the excitation mode of baryons? Can they be described by a three-quark or a quark-diquark structure? How important is the dynamics in baryon-meson systems? There is an intense worldwide effort ongoing to study the nucleon and Δ excitation spectrum with photo-induced reactions. Much less is known on the excitation spectrum of double and triple strange baryons Ξ and Ω. The study of Ξ baryons will provide independent information on the excitation pattern and allows to verify if the picture deduced from the excited nucleon and Δ states is correct and has a general validity for baryons in the light quark sector. The $\overline{P}$ANDA detector will give simultaneous access to the excited states of baryons and anti-baryons in $\bar{p}p \rightarrow \Xi\bar{\Xi}$ + meson(s). The large cross section for these reactions in the order of μb allows high statistics data to be collected in reasonable time.

2. The $\bar{P}$ANDA Experiment

$\overline{P}$ANDA has a wide physics program which, besides baryon and meson spectroscopy, includes studies of QCD dynamics, nucleon structure and electromagnetic processes as well as the study of hadrons in nuclei and hypernuclear physics [1]. The $\overline{P}$ANDA detector consists of two parts,

Content from this work may be used under the terms of the Creative Commons Attribution 3.0 licence. Any further distribution of this work must maintain attribution to the author(s) and the title of the work, journal citation and DOI.
Published under licence by IOP Publishing Ltd

the Target Spectrometer which has a cylindrical geometry and is placed around the interaction region and the Forward Spectrometer for the reconstruction of high-energy, forward-boosted particles. The specific design allows the $\overline{\text{P}}$ANDA detector to provide almost 4π acceptance for both charged and neutral particles.
A property of hyperons is their long life time, and hence long decay length. For the reconstruction of these particles a central tracking system is used. This central tracking system consists of the Micro Vertex Detector (MVD), the Straw Tube Tracker (STT) and the Gas Electron Multiplier (GEM) stations. The MVD is a silicon based high precision vertex detector and the innermost sub-detector enclosing the interaction point. The STT consists of gas filled drift tubes arranged in a cylindrical geometry around the beam axis. The GEM stations are also gas based detectors with planar geometry located directly downstream of the STT, and are thus important to reconstruct the trajectories of particles emitted in the forward direction within the Target Spectrometer. This central tracking system provides a high tracking resolution. In addition, the information from these detectors is combined with other components of $\overline{\text{P}}$ANDA to provide a good particle identification.

3. Simulation of $\Xi\,(1820)^-$

In the following we focus on the channel $\bar{\text{p}}\text{p} \rightarrow \Xi\,(1820)^-\ \bar{\Xi}^+$ and its charge conjugate. The full decay tree is shown in figure 1.

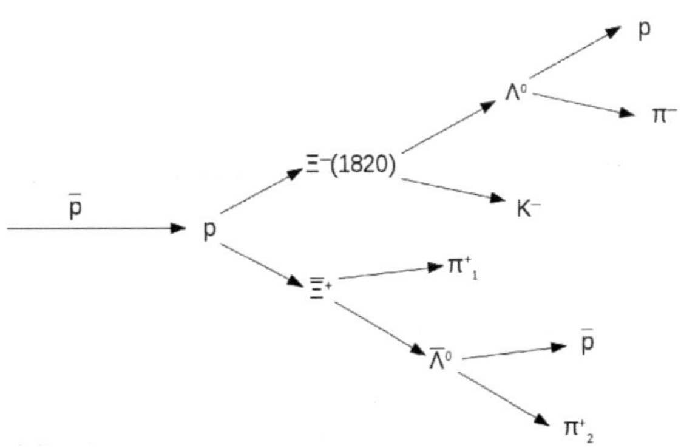

Figure 1. Reaction and decay tree for the event generation.

3.1. Event Generation

For this study 1.5 million signal events were generated using the event generator EvtGen [2]. For the charge conjugate channel another 1.5 million events were generated. If not otherwise specified, the charge conjugate process is implicitly included in the following description of the reconstruction strategy. For the production reaction $\bar{\text{p}}\text{p} \rightarrow \Xi\,(1820)^-\ \bar{\Xi}^+$ the "Phase Space" model, generating an isotropic angular distribution, was used, because a more realistic treatment has not yet been implemented in EvtGen. This simplification does not affect the strategy used in this study. The chosen beam momentum $p_{\bar{p}} = 4.6\,\text{GeV/c}$ corresponds to a center-of-mass energy of 100 MeV above the threshold to produce $\Xi\,(1820)^-$ and $\bar{\Xi}^+$. The production cross section $\sigma(\bar{\text{p}}\text{p} \rightarrow \Xi\,(1820)^-\ \bar{\Xi}^+)$ is unknown. However, for the ground state $\Xi^-\ \bar{\Xi}^+$ production in $\bar{\text{p}}\text{p}$ collisions, a cross section $\sigma \simeq 2\mu\text{b}$ has been measured at $p = 3\,\text{GeV/c}$ [3] and the comparison of ground state and excited single strange hyperon production shows similar cross sections for both species [4]. We therefore assume the cross section $\sigma(\bar{\text{p}}\text{p} \rightarrow \Xi\,(1820)^-\ \bar{\Xi}^+)$ to be of the order of $1\,\mu\text{b}$. The generated mass distribution for $\Xi\,(1820)^-$ is shown in figure 2. The properties of $\Xi\,(1820)^-$

Journal of Physics: Conference Series **742** (2016) 012028 doi:10.1088/1742-6596/742/1/012028

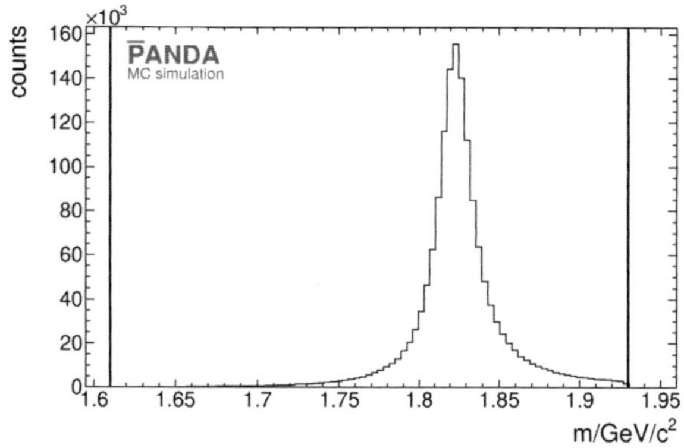

Figure 2. Generated mass distribution of $\Xi\,(1820)^-$. The black vertical lines denote the limits of the kinematically allowed region.

listed below are taken from Ref. [5]. The mass of $\Xi\,(1820)^-$ is $M = (1.823 \pm 0.005)$ GeV/c^2, the width is $\Gamma = (0.024 \pm 0.006)$ GeV/c^2. Spin and parity are given as $3/2^-$, however this assignment is still labeled as uncertain.

3.2. Reconstruction

For the reconstruction of the $\Xi\,(1820)^-$ one starts with the final state particles and proceeds backwards through the decay chain.

3.2.1. Final state particles The selected final state particles are protons, antiprotons, π^-, π^+, K^- and K^+ mesons. For the reconstruction of these particles the so-called "ideal pattern recognition" was used. This means that the origin of the hit is extracted from the propagator and all hits are grouped by using this information. The usage of a realistic pattern recognition is foreseen for a more detailed future analysis. To achieve a more realistic reconstruction efficiency, only particles with a minimum number of 4 hits in at least one inner tracking detector (MVD, STT or GEM) are selected. The particle identification (PID) is also "ideal" meaning that the true particle species is selected according to the Monte Carlo (MC) truth value. The reconstruction efficiency for the final state particles is shown in table 1. All reconstruction efficiencies are calculated with the MC matched particles and do not include branching ratios in the decay chain.

Table 1. Reconstruction efficiency for $\bar{p}p \to \Xi\,(1820)^-\,\bar{\Xi}^+$ (left) and $\bar{p}p \to \bar{\Xi}\,(1820)^+\,\Xi^-$ (right). The error of the efficiency $\sigma_\epsilon \simeq 0.1\%$ is purely statistical.

final state	ϵ [%]
π^-	83.5
π_1^+ ($\bar{\Xi}^+$)	80.9
π_2^+ ($\bar{\Lambda}$)	83.1
K^-	78.6
p	84.4
$\bar{p}$	78.3

final state	ϵ [%]
π^+	83.0
π_1^- (Ξ^-)	80.4
π_2^- (Λ)	82.7
K^+	83.3
p	80.7
$\bar{p}$	80.9

3.2.2. Reconstruction of Λ and $\bar{\Lambda}$ For the reconstruction of Λ hyperons a proton and a π^- meson are combined. The same is done with an antiproton and a π^+ meson for the reconstruction

of $\bar{\Lambda}$ hyperons. After combining the daughter particles, those candidates outside a symmetric invariant mass window around the nominal Λ mass of $0.3\,\mathrm{GeV/c^2}$ width are rejected. A kinematic vertex constraint fit is performed on the selected candidate. This means that the tracks of the daughter particles are fitted by taking account of the constraint that the tracks from the daughter particles come from a common vertex point. After the vertex constraint fit, a kinematic mass constraint fit is performed on the fitted candidate. Only those particles which have a probability of χ^2 value larger than 1% in both fits are selected. If more than one candidate fulfilling this condition is found, the candidate with the smallest χ^2 value in the vertex fit is chosen.
For the reconstruction of the Λ and $\bar{\Lambda}$ mass, a double-Gaussian fit is performed on the mass distribution of the selected Λ and $\bar{\Lambda}$ candidates. The peak position of the inner component of the double-Gaussian fit is taken as the value of the reconstructed mass. Both for Λ and $\bar{\Lambda}$ a peak position $M_\Lambda = 1.116\,\mathrm{GeV/c^2}$ in agreement with the input, and a width $\sigma_\Lambda = (2.1 \pm 0.1)\ \mathrm{MeV/c^2}$ is obtained.

3.2.3. Reconstruction of Ξ^- and $\bar{\Xi}^+$ In contrast to the Λ ($\bar{\Lambda}$) decay, the Ξ^- ($\bar{\Xi}^+$) decay contains a neutral particle in the final state. To include this case, an extension of the kinematic vertex fitter was developed. Apart from this, the reconstruction of Ξ^- ($\bar{\Xi}^+$) follows the same procedure as the reconstruction of Λ ($\bar{\Lambda}$). For the $\bar{\Xi}^+$ baryon $\bar{\Lambda}$ and π^+ are combined, for Ξ^- in the charge conjugate channel Λ and π^-. In case of a correct selection for both π^- and π^+ as daughter particle of the Λ and $\bar{\Lambda}$, respectively, only one π^- or π^+ candidate remains within the primary particles of the reaction chain, which must be the daughter particle of Ξ^- or $\bar{\Xi}^+$, respectively. The correct selection of Λ and $\bar{\Lambda}$ daughter pions is assured by the choice of the best fitted Λ and $\bar{\Lambda}$ candidate, as described in the previous section. The pions associated to the Λ and $\bar{\Lambda}$ decay are removed from the pion candidate lists used for the reconstruction of Ξ^- and $\bar{\Xi}^+$.
The result of the mass fit for $\bar{\Xi}^+$ reads $M = 1.322\,\mathrm{GeV/c^2}$, $\sigma = (3.69 \pm 0.01)\ \mathrm{MeV/c^2}$ and for Ξ^- reads $M = 1.322\,\mathrm{GeV/c^2}$, $\sigma = (4.00 \pm 0.01)\ \mathrm{MeV/c^2}$.

3.2.4. Reconstruction of $\Xi(1820)^-$ and $\bar{\Xi}(1820)^+$ In order to reconstruct the excited Ξ^- state, Λ and K^- mesons are combined for $\Xi(1820)^-$ and $\bar{\Lambda}$ and K^+ for $\bar{\Xi}(1820)^+$, using the best candidates of Λ and $\bar{\Lambda}$.
The reconstruction scheme is as for Ξ^- and $\bar{\Xi}^+$, but no mass constraint fit is performed. Figure 3 shows the mass distribution of $\Xi(1820)^-$ after different cuts. The resulting mass and width obtained from the double-Gaussian fit are $M_{\Xi^*} = 1.823\,\mathrm{GeV/c^2}$ and $\sigma_{\Xi^*} = (14.2 \pm 0.1)\ \mathrm{MeV/c^2}$ for $\Xi(1820)^-$ and $M_{\bar{\Xi}^*} = 1.823\ \mathrm{GeV/c^2}$ and $\sigma_{\bar{\Xi}^*} = (14.2 \pm 0.1)\ \mathrm{MeV/c^2}$ for $\bar{\Xi}(1820)^+$. The mass values are consistent with the input value.

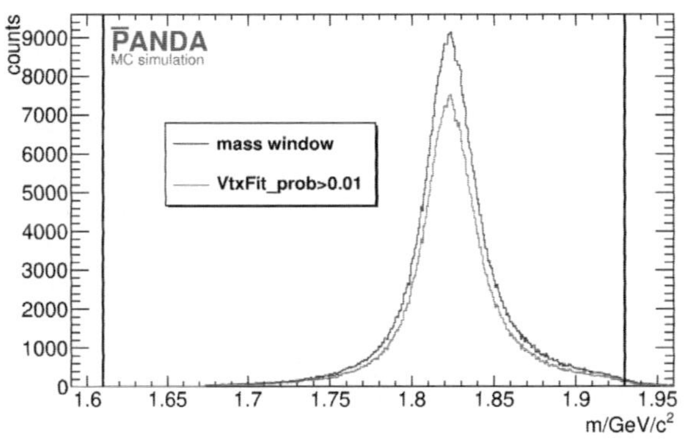

Figure 3. Mass distribution for $\Xi(1820)^-$ after the mass cut in blue and after the vertex fit probability cut in red. The black vertical lines denote the limits of the kinematically allowed region.

3.2.5. Reconstruction of the full decay chain To reconstruct the full reaction chain $\Xi(1820)^-$ and $\bar{\Xi}^+$ candidates are combined to the initial $\bar{p}p$ system. This is also done with the $\bar{\Xi}(1820)^+$ and Ξ^- for the charge conjugate channel. The resulting four-momentum vector of both daughter particles – here $\Xi(1820)^-$ and $\bar{\Xi}^+$ and their charge conjugate particles – is fitted with the constraint to match to the initial four-momentum vector of the $\bar{p}p$ entrance channel. After the fit only those candidates are selected which have a probability of more than 1 %.

4. Background

For background studies 15 million events have been simulated with the Dual Parton Model based generator DPM [6]. Since the ratio of the number of simulated background and signal events does not reflect the ratio of the respective cross section, a scaling factor is needed. The scaling factor B is calculated with the number of generated events and the cross section of signal and background as

$$B = \frac{N_{\text{sig}}^{\text{gen}}/\sigma_{\text{sig}}}{N_{\text{bg}}^{\text{gen}}/\sigma_{\text{bg}}}, \tag{1}$$

where $N_{\text{sig}}^{\text{gen}}$ is the number of generated signal events and $N_{\text{bg}}^{\text{gen}}$ the number of generated background events. The signal and background cross sections are given by $\sigma_{\text{sig}} = 1\,\mu\text{b}$ and $\sigma_{\text{bg}} = 60\,\text{mb}$ [5], respectively. The scaling factor is then $B = 6000$. All background events are subject to the same reconstruction procedure including identical cuts as for the signal events. The number of signal events is 8,614, including the branching ratios: BR($\Lambda \to \pi^-$ p)= 0.639, BR($\Xi^- \to \Lambda\ \pi^-$)= 0.999 and BR($\Xi(1820)^- \to \Lambda\ \text{K}^-$)= 0.3 [5] for the decay tree, but no background events survived the applied cuts.
This means that only a lower limit for the significance can be given as

$$S = \frac{N_{\text{sig}}}{\sqrt{N_{\text{sig}} + N_{\text{bg}} \cdot B}}. \tag{2}$$

In this first study, still based on several simplifications, we would then obtain a significance of 71 for a single background event and a signal-to-background ratio of 1.44:1. To achieve a more realistic estimate of the signal-to-background ratio, the used "ideal" pattern recognition and "ideal" particle identification need to be replaced by realistic ones and the statistical uncertainties must be reduced by generating and analyzing a much larger background sample.

5. Summary and Outlook

Each final state particle has a reconstruction efficiency of nearly 80 %. The reconstructed mass for $\Xi(1820)^-$ and $\bar{\Xi}(1820)^+$ are in good agreement with the values in the literature [5]. The topology of the decay chain suppresses the background efficiently. In future studies, the "ideal" particle identification and "ideal" pattern recognition will be replaced by more realistic algorithms. In addition a larger background sample will be generated and analyzed.

References

[1] Erni W., *et al.* 2009 *arXiv preprint* arXiv: 0903.3905
[2] Lange David J., 2001 *Nucl. Instr. Meth.* A **462**, p. 152
[3] Musgrave B., *et al.* 1965 *Nuovo Cimento* **35**, p. 735
[4] Flaminio V., Moorhead W. G., Morrissson D. R. O. and Rivoire N. 1984 **CERN-HERA84-01**(unpublished)
[5] Olive K. A. *et al.* [Particle Data Group Collaboration] 2014 *Chin. Phys.* C **38** 090001.
[6] Capella A., Sukhatme U., Tan C. I., & Van J. T. T. 1994 *Phys. Rept.* **236** p. 225.

Journal of Physics: Conference Series **742** (2016) 012029 doi:10.1088/1742-6596/742/1/012029

Pion induced reaction with carbon and polyethylene targets obtained by HADES-GSI in 2014

Pablo Rodríguez-Ramos[1,2] **for the HADES collaboration**

[1]Nuclear Physics Institute, Academy of Sciences of the Czech Republic, Řež, Czech Republic.
[2]Czech Technical University in Prague, Faculty of Nuclear Sciences and Physical Engineering, Prague, Czech Republic

E-mail: ramos@ujf.cas.cz

Abstract. In the summer of 2014, HADES was conducting measurements with secondary pion-beam using different targets. The program was devoted to measure dilepton radiation from baryonic resonances. In particular we investigated a sub-threshold coupling of ρ to baryonic resonances in the second resonance region (N(1520), N(1535)). Most of the beam time was dedicated to measurement of e^+e^- production from Polyethylene target at pion beam momentum of 0.69 GeV/c. In addition we run part of the time with pure carbon target. This allow us to study exclusive $\pi^- + p \to ne^-e^+$ channel. The normalization of spectra has been done using elastic scattering of pion on proton and carbon. The simulations of dilepton yields for π^0, Δ and N(1520) Dalitz decay using PLUTO was carried out.

1. Introduction

The HADES spectrometer [1] as shown in Fig. 1 is installed at SIS18 synchrotron in GSI Darmstadt, Germany. It is designed to measure systematically the production of electron-positron pairs in elementary and heavy-ion collisions at SIS18 energy range. It consists of 6 identical sectors covering the full azimuthal range and polar angles from $18° - 85°$ with respect to the beam direction. Each sector contains: A Ring Imaging CHerenkov (RICH) detector used for electron identification; two Mini-Drift Chambers (MDC) placed in front and two (MDC) placed behind the toroidal magnetic field used to determine momenta of charged particles; Time-Of-Flight detectors (TOF+RPC) and Pre-Shower detector improving the electron identification. The first level trigger is obtained by a fast multiplicity signal coming from the TOF wall, combined with a reaction signal from the START detector.

Content from this work may be used under the terms of the Creative Commons Attribution 3.0 licence. Any further distribution of this work must maintain attribution to the author(s) and the title of the work, journal citation and DOI.
Published under licence by IOP Publishing Ltd

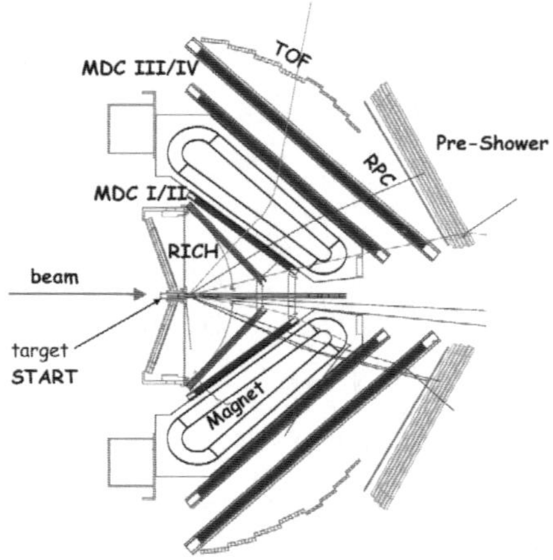

Figure 1. Schematic layout of the HADES detector.

2. Motivation

The addition of a secondary pion beam to the already available proton and heavy ion (HI) beams will allow to study properties of hadrons using the e^+e^- probe with the same detector system at different nuclear matter densities produced by pion, proton and HI beams. At ultra relativistic beam energies, the radiation in the spectral below the vector meson pole masses can be explained as $\pi^+\pi^-$ annihilation in the s-channel, at low beam energies the radiation of the fireball is better understood as electromagnetic decays of baryonic resonances. [2] The goal of the run was to investigate couplings of vector meson to baryon resonance, which play an important role in the description of in-medium ρ properties.

3. Experimental Layout

The π^- beam is generated by a primary ^{14}N beam, provided by the SIS18 synchrotron, with an intensity close to the space-charge limit of 0.8 -1.0 10^{11} ions/spill. The pions are transported to the HADES target, located 33m downstream of the production point by a beam-line composed of a lattice of 7 quadrupole and 2 dipole magnets, see Fig. 2. We used polyethylene (C_2H_4) as well as Carbon targets at pion beam momentum of 656 MeV/c, 690 MeV/c, 748 MeV/c, 800 MeV/c, see Table 1. The outgoing spectrum of pions is widely open both. Properties of the targets are given in Table 2.

Collected statistics in beam time AUG14/SEP14			
Target	**p**(MeV/c)	**Sum of events** (10^6)	**Data** (h)
PE	690	774.7	175.56
PE	748	76.5	11.61
PE	656	42.4	14,08
PE	800	52.4	7,48
C	690	115,7	13,06
C	800	41,2	6,27
C	748	42,2	6,8
C	656	41,9	14,75

Table 1. Collected statistics in beam time Aug14/Sep14

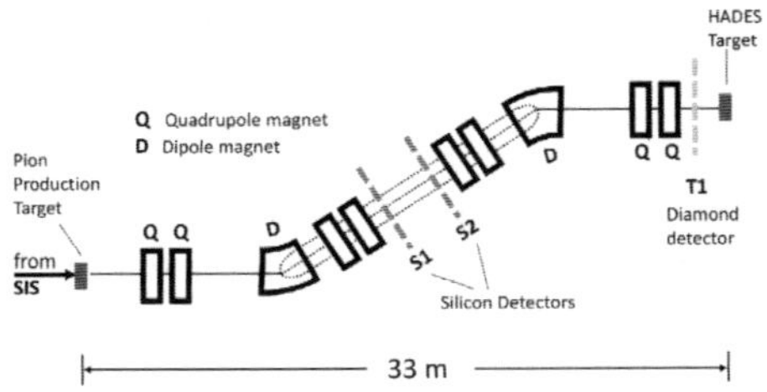

Figure 2. Schematic drawing of the pion beam-line

Target	PE=C_2H_4	C
Diameter (cm)	1.2	1.2
Length cm	4.6	2.52
M (g/mol)	28	12
Density(g/cm^3)	0.93	1.85
No. of carbon atoms	$18.4 \cdot 10^{22}$	$23.4 \cdot 10^{22}$

Table 2. Properties of the targets

3.1. Subtraction of Carbon contribution from PE target

First of all, we have to normalize the signal from both targets with the number of pions that went through the START detector. $Signal - scaled = \frac{Inv.Mass-CB}{Livetime}$ where $Livetime = \frac{M2\&\&START}{M2\&\&START-rate}$, where $M2\&\&START$ is the number of triggers, and $M2\&\&START - rate$ with the correction of dead time. The yield will be finally normalized by Signal-scaled/START. Finally using the corresponding factor $f = \frac{No.\ C\ atoms\ in\ PE}{No.\ C\ atoms\ in\ C} = 0.78$ taken from Table 2, we can subtract the C contribution from the PE using the formula: $H = PE - f \cdot C$. To check the consistency of the procedure, we normalized by the same way the number of $\pi^- + p$ and $\pi^- + C$ elastic scattering events collected during the experiment, see Fig. 3.

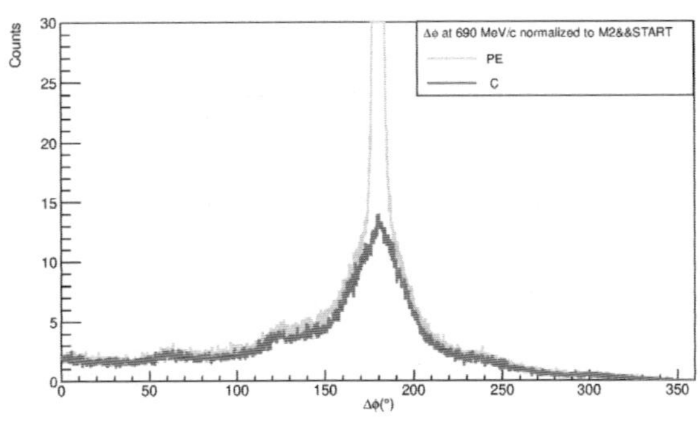

Figure 3. $\Delta\phi$ Spectra of elastically scattered pions scaled by the normalization factor

4. Simulations

In the analysis of simulations events were generated following theoretical distributions, using a comprehensive and modular ROOT-based event generator called PLUTO [2] developed by the HADES collaboration. For the simulations of $\pi^- + C$ we used a quasi-free model normalized by a factor 2, the list of channels simulated is shown in Table 3. No form factors are used for the treatment of the Dalitz decay of N(1520) and $\Delta(1232)$.

Channel	$\sigma\,(mb)$	**Branching Ratio**(%)	**Data source**
$\pi^- + p \rightarrow n\pi^0$	9.2	0.012	1
$\pi^- + p \rightarrow n\pi^0\pi^0 \quad \pi^- + p \rightarrow p\pi^0\pi^0\ (\pi^0 \rightarrow \gamma e^+ e^-)$	7.4	$4 \cdot 10^{-5}$	2
$\pi^- + p \rightarrow \Delta(\Delta \rightarrow Ne^+e^-)$	8.4	$4 \cdot 10^{-5}$	3
$\pi^- + p \rightarrow N\,(1520)^0\,(N\,(1520) \rightarrow ne^+e^-)$	20.5	$4 \cdot 10^{-5}$	4
$\pi^- + p \rightarrow n\eta(\eta \rightarrow \gamma e^+e^-)$	$0.3(p)\,0.7\,(C)$	0.006	5
$\pi^- p \rightarrow n\rho(\rho \rightarrow e^+e^-)$	4	$4 \cdot 10^{-5}$	6

Table 3. Cocktail source. 1: Landolt-Börnstein (L-B) for p $\in (0.6 - 0.72)\,MeV/c$ [4], 2: Crystall Ball for $\sqrt{s} = 1.461\ GeV/c^2$ with 20% reduction [5], 3: From single and double pion production cross sections, 4: From single and double pion production cross sections, 5: Parametrization from L-B data, 6: Manley' s analysis [6].

4.1. Inclusive e^+e^- cocktail

The inclusive electron pair differential cross section as a function of invariant mass and missing mass preliminary of the dielectron with respect to the pion-nucleon system were compared with the expected contributions for $\pi^- + C$ and $\pi^- + p$ simulations obtained from experiment by PLUTO event generator. The result of simulation are shown in Fig. 3.

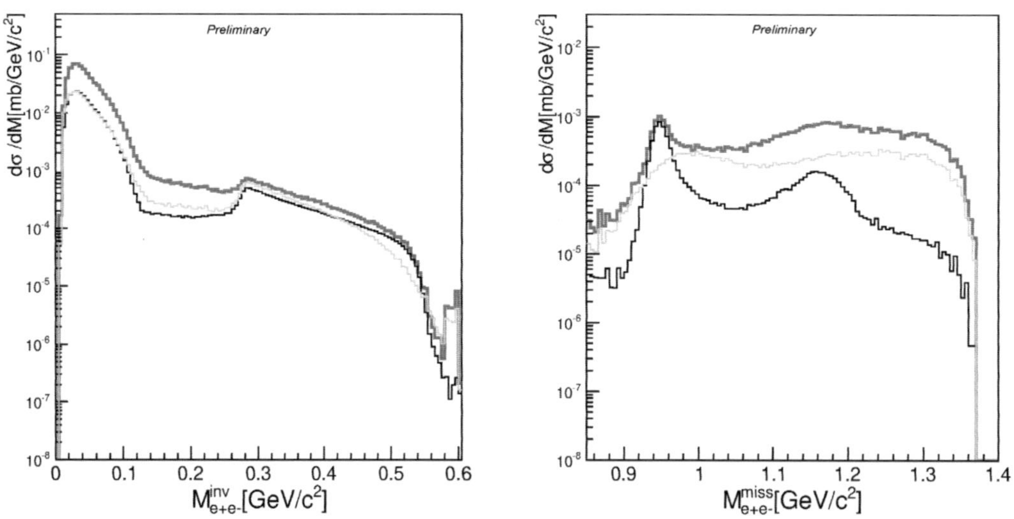

Figure 4. Left: e^+e^- Invariant mass distribution. **Right**: e^+e^- Missing mass distribution. Black line: Sum of $\pi^- + p$ simulations, Green line: Sum of $\pi^- + C$ simulations, Red line: Sum of $\pi^- + PE$ simulations

4.2. Exclusive e^+e^- cocktail

The exclusive electron pair differential cross sections as a function of invariant mass and missing mass with the proper cuts to select the $\pi^- + p \rightarrow ne^+e^-$ channel decays for $\pi^- + p$ simulations obtained by PLUTO event generator are shown below. It can be seen that the η contribution is very efficiently suppressed. The N(1520) Dalitz decay contribution falls off very rapidly at large invariant masses, due to the absence of form factor. The ρ contribution is dominating for invariant masses larger than 0.25 GeV/c^2 and consists of both resonant (i.e. via N(1520) decay) and non-resonant processes).

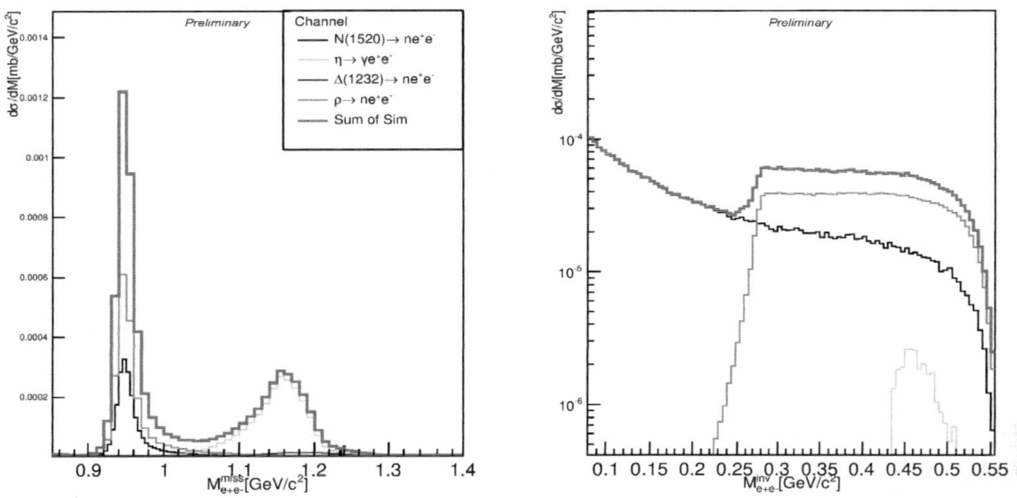

Figure 5. Left: Missing mass for $M^{inv}_{e^+e^-} > 0.12\ GeV/c^2$. **Right**: Invariant Mass for $0.9 < M^{miss}_{e^+e^-} < 1.0\ GeV/c^2$

5. Conclusion

The differential distributions obtained in the experiment will be compared to the e^+e^- cocktail from η, ρ, $\Delta(1232)$ and N(1520). It is expected that e^+e^- yield of invariant mass above 0.25 GeV/c^2 region corresponding to the missing mass between $[0.9 - 1.0]$ GeV/c^2 will be consisted with $N(1520) \to n\rho \to ne^+e^-$. Models associate the excess of dilepton measured in HI reactions with the excitation and decay of baryonic resonances into dileptons via intermediate ρ meson.

6. Acknowledgements

The collaboration gratefully acknowledges to: INFN-LNS Catania(Italy); LIP Coimbra(Portugal):PTDC/FIS/113339/2009; SIP JUC Cracow(Poland):NN202198639; GSI Darmstadt(Germany): Helmholtz Alliance HA216/EMMI; TU Darmstadt (Germany): VH-NG-823, Helmholtz Alliance HA216/EMMI; HZDR, Dresden(Germany):283286, 05P12CRGHE; Goethe University, Frankfurt (Germany):Helmholtz Alliance HA216/EMMI, HIC for FAIR (LOEWE), GSI FE, BMBF 06FY9100I, HGS-Hire, H-QM research school TU Munchen, Garching (Germany): BMBF 06MT7180; JLU Giessen (Germany):BMBF:05P12RGGHM; University Cyprus, Nicosia (Cyprus):UCY/3411-23100; IPN Orsay, Orsay Cedex (France): CNRS/IN2P3; NPI AS CR, Rez, (Czech Republic): GACR 13-067595 and LM201549 of MSMT of CR.

References

[1] G. Agakishiev *et al.* [HADES Collaboration], "The High-Acceptance Dielectron Spectrometer HADES," Eur. Phys. J. A **41**, 243 (2009) doi.10.1140/epja/i2009-10807-5 [arXiv:0902.3478 [nucl-ex]].
[2] The HADES Pion Beam Facility, Nuclear Physics News, Vol. 25, No. 2, 2015.
[3] I. Frohlich *et al.*, "Pluto: A Monte Carlo Simulation Tool for Hadronic Physics," PoS ACAT **2007**, 076 (2007)
[4] Baldini A, Flaminio V, Moorhead W G and Morrison D R O 1988 Total Cross-Sections for Reactions of High Energy Particles 1st ed (Landolt-Bornstein: New Series - Elementary Particles, Nuclei and Atoms vol 12)
[5] Prakhov S et al. (Crystal Ball) 2005 Phys. Rev. C 72(1) 015203
[6] Manley D M, Arndt R A, Goradia Y and Teplitz V L 1984 Phys. Rev. D 30(5) 904936

Journal of Physics: Conference Series **742** (2016) 012030 doi:10.1088/1742-6596/742/1/012030

Large-scale configuration interaction description of the structure of nuclei around ^{100}Sn and ^{208}Pb

Chong Qi

Department of Physics, KTH Royal Institute of Technology, SE-10691 Stockholm, Sweden

E-mail: chongq@kth.se; https://www.kth.se/profile/chongq/

Abstract. In this contribution I would like to discuss briefly the recent developments of the nuclear configuration interaction shell model approach. As examples, we apply the model to calculate the structure and decay properties of low-lying states in neutron-deficient nuclei around ^{100}Sn and ^{208}Pb that are of great experimental and theoretical interests.

1. Introduction

The FAIR-NUSTAR facility aims at addressing fundamental nuclear physics questions including: How are complex nuclei built from their basic constituents? What are the limits for existence of nuclei? How to explain collective phenomena from individual motion? Nuclear theory plays a critical role in explaining those emerging phenomena and new data from challenging measurements on radioactive beam facilities as well as in predicting the structure and decay properties of nuclei in regimes that are not accessible in the laboratory. Nuclear theory also has mutually beneficial interplay with other many-body physics including atomic physics and condensed matter physics as well as astrophysics and cosmology. The so-called *ab initio* approaches (in the sense that realistic nucleon-nucleon interaction are applied without much *ad hoc* adjustment), the nuclear shell model defined in a finite model space and the density functional theory are among the most commonly employed nuclear models that have been developed. Since all audience are not familiar with nuclear structure theory, firstly I would like to give an introduction to it.

The guiding principle for microscopic nuclear theory is that the building blocks of the nucleus, protons and neutrons, can be approximately treated as independent particles moving in a mean field that represents the average interaction between all particles. The single-particle motion provides a zeroth-order picture of the nucleus on top of which one has to consider the residual interaction between different particles. Within the *ab initio* family, the no-core shell model approach aims at considering the residual correlation between all nuclei in a large space defined by the harmonic oscillator. As a result, only light nuclei below ^{16}O can be evaluated (see, Fig. 1). The nuclear shell model, as we call it, is a full configuration interaction approach. It considers the mixing effect of all possible configurations within a given model space. The model space is usually defined by taking a few single-particle orbitals near the Fermi surface. The number of orbitals one can include is highly restricted due to computation limitation. As an example, the dimension for the Pb isotopes are given in the right panel of Fig. 1. Despite of this challenge, the nuclear shell model is by far the most accurate and precise theory available on the market. State-of-the-art configuration interaction algorithms are able to diagonalize matrices with dimension up to 2×10^{10} ($\sim 10^9$ with the inclusion of three-body interaction or if only identical particles are considered).

Below I will give a brief review on the challenges and recent developments of the nuclear configuration interaction shell model approach at our group. I will also mention a few simple

Content from this work may be used under the terms of the Creative Commons Attribution 3.0 licence. Any further distribution of this work must maintain attribution to the author(s) and the title of the work, journal citation and DOI.
Published under licence by IOP Publishing Ltd

but efficient truncation schemes. Shell-model calculations have been shown to be very successful in describing nuclei below ^{100}Sn. Now we aim at giving a microscopic description of heavier nuclei, in particular those around the shell closures marked by arrows in Fig. 1, which may be possible with the help of Petaflops supercomputer. I will show some results we obtained for intermediate-mass and heavy nuclei around ^{100}Sn and ^{208}Pb. I will explain the structure and decay studies of those nuclei, regarding both experimental and theoretical opportunities. It may be interesting to mention that a proper description of the $N = 126$ isotones is important for our understanding of the astrophysical r process and its third peak.

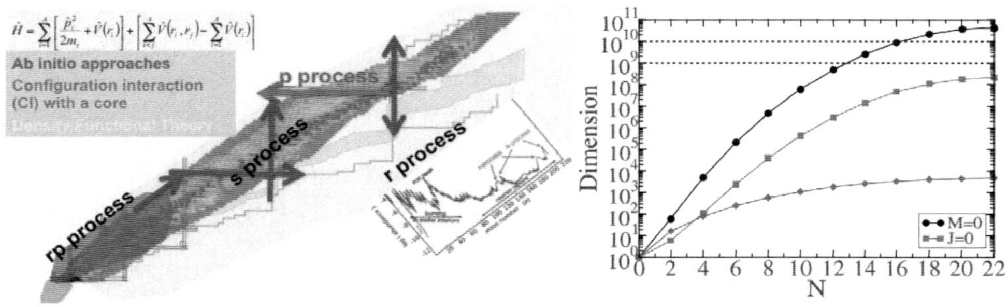

Figure 1. Left: Schematic plot for the territories of different nuclear models. Heavier nuclei around the arrows can also be described within the shell model approach. Right: Numbers of $M^\pi = 0^+$ and $I^\pi = 0^+$ shell-model states in even-even Pb isotopes as a function of valence neutron numbers N in the model space defined by orbitals $2p_{1/2,3/2}$, $1f_{5/2,7/2}$, $0i_{13/2}$, $0h_{9/2}$.

2. Configuration interaction shell model and the effective interaction

The residual interaction between valence particles around the Fermi surface is mostly supposed to be of two-body nature. A common practice is to express the effective Hamiltonians in terms of single-particle energies and two-body matrix elements as

$$H_{eff} = \sum_{\alpha} \varepsilon_\alpha \hat{N}_\alpha + \frac{1}{4} \sum_{\alpha\beta\delta\gamma JT} \langle \alpha\beta|V|\gamma\delta\rangle_{JT} A^\dagger_{JT;\alpha\beta} A_{JT;\delta\gamma}, \quad (1)$$

where we have assumed isospin symmetry in the effective Hamiltonian, $\alpha = \{nljt\}$ denote the single-particle orbitals and ε_α stand for the corresponding single-particle energies. $\hat{N}_\alpha = \sum_{j_z,t_z} a^\dagger_{\alpha,j_z,t_z} a_{\alpha,j_z,t_z}$ is the particle number operator. $\langle \alpha\beta|V|\gamma\delta\rangle_{JT}$ are the two-body matrix elements coupled to spin J and isospin T. A_{JT} ($A^\dagger_{JT}$) is the fermion pair annihilation (creation) operator. The two-body matrix elements can be calculated from realistic nucleon-nucleon potential where one has to consider the effect of its short range repulsion and the core polarization effects induced by the assumed inert core. Moreover, an optimization of the monopole interaction is necessary in most cases due to the neglect of three-body and other effects. The total energy of the state i is calculated to be

$$E_i^{\rm tot} = C + N\varepsilon_0 + \frac{N(N-1)}{2} V_1 + [T(T+1) - \frac{3}{4}N] V_0 + E_i^{\rm SM}, \quad (2)$$

where the constant C denotes the (negative) binding energy of the core and $E_i^{\rm SM}$ is the shell model energy. ε_0 is a mean single-particle energy. The relative value of the $T = 0$ and $T = 1$ monopole interaction $V_{0,1}$ determines the relative position of the nuclear states with different total isospin T. One may rewrite the Hamiltonian as $H_{eff} = H_m + H_M$ where H_m and H_M denote the (diagonal) monopole and Multipole Hamiltonians, respectively. The shell model energies can be written as

$$E_i^{\rm SM} = \sum_{\alpha} \varepsilon_\alpha < \hat{N}_\alpha > + \sum_{\alpha\leq\beta} V_{m;\alpha\beta} \left\langle \frac{\hat{N}_\alpha(\hat{N}_\beta - \delta_{\alpha\beta})}{1+\delta_{\alpha\beta}} \right\rangle + \langle \Psi_i|H_M|\Psi_i\rangle,$$

where $\sum_\alpha < \hat{N}_\alpha >= N$, Ψ_i is the calculated shell-model wave function of the state i.

Truncations often have to be applied in order to reduce the size of the shell-model bases. The simplest way of truncation is to restrict the maximal/minimal numbers of particles in different orbitals. This method is applied both to no-core and empirical shell model calculations. In Ref. [1] we studied the structure and electromagnetic transition properties of light Sn isotopes within the large $gdsh_{11/2}$ model space by restricting the maximal number of four neutrons that can be excited out of the $g_{9/2}$ orbital. However, the convergence can be very slow if there is no clear shell or subshell closure or if single-particle structure are significantly modified by the monopole interaction, as it happens in neutron-rich light nuclei (see, e.g., Ref. [2]).

For systems involving the same kind of particles, the low-lying states can be well described within the seniority scheme [3]. This is related to the fact that the $T = 1$ two-body matrix elements in Eq. (1) is dominated by monopole pairing interactions with $J = 0$. The seniority quantum number is related to the number of particles that are not paired to $J = 0$. Our recent studies on the seniority coupling scheme may be found in Refs. [4, 5, 6, 7]. One can also derive the exact solution of the pairing Hamiltonian by diagolizing the matrix spanned by the seniority $v = 0$, spin $I = 0$ states which represent only a tiny part of the total wave function. This is applied in Ref. [2, 8, 9]. The seniority coupling will be broken if both protons and neutrons are present where neutron-proton (np) coupling may be favored instead. There has been a long quest for the possible existence of np pairing in $N \sim Z$ nuclei for which there is still no conclusive evidence after long and extensive studies (see, recent discussions in Refs. [10, 11, 12, 13, 14]).

One can evaluate the importance of a given basis vector ψ_i within a partition through a perturbation measure

$$R_i = \frac{|\langle\psi_i|H_{eff}|\psi_c\rangle|}{\epsilon_i - \epsilon_c} \tag{3}$$

where ψ_c is the chosen reference with unperturbed energy ϵ_c. It is expected that the basis vectors with larger R_i should play larger role in the given state dominated by the reference basis ψ_c, from which truncation scheme can be defined. The off-diagonal matrix elements $\langle\psi_i|H_{eff}|\psi_c\rangle$ are relatively weak in comparison to the diagonal ones. The most important configurations may be selected by considering the difference of unperturbed energy difference as $r_i = \epsilon_i - \epsilon_c$. An truncated model space can thus be defined by taking those with smallest r_i. The challenge here is that the truncated bases may not conserve angular momentum. An angular momentum conserved correlated basis truncation approach is introduced in Ref. [15]. Alternatively, one may consider an importance truncation based on the total monopole energy as [16]

$$E_P^{\mathrm{m}} = \sum_\alpha \varepsilon_\alpha N_{P;\alpha} + \sum_{\alpha\leq\beta} V_{m;\alpha\beta} \frac{N_{P;\alpha}(N_{P;\beta} - \delta_{\alpha\beta})}{1+\delta_{\alpha\beta}},$$

where $N_{P;\alpha}$ denotes the particle distributions within a given partition P. One can order all partitions according to the monopole energy E_P^{m} and consider the lowest ones for a given truncation calculation. The idea behind is that the Hamiltonian is dominated by the diagonal monopole channel. The monopole interaction can change significantly the (effective) mean field and drive the evolution of the shell structure.

3. Selected results

We have been evaluating the structure and decay properties of nuclei following the arrows as indicated in Fig. 1. The robustness of the $N = Z = 50$ shell closures has been studied extensively recently, which has fundamental influence on our understanding of the structure of nuclei around the presumed doubly magic nucleus ^{100}Sn. It was argued that ^{100}Sn may be a soft core in analogy to the soft $N = Z = 28$ core ^{56}Ni. It seems such a possibility can be ruled out based on indirect information from recent measurements in this region [1, 17, 18, 19, 20]. It is still difficult to measure the single-particle states outside the ^{100}Sn core. The neutron single-particle states $d_{5/2}$ and $g_{7/2}$ orbitals in ^{101}Sn are very close to each other. A flip between the $g_{7/2}$ and $d_{5/2}$ orbitals from ^{103}Sn to ^{101}Sn was suggested in Ref. [21]. This result was used in the construction of the effective Hamiltonian [22] where the monopole interaction was optimized by fitting to all low-lying states in Sn isotopes using a global optimization method. The effective interaction has been shown to be successful in explaining many properties of nuclei in this region. The asymmetric electric quadrupole (E2) transition shape in Sn isotopes is suggested to be induced by the Pauli blocking effect [1]. A systematic study on the E2 transition in Te isotopes is done in Ref. [20].

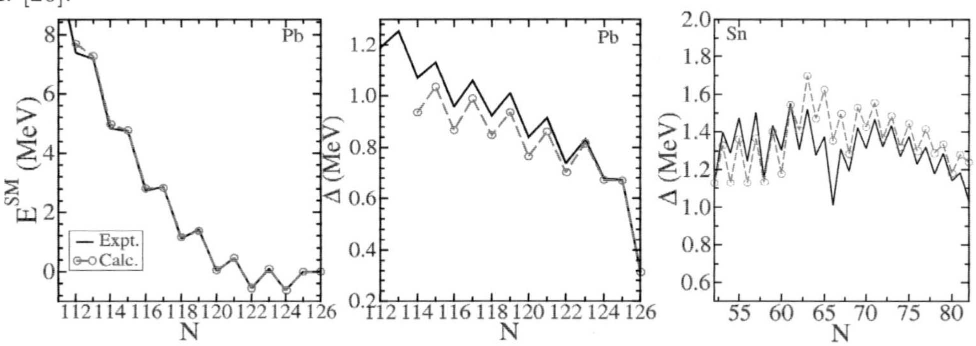

Figure 2. Shell-model energies for Pb isotopes with neutron number $N < 126$ (left) and neutron odd-even mass staggering (empirical pairing gaps) in Pb (middle) and Sn (right) isotopes extracted from the experimental and calculated binding energies.

We have done truncation calculations for two isotopes ^{200,194}Pb by considering the relative importance as defined by the monopole Hamiltonian and monopole+diagonal pairing Hamiltonian. Convergence can be reached with a small portion (around 10%) of the total M-scheme wave function in both cases. We have also done pair-truncated shell-model calculations with collective pairs as building blocks in Refs. [13, 16, 23] for both the standard shell model and continuum shell model in the complex energy plane. One example is the proton-unbound nucleus ^{109}I [24] for which the level structure and E2 transition properties are very similar to those of ^{108}Te [17] and ^{109}Te [25], indicating that the odd proton in ^{109}I is weakly coupled to the ^{108}Te daughter nucleus like a spectator.

In Fig. 2 we plotted the calculated shell-model energy for Pb isotopes and compared them with experimental data. Those energies are defined in the hole-hole channel relative to the assumed core ^{208}Pb. For nuclei heavier than ^{196}Pb, the difference between theory and experiment is less than 100 keV. The largest deviation appears in the case of ^{194}Pb for which the calculation overestimate E^{SM} by 300 keV. The empirical pairing gaps can be extracted from the binding energy by using the simple three-point formula, which carry important information on the two-nucleon pair clustering as well as α clustering in the nuclei involved [26, 27]. In nuclear systems the pairing collectivity manifests itself through the coherent contribution of many shell-model configurations, which lead to large pairing gaps. The results for Pb and Sn isotopes are shown as a function of the neutron number in the middle and right panels of Fig. 2. The overall agreement between experiments and calculations on the pairing gaps are quite satisfactory. Noticeable differences are only seen for mid-shell nuclei $^{196-198}$Pb and mid-shell Sn isotopes.

Our shell-model calculations can reproduce well the excitation energies of the low-lying 0^+ and 2^+ states in isotopes $^{198-206}$Pb. The excitation energies of the first 2^+ isotopes show a rather weak parabolic behavior In the lighter Pb isotopes the excitation energy of the second 0^+ state decreases rapidly with decreasing neutron number. It even becomes the first excited state in $^{184-194}$Pb. Within a shell-model context, those low-lying 0^+ states may be interpreted as coexisting deformed states which are induced by proton pair excitations across the $Z = 82$ shell gap. The energy of those core-excited configurations get more favored in mid-shell Pb isotopes in relation to the stronger neutron-proton correlation in those nuclei.

Another interesting phenomena is the nearly linear behavior of quadrupole moments in Cd, Sn as well as Pb isotopes for states involving the $h_{11/2}$ and $i_{13/2}$ orbitals, which can be explained in terms of shell occupancy within the seniority coupling scheme. As the occupancy increases, the quadrupole moments follow a linear decreasing trend and eventually vanish around half-filling.

There has already been a long effort answering the question whether the formation probabilities of neutron-deficient $N \sim Z$ isotopes are larger compared to those of other nuclei [28, 29]. The α decays from $N \sim Z$ nuclei can provide an ideal test ground for our understanding of the np correlation. We have evaluated within the shell-model approach the nn and pp two-body clustering in ^{102}Sn and ^{102}Te and then evaluated the correlation angle between the two pair by switching on and off the np correlation [14]. If the np correlation is switched on, in particular if a large number of levels is included, there is significant enhancement of the four-body clustering at zero angle. This is eventually proportional to the α formation probability. It should be mentioned that, one need large number of orbitals already in heavy nuclei in order to reproduce properly the α clustering at the surface. The inclusion of np correlation will make the problem even more challenging due to the huge dimension.

4. Summary

We present briefly our recent works on the configuration interaction shell model calculations. A simple truncation scheme can be established by considering configurations with lowest monopole energies. Good convergence for Pb isotopes is reached for both the energy and wave function. Large scale calculations are carried out to study the spectroscopic and transition properties of nuclei around ^{100}Sn and ^{208}Pb that cannot be reached by standard shell model calculations. Both the ground state binding energies and excitation energies of low-lying states of the Sn and Pb isotopes can be reproduced very well.

Acknowledgement

This work is supported by the Swedish Research Council (VR) under grant Nos. 621-2012-3805, and 621-2013-4323. The computations were partly performed on resources provided by the Swedish National Infrastructure for Computing (SNIC) at PDC, KTH, Stockholm.

References

[1] Back T, Qi C, Cederwall B, Liotta R, Moradi F G, Johnson A, Wyss R and Wadsworth R 2013 *Phys. Rev. C* **87** 031306
[2] Xu Z X and Qi C 2013 *Physics Letters B* **724** 247 – 252
[3] Talmi I 1993 *Simple Models of Complex Nuclei: The Shell Model and Interacting Boson Model* (Harwood Academic Publishers)
[4] Qi C, Wang X B, Xu Z X, Liotta R J, Wyss R and Xu F R 2010 *Phys. Rev. C* **82** 014304
[5] Qi C 2011 *Phys. Rev. C* **83** 014307
[6] Qi C, Xu Z X and Liotta R J 2012 *Nucl. Phys. A* **884** 21–35
[7] Qi C 2012 *Phys. Lett. B* **717** 436–440
[8] Qi C and Chen T 2015 *Phys. Rev. C* **92** 051304
[9] Changizi S, Qi C and Wyss R 2015 *Nucl. Phys. A* **940** 210 – 226
[10] Cederwall B *et al.* 2011 *Nature* **469** 68–71
[11] Frauendorf S and Macchiavelli A 2014 *Prog. Part. Nucl. Phys.* **78** 24 – 90
[12] Qi C, Blomqvist J, Bäck T, Cederwall B, Johnson A, Liotta R J and Wyss R 2011 *Phys. Rev. C* **84** 021301
[13] Xu Z X, Qi C, Blomqvist J, Liotta R J and Wyss R 2012 *Nucl. Phys. A* **877** 51–58
[14] Qi C and Wyss R 2016 *Physica Scripta* **91** 013009
[15] Jiao L F, Sun Z H, Xu Z X, Xu F R and Qi C 2014 *Phys. Rev. C* **90** 024306
[16] Qi C, Jia L Y and Fu G J *to be published*
[17] Bäck T *et al.* 2011 *Phys. Rev. C* **84** 041306
[18] Hinke C B *et al.* 2012 *Nature* **486** 341–345
[19] Guastalla G *et al.* 2013 *Phys. Rev. Lett.* **110**(17) 172501
[20] Doncel M *et al.* 2015 *Phys. Rev. C* **91**(6) 061304
[21] Darby I G *et al.* 2010 *Phys. Rev. Lett.* **105**(16) 162502
[22] Qi C and Xu Z X 2012 *Phys. Rev. C* **86**(4) 044323
[23] Jiang H, Qi C, Lei Y, Liotta R, Wyss R and Zhao Y M 2013 *Phys. Rev. C* **88**(4) 044332
[24] Procter M G *et al.* 2011 *Phys. Lett. B* **704** 118 – 122
[25] Procter M G *et al.* 2012 *Phys. Rev. C* **86** 034308
[26] Andreyev A N *et al.* 2013 *Phys. Rev. Lett.* **110** 242502
[27] Qi C, Andreyev A, Huyse M, Liotta R, Duppen P V and Wyss R 2014 *Phys. Lett. B* **734** 203 – 206
[28] Seweryniak D *et al.* 2006 *Phys. Rev. C* **73** 061301
[29] Liddick S N *et al.* 2006 *Phys. Rev. Lett.* **97** 082501

Journal of Physics: Conference Series **742** (2016) 012031 doi:10.1088/1742-6596/742/1/012031

Verification of passive cooling techniques in the Super-FRS beam collimators

C A Douma, J Gellanki, M A Najafi, H Moeini, N Kalantar-Nayestanaki, C Rigollet, O J Kuiken, M F Lindemulder, H A J Smit and H J Timersma.

KVI-CART, University of Groningen, Zernikelaan 25, 9747 AA Groningen, The Netherlands.

E-mail: c.a.douma@rug.nl

Abstract. The Super FRagment Separator (Super-FRS) at the FAIR facility will be the largest in-flight separator of heavy ions in the world. One of the essential steps in the separation procedure is to stop the unwanted ions with beam collimators. In one of the most common situations, the heavy ions are produced by a fission reaction of a primary ^{238}U-beam (1.5 GeV/u) hitting a ^{12}C target (2.5 g/cm^2). In this situation, some of the produced ions are highly charged states of ^{238}U. These ions can reach the collimators with energies of up to 1.3 GeV/u and a power of up to 500 W. Under these conditions, a cooling system is required to prevent damage to the collimators and to the corresponding electronics. Due to the highly radioactive environment, both the collimators and the cooling system must be suitable for robot handling. Therefore, an active cooling system is undesirable because of the increased possibility of malfunctioning and other complications. By using thermal simulations (performed with NX9 of Siemens PLM), the possibility of passive cooling is explored. The validity of these simulations is tested by independent comparison with other simulation programs and by experimental verification. The experimental verification is still under analysis, but preliminary results indicate that the explored passive cooling option provides sufficient temperature reduction.

1. Introduction

The Super FRagment Separator (Super-FRS) will be the largest in-flight separator of exotic nuclei in the world [1]. It produces beams of exotic nuclei to be used by the experiments of the new FAIR accelerator facility [2]. All kinds of exotic nuclei up to uranium can be produced and separated [1].

To produce the exotic nuclei, a high-luminosity ^{238}U-beam is used. This beam is produced by the primary SIS-ring [2] with an energy of 1.5 GeV/u and is collided with a fixed ^{12}C target of 2.5 g/cm^2 to produce exotic nuclei by fission and fragmentation reactions [1]. The produced exotic nuclei are extracted from the target by their forward momentum and then guided through a series of dipole and quadrupole magnets and beam collimators to remove the unwanted ions.

Some of these exotic nuclei can hit the first beam collimator with an energy of up to 1.3 GeV/u and a

Content from this work may be used under the terms of the Creative Commons Attribution 3.0 licence. Any further distribution of this work must maintain attribution to the author(s) and the title of the work, journal citation and DOI.
Published under licence by IOP Publishing Ltd

power of up to 500 W [3]. These energies will activate the first collimators and their surroundings up to a level that is dangerous for humans, even when the beam is switched off [1]. After more unwanted ions are stopped by previous collimators, the incident power on the next collimator decreases and the activation is less severe. Hence, it is only dangerous for humans to be around the first few collimators. This area is called the pre-separator [1], where every action is handled by robots.

The next stage of the Super-FRS is called the main separator. Humans can safely enter this area when the beam is switched off [1]. After the main separator area, the fragment separator splits up into three different branches that can guide the exotic beam to different experimental areas [1]. A schematic overview of the different experimental areas in the Super-FRS is given in figure 1.

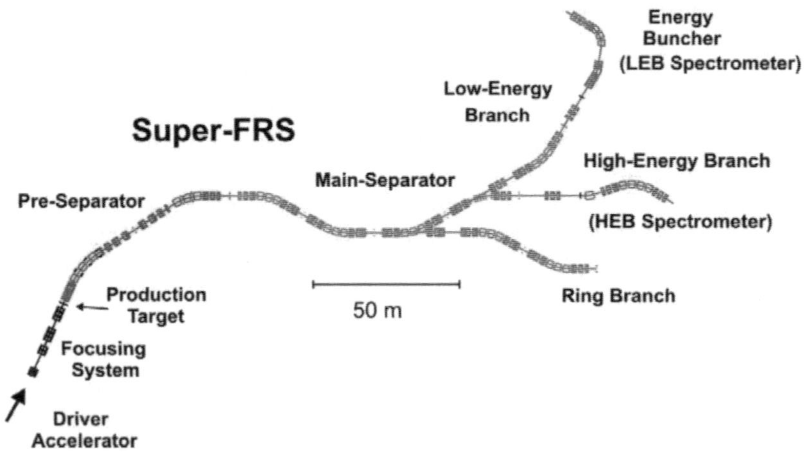

Figure 1. Schematic overview of the different experimental areas in the Super-FRS. The beam from the Driver accelerator is the ^{238}U-beam from the SIS-ring. The pre-separator area is drawn in red. The main separator area is drawn in blue [1].

Beam collimators that stop unwanted ions on the left and the right of the beam center are denoted as X-slits (since they move along the x-axis). Beam collimators that stop unwanted ions at the top and bottom of the beam center are denoted as Y-slits (since they move along the y-axis). In this paper, we focus on the thermal cooling of the first X-slit after the target. In the worst-case scenario this collimator receives $^{238}U^{90+}$ ions with an energy of 1.3 GeV/u and a power of 500 W [3]. All of these ions are stopped by the collimator. This means that the X-slit system should be sufficiently cooled to handle this 500 W.

2. Overview of the X-slit system

The actual collimation in the X-slit system is done by two blocks of Densimet. Densimet is an alloy of 97% Tungsten, 2% Nickel and 1% Iron [3]. These blocks are hanging in vacuum by two stainless steel rods. The movement of these blocks is controlled and read out by stepping motors and electronics located on the other side of the vacuum wall (in air). See figure 2.

On top of the X-slit system is a robot mount. This allows the robot to lift the entire X-slit system out of the beam line and replace it with a new one. This is a prerequisite, since some of the X-slit systems are located inside the pre-separator [1], [3]. Hence, replacement of the X-slit system must be possible in case of a malfunction. During the replacement procedure the Densimet blocks are disconnected by the robot and then reconnected to the new system.

3. Cooling options for the X-slit system

Our simulations (performed with Siemens NX 9.0 [4]) indicate that without cooling, the Densimet blocks can locally heat up to 700 °C. Due to conduction and radiation, the top flange of the X-slit system (part of the wall of the vacuum chamber) will then heat up to about 100 °C. This might cause malfunction in

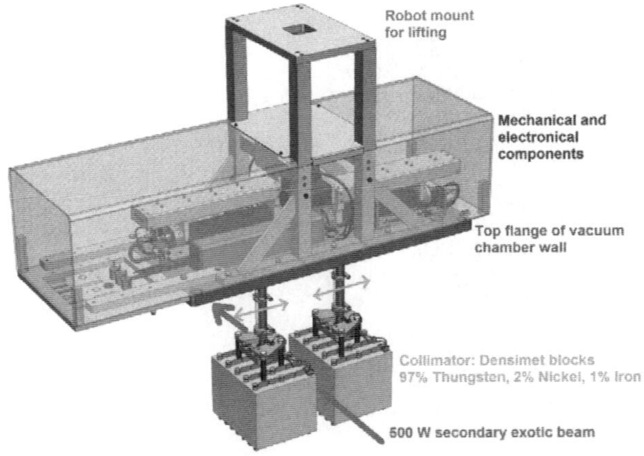

Figure 2. Computer drawing of the complete X-slit system with all of its components, as designed at KVI-CART.

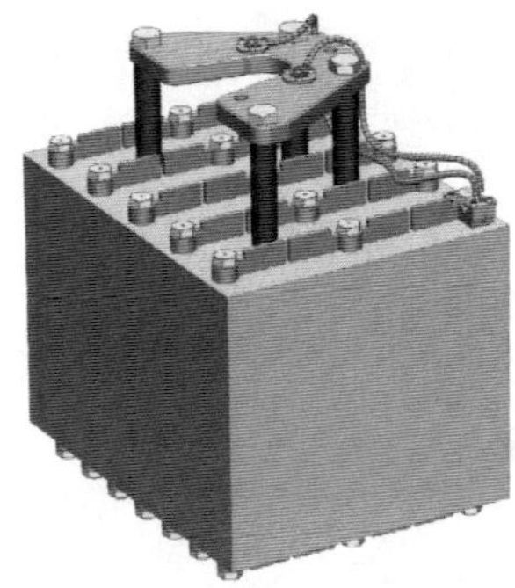

Figure 3. Computer drawing of a Densimet block of the X-slit system with stainless steel ribs on top. The dimensions of the block are 250 mm along the beam axis, see figure 2, 180 mm horizontal and 196 mm vertical.

the electronics located on the other side of the flange. Hence, cooling is required. For this cooling, one basically has two options: active cooling and passive cooling.

Our simulations also show that active water cooling on the Densimet blocks can easily reduce the temperature of the blocks from 700 °C to less then 200 °C and can, therefore, prevent the electronics from heating up too much. However, an active water cooling system adds a lot of complexity to the design of the X-slit system. Since this system has to run inside the pre-separator area where humans cannot enter, a complex system is undesirable since repair work as a result of a malfunction is extremely difficult, if not impossible. Especially (radioactive) water leaks can cause a lot of trouble since the pre-separator area is difficult to clean. Therefore, active cooling is undesirable.

Passive cooling means that one artificially increases the infrared emission of the Densimet blocks. Since tungsten has an emissivity below $\epsilon = 0.04$, [5], the emissivity of Densimet will also be very low. Hence, a lot of cooling can be done by increasing this emissivity. Our solution is to put ribs of stainless steel on the top and the bottom of the Densimet blocks (see figure 3). Since stainless steel has an emissivity of about $\epsilon = 0.65$ [4], [3], this provides an easy and effective cooling method without additional possibilities of malfunction.

Our simulations show that with the stainless steel ribs the Densimet will only heat up to 550 °C and the top flange of the X-slit system only heats up to about 35 °C [3]. This suggests that passive cooling with stainless steel ribs is enough to prevent damage to the mechanical and electronical components of the X-slit system (see figure 2). The question that then remains is how reliable our simulations are.

4. Simulation verification

In order to verify our simulation results, our procedure with Siemens NX 9.0 is compared to a simulation with Comsol [6]. For this comparison a simple test scenario is used where a plain Densimet block floats in the vacuum and is irradiated with a ^{238}U-beam of 500 W and 1.5 GeV/u. The beam profile is Gaussian with $\sigma = 5$ mm in both the x- and y-direction. The beam hits the block at 60 mm from the edge. A Densimet emissivity of $\epsilon = 0.07$ was used [3]. Results for a non-uniform triangular mesh of 0.3 mm around the beam spot are displayed in figures 4 - 5.

The overall difference between the temperature distributions produced by the NX and Comsol simulations is no more than 0.74 °C. This suggests that both simulation procedures can be regarded as reliable.

Figure 4. NX simulation results for the test scenario on a non-uniform triangular mesh.

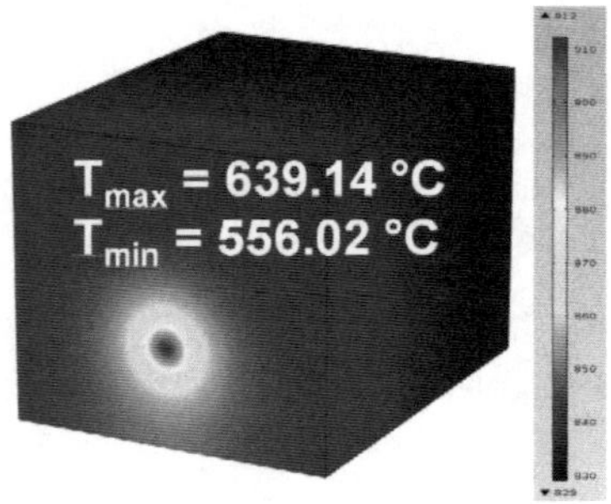

Figure 5. Comsol simulation results for the test scenario on a non-uniform triangular mesh.

5. Experimental verification

Our simulation results are further tested with an experimental verification. For this procedure a small test setup with Densimet blocks of $30 \times 30 \times 50$ mm^3 is used. The blocks are irradiated by a $^{20}Ne^{5+}$-beam of 30 MeV/u (21.6 W). The experiment is performed with the AGOR cyclotron at KVI-CART [7]. During irradiation, the temperatures of the Densimet blocks are measured with K-type thermocouples. An overview of the setup is given in figure 6.

Figure 6. The constructed experimental setup used to verify the X-slit heat simulations.

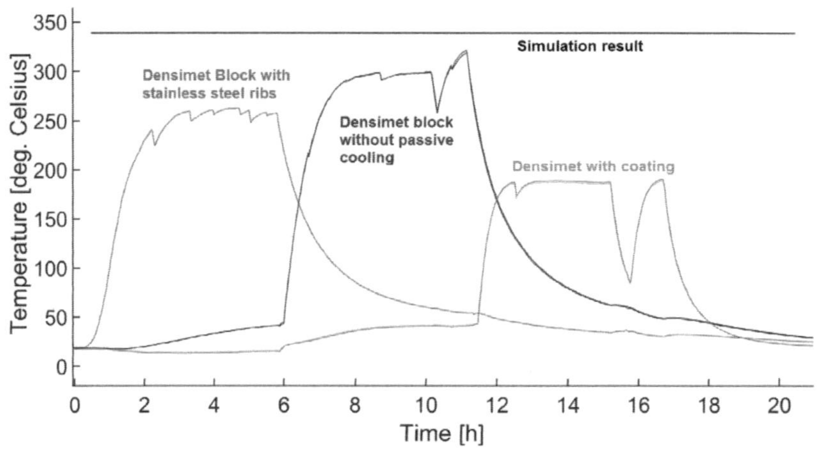

Figure 7. Temperature measurements of the Densimet blocks by two K-type thermocouples (linearly calibrated on the steam point and the ice point). The kinks in the graphs match the short beam failures. The high spike in the two blue graphs (two thermocouples) corresponds to a temporary increase of beam power to 30 W.

Three different Densimet blocks are irradiated one by one. One block is without any cooling (middle block in figure 6), another block is provided with passive cooling by stainless steel ribs (the bottom block) and the third block is provided with another type of passive cooling: a special coating with $\epsilon > 0.9$ [3]. This third block also had slightly different dimensions of $25 \times 25 \times 50$ mm^3. This allows us to both verify our simulation procedure and study the effect of passive cooling methods. The results are displayed in figure 7.

The black horizontal line in figure 7 corresponds to the temperature of the uncooled block in our latest NX simulation. So far, we have run only steady state simulations, but this work is still ongoing. So far, the difference between experiment and simulation is about 40 °C. We would like to point out that for any reasonable choice of parameters, the simulated results have always been higher than the experimental results.

It is also clear from figure 7 that the stainless steel ribs do provide passive cooling with respect to the uncooled block. The coating provides better passive cooling than the stainless steel ribs. However, after removing the setup from the beam-line, the coating sustained some damage while the stainless steel ribs appeared to be undamaged. Given the fact that the X-slit system should run inside the pre-separator area for many years, passive cooling by stainless steel ribbons does seem like the better option.

6. Conclusion

According to our results, passive cooling by connecting stainless steel ribs to the Densimet blocks seems to be the most suited cooling method for the X-slit system. Our NX simulations show that with this cooling the electrical and mechanical components will have a temperature below 35 °C (see section 3). Our simulation procedure has been tested against another simulation procedure and bench-marked by a measurement. The difference between our NX and Comsol simulation was no more then 0.74 °C, which suggests that our procedure is reliable. The steady state difference between our NX simulation and our experimental data is about 40 °C. This difference is still unresolved. The fact that each simulation shows a higher temperature than our experimental data might suggest that also in reality the electrical and mechanical components will have a temperature below 35 °C. However, the 40 °C difference should first be resolved and transient simulations should be run before any definitive statement can be made.

Acknowledgments

We would like to acknowledge C Nociforo, C Karagiannis and the NUSTAR Super-FRS Collaboration for their helpful input.

References

[1] Geissel H *et al* 2009 *Technical design report on the Super-FRS* Tech. rep. GSI and Collaborators
[2] Gutbrod H H, Augustin I, Eickhoff H, Gro K D, Henning WF, Kramer D, Walter G 2006 *FAIR baseline technical report* Tech. rep. GSI
[3] Gellanki J *et al* 2015 *Super-FRS slit system and possible passive cooling techniques* (*GSI Report* vol 2015-1) (Darmstadt: GSI Helmholtzzentrum für Schwerionenforschung) http://repository.gsi.de/record/184207
[4] Siemens PLM Software 2015 NX CAE http://www.plm.automation.siemens.com/en_us/products/nx/for-simulation/cae/
[5] Weast R C and Chemical Rubber Company (Cleveland, Ohio) 1975 *CRC Handbook of Chemistry and Physics: A Ready-reference Book of Chemical and Physical Data* (CRC Press)
[6] COMSOL, Inc 2015 Comsol: Modelinga nd Simulation for Everyone http://www.comsol.com/
[7] Brandenburg S, Ostendorf R, Hofstee M, Kiewiet H and Beijers H 2007 *Nuclear Instruments and Methods in Physics Research Section B: Beam Interactions with Materials and Atoms* **261** 82 – 85 ISSN 0168-583X
[8] Geissel H *et al* 2008 *Nuclear Instruments and Methods in Physics Research Section B: Beam Interactions with Materials and Atoms* **266** 4183 – 4187 ISSN 0168-583X
[9] Mathworks 2015 Matlab: The Language of Technical Computing http://nl.mathworks.com/products/matlab/
[10] Orona L M, Weick H, Mattila J, Amjad F, Kozlova E, Karagiannis C, Behr K H and Winkler M 2013 *International Journal of Advanced Robotic Systems* **10** 9

Journal of Physics: Conference Series **742** (2016) 012032 doi:10.1088/1742-6596/742/1/012032

Leptonic decays of D-wave vector quarkonia

A Krassnigg[1], **M Gomez-Rocha**[2] **and T Hilger**[1]

[1]Institute of Physics, University of Graz, NAWI Graz, A-8010 Graz, Austria

[2]ECT*, Villa Tambosi, 38123 Villazzano (Trento), Italy

E-mail: andreas.krassnigg@uni-graz.at

Abstract. We give a short and basic introduction to our covariant Dyson-Schwinger-Bethe-Salpeter-equation approach using a rainbow-ladder truncated model of QCD, in which we investigate the leptonic decay properties of heavy quarkonium states in the pseudoscalar and vector channels. Comparing the magnitudes of decay constants, we identify radial 1^{--} excitations in our calculation with experimental excitations of J/Ψ and Υ. Particular attention is paid to those states regarded as D-wave states in the quark model. We predict e^+e^--decay width of the $\Upsilon(1^3D_1)$ and $\Upsilon(2^3D_1)$ states of the order of ≈ 15 eV or more. We also provide a set of predictions for decay constants of pseudoscalar radial excitations in heavy quarkonia.

1. Introduction

The Dyson-Schwinger-Bethe-Salpeter-equation (DSBSE) approach to hadrons is a modern nonperturbative framework based on QCD as a continuum quantum field theory [1]. Phenomenological DSBSE studies mostly make use of the rainbow-ladder (RL) setup, since this provides excellent value for money or, in other words, an optimal balance of richness and feasibility, e. g., [2–16]. Beyond RL one can explore the infinite set of Dyson-Schwinger equations (DSEs) in symmetry-preserving truncation schemes, see [17–19] and references therein.

2. DSBSE Setup

The Dyson-Schwinger equation (DSE) for the dressed quark propagator S describes how $S(p)$ behaves at quark momentum p depending on the dressed gluon propagator $D(q)$ and the dressed quark-gluon vertex (QGV) $\Gamma(p,q)$ (the bare QGV is denoted by γ), which is depicted in the left panel of Fig. 1. The right panel of this figure shows the definition of the quark self energy $\Sigma(p)$ [20, 21]. S_0 denotes the free quark propagator, which serves as a driving term in this inhomogeneous integral equation and is also used to formulate a renormalization condition for the regularized integral [22]. In all figures, blobs depict dressed quantities, while dots depict the bare counterparts. In particular, the scalar part of S_0 is the current-quark mass m_q. Here and

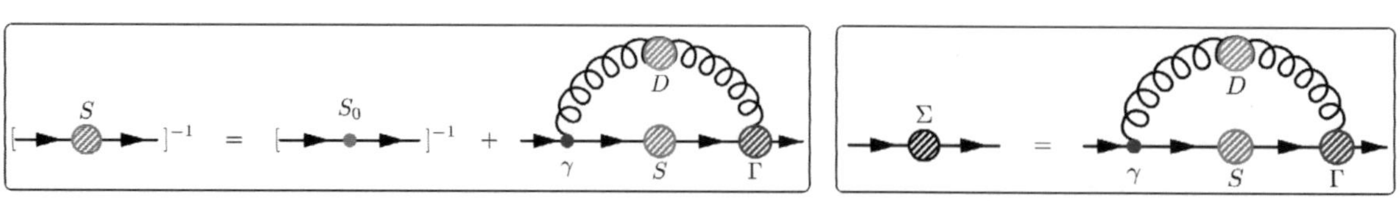

Figure 1. *Left panel:* quark DSE; *Right panel:* quark self energy Σ.

Content from this work may be used under the terms of the Creative Commons Attribution 3.0 licence. Any further distribution of this work must maintain attribution to the author(s) and the title of the work, journal citation and DOI.
Published under licence by IOP Publishing Ltd

Journal of Physics: Conference Series **742** (2016) 012032 doi:10.1088/1742-6596/742/1/012032

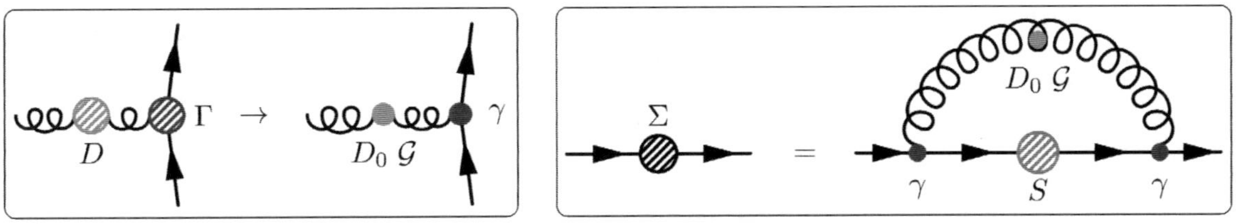

Figure 2. *Left panel:* RL *Ansatz*; *Right panel:* rainbow approximation of Σ in the quark DSE.

in the following, we assume all quantities to be renormalized and omit indices for simplicity. Note that, due to the appearance of the inverses of S and S_0, the quark DSE is nonlinear.

In RL truncation one makes the *Ansatz* depicted in the left panel of Fig. 2 to replace the product of the dressed gluon propagator and dressed QGV by the product of their bare counterparts multiplied with an effective interaction $\mathcal{G}$ [23]. This function of the gluon momentum squared is then modeled and used to conduct sophisticated covariant studies of hadrons. The right panel of Fig. 2 shows the effect of this *Ansatz* in the quark self energy as defined in the right panel of Fig. 1. Basically, one is left with known quantities and can solve the equation after making an informed choice for $\mathcal{G}$.

The two-body bound-state equation in a quantum field theory is the Bethe-Salpeter equation (BSE), which is depicted in the left panel of Fig. 3. Ingredients are the dressed quark propagator, which one gets from solving its DSE, and the amputated quark-antiquark scattering kernel K, which is unknown a priori [24]. The RL truncation, inspired and guided by the axial-vector Ward-Takahashi-identity (AVWTI), leads to an effective-gluon exchange ladder kernel as shown in the center panel of Fig. 3. The AVWTI is the manifestation of chiral symmetry in the QCD context, and respecting it guarantees a correct in-principle as well as phenomenological setup of the framework with respect to chiral symmetry and its dynamical breaking. Concretely, RL truncation satisfies the AVWTI and thus one automatically has a massless pion in the chiral limit, if chiral symmetry is dynamically broken by the effective interaction [25, 26].

The solution of the BSE is the Bethe-Salpeter amplitude (BSA) denoted in the left panel of Fig. 3 for the pion by Γ_π. It is the covariant analogue of a wave function and contains all information about the state [27–29]. It is found at a particular value of one of the arguments of Γ_π, the total momentum squared P^2, which gives the bound-state's mass M via $M^2 = -P^2$.

The ladder-truncated BSE kernel again contains only known objects such that a solution is straightforward. The quark DSE and meson BSE are coupled, since the DSE solution appears as input in the BSE. We solve this coupled set of equations numerically in Euclidean momentum space. The appropriate tools are all at hand and can be found in [30–35]. In order to calculate leptonic decay constants of mesons, one has to compute a diagram as shown in the right panel of Fig. 3, where the appropriate axial-vector current for f_π is depicted as a black-yellow blob and denoted by $\gamma_5\gamma_\mu$ for the pseudoscalar case; the vector decay constant is obtained via γ_μ.

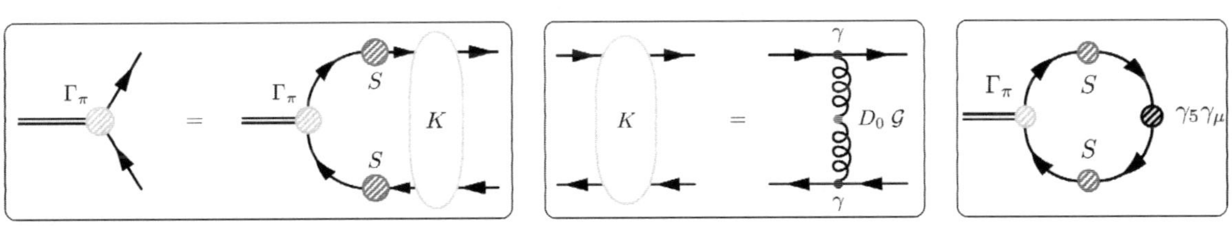

Figure 3. *Left panel:* meson BSE; *Center panel:* ladder-truncated BSE kernel; *Right panel:* diagram to compute the pseudoscalar leptonic decay constant.

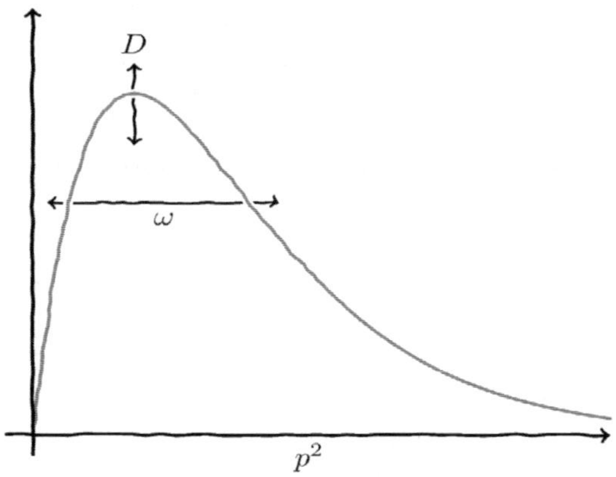

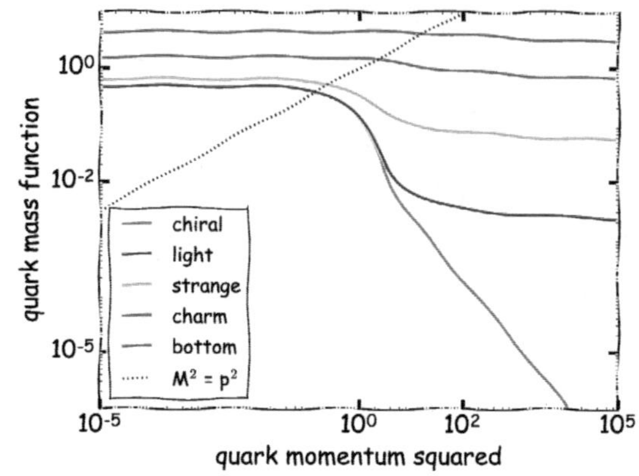

Figure 4. Sketches of: *Left panel:* effective coupling $\mathcal{G}$; *Right panel:* typical set of solutions of the gap equation: mass functions from the chiral limit to b quark.

Table 1. Calculated pseudoscalar and vector meson decay constants in MeV compared to experimental data [40, 41], where available. *Left*: charmonium; *right*: bottomonium.

State	J^{PC}	Calc. I	II	Exp.	State	J^{PC}	Calc. I	II	Exp.
Pseudoscalar					Pseudoscalar				
η_c	0^{-+}	401	378	339(14)	η_b	0^{-+}	773	756	–
$\eta_c(2S)$	0^{-+}	244(12)	82	189(50)	$\eta_b(2S)$	0^{-+}	419(8)	285	–
$\eta_c(3S)$	0^{-+}	145(145)	206	–	$\eta_b(3S)$	0^{-+}	534(57)	333	–
$\eta_c(4S)$	0^{-+}	–	87	–	$\eta_b(4S)$	0^{-+}	–	40(15)	–
Vector					Vector				
J/Ψ	1^{--}	450	411	416(5)	Υ	1^{--}	768	707	715(5)
$\Psi(2S)$	1^{--}	30(3)	155	294(4)	$\Upsilon(2S)$	1^{--}	467(17)	393	497(4)
$\Psi(3770)$	1^{--}	118(91)	45	99(3)	$\Upsilon(1^3D_1)$	1^{--}	41(7)	371(2)	–
$\Psi(4040)$	1^{--}	–	188	187(8)	$\Upsilon(3S)$	1^{--}	–	9(5)	430(4)
$\Psi(4160)$	1^{--}	–	1	142(34)	$\Upsilon(2^3D_1)$	1^{--}	–	165(50)	–
$\Psi(4415)$	1^{--}	–	262	161(10)	$\Upsilon(4S)$	1^{--}	–	20(15)	341(18)

3. Results and Discussion

A very successsful and typical sophisticated effective interaction was given in [36], which we use also herein. The high-momentum behavior of the interaction is that of perturbative QCD; its shape at intermediate momenta is determined by essentially two parameters: an inverse effective range ω and an overall strength D [37]. Variations of these parameters change the interaction's shape as shown in the left panel of Fig. 4 [38]. The right panel of Fig. 4 shows a characteristic set of solutions of the quark DSE in terms of the quark mass function, where m_q assumes typical values for the masses of the up/down (light), strange, charm, and bottom quarks [39]. In addition to those, we also show the solution in the chiral limit, i. e., with $m_q = 0$ [35].

The most notable feature of this set of mass functions is the transition from the perturbative domain at the right of the plot, where one essentially deals with current quarks, to the left of the plot, where dynamical chiral symmetry breaking (DχSB) is clearly visible in that the quarks show masses typical for corresponding constituent quarks. The border between these two domains is ≈ 1 GeV^2, a scale set by the model parameters responsible for DχSB.

After detailed studies of quarkonium spectroscopy in [39, 42] we focus on leptonic decay constants herein. Those were calculated for pseudoscalar and vector mesons already elsewhere, e. g., [42–44], but never systematically or comprehensively for radially excited states. Our results are given in Tab. 1 for two different sets of model parameters and compared to experimental data where available. Set I provides the values obtained for the results as given in [39], i. e., $\omega = 0.7$ GeV and $D = 0.5$ GeV^2 for charmonium, and $\omega = 0.7$ GeV and $D = 1.3$ GeV^2 for bottomonium, which were fitted separately for $\bar{c}c$ and $\bar{b}b$ to ground- and low-lying excited quarkonium states [39]. In order to have a common parameter set for both as well as to avoid technical complications [45] which result in a limited number of states in set I, we present also set II, which is uniformly obtained via $\omega = 0.3$ GeV and $D = 1.3$ GeV^2.

We found that our results for decay constants do not change much from one parameter set to another, so they are robust in terms of their order of magnitude. However, some changes occur. For example, excitations can switch places as seen for the first and second J/Ψ excitation in sets I and II. Also, the smaller the numbers get, the more sensitive they tend to be to parameter changes. Still, on the whole one can identify the relevant patterns and we turn to set II for further analysis. By the size of the decay constants we match our third radial Υ excitation to the missing $\Upsilon(1^3D_1)$ and the fifth to the $\Upsilon(2^3D_1)$, thus reasonably reordering our results.

Then, our prediction for the $\Upsilon(1^3D_1)$ state, at a mass of 10.2 ± 0.1 GeV [39], is a decay width $\Gamma(\Upsilon(1^3D_1) \to e^+e^-) \approx 15$ eV. For the $\Upsilon(2^3D_1)$ state we obtain $\Gamma(\Upsilon(2^3D_1) \to e^+e^-) \approx 75$ eV. These predictions for the decay widths are about one order of magnitude larger than those from other approaches, see [46, 47] and references therein. A more in-depth comparison is thus necessary and will be published elsewhere. We also note that in terms of experimental sensitivity, the upper limits for these decays are, in fact, of the order of 40 eV or less [48].

To check the S/D-wave assignment, we investigated the contributions from various covariant structures in the vector-meson BSA to the state's canonical norm. While l is not a Lorentz invariant, we can investigate the situation in the meson rest frame in which we solve the BSE [49]. Upon inspection, the picture holds, i. e., those states with larger leptonic decay constants receive dominant contributions to the norm from S-wave BSA components. In those cases which we compare to D-wave states, the dominant contributions are shifted to D-wave BSA components with small additional contributions from S-wave. We defer a more detailed analysis and discussion of this matter to the future as well.

Acknowledgments

A. K. would like to thank the organizers of the FAIRNESS2016 workshop for their hospitality and the stimulating and productive atmosphere. This work was supported by the Austrian Science Fund (FWF) under project no. P25121-N27.

References

[1] Gomez-Rocha M, Hilger T and Krassnigg A 2016 *Phys. Rev.* D **93** 074010
[2] Alkofer R, Kloker M, Krassnigg A and Wagenbrunn R F 2006 *Phys. Rev. Lett.* **96** 022001
[3] Eichmann G, Krassnigg A, Schwinzerl M and Alkofer R 2008 *Annals Phys.* **323** 2505–2553
[4] Holl A, Krassnigg A and Roberts C D 2005 *AIP Conf. Proc.* **768** 378
[5] Eichmann G, Alkofer R, Krassnigg A and Nicmorus D 2008 *PoS* **CONFINEMENT8** 077
[6] Eichmann G, Alkofer R, Krassnigg A and Nicmorus D 2010 *Phys. Rev. Lett.* **104** 201601
[7] Alkofer R, Eichmann G, Krassnigg A and Nicmorus D 2010 *Chinese Physics* C **34** 1175
[8] Dorkin S M, Hilger T, Kämpfer B and Kaptari L P 2010 (*Preprint* 1012.5372)
[9] Blank M and Krassnigg A 2010 *Phys. Rev.* D **82** 034006
[10] Blank M, Krassnigg A and Maas A 2011 *Phys. Rev.* D **83** 034020

[11] Mader V, Eichmann G, Blank M and Krassnigg A 2011 *Phys. Rev.* D **84** 034012
[12] Sanchis-Alepuz H and Fischer C S 2014 *Phys. Rev.* D **90** 096001
[13] Fischer C S, Kubrak S and Williams R 2014 *Eur. Phys. J.* A **50** 126
[14] Fischer C S, Kubrak S and Williams R 2015 *Eur. Phys. J.* A **51** 10
[15] Leitao S *et al.* 2016 *EPJ Web Conf.* **113** 03013
[16] Raya K *et al.* 2016 *Phys. Rev.* D **93** 074017
[17] Bhagwat M S *et al.* 2004 *Phys. Rev.* C **70** 035205
[18] Gomez-Rocha M, Hilger T and Krassnigg A 2015 *Few Body Syst.* **56** 475–480
[19] Gomez-Rocha M, Hilger T and Krassnigg A 2015 *Phys. Rev.* D **92** 054030
[20] Krassnigg A and Maris P 2005 *J. Phys. Conf. Ser.* **9** 153–160
[21] Eichmann G *et al.* 2009 *Phys. Rev.* C **79** 012202(R)
[22] Holl A, Krassnigg A and Roberts C D 2003 (*Preprint* nucl-th/0311033)
[23] Eichmann G *et al.* 2008 *Phys. Rev.* C **77** 042202(R)
[24] Krassnigg A 2009 *Phys. Rev.* D **80** 114010
[25] Holl A, Krassnigg A and Roberts C D 2004 *Phys. Rev.* C **70** 042203(R)
[26] Holl A, Krassnigg A, Maris P, Roberts C D and Wright S V 2005 *Phys. Rev.* C **71** 065204
[27] Dorkin S M, Hilger T, Kaptari L P and Kämpfer B 2011 *Few Body Syst.* **49** 247–254
[28] Hilger T U 2012 *Medium Modifications of Mesons* Ph.D. thesis TU Dresden
[29] Hilger T, Gomez-Rocha M and Krassnigg A 2015 *Phys. Rev.* D **91** 114004
[30] Bhagwat M S *et al.* 2007 *Few-Body Syst.* **40** 209–235
[31] Krassnigg A 2008 *PoS* **CONFINEMENT8** 075
[32] Blank M and Krassnigg A 2011 *Comput. Phys. Commun.* **182** 1391
[33] Blank M and Krassnigg A 2011 *AIP Conf. Proc.* **1343** 349
[34] Dorkin S M, Kaptari L P, Hilger T and Kämpfer B 2014 *Phys. Rev.* C **89** 034005
[35] Hilger T 2016 *Phys. Rev.* D **93** 054020
[36] Maris P and Tandy P C 1999 *Phys. Rev.* C **60** 055214
[37] Popovici C, Hilger T, Gomez-Rocha M and Krassnigg A 2015 *Few Body Syst.* **56** 481–487
[38] Hilger T, Gomez-Rocha M and Krassnigg A 2015 (*Preprint* 1508.07183)
[39] Hilger T, Popovici C, Gomez-Rocha M and Krassnigg A 2015 *Phys. Rev.* D **91** 034013
[40] Olive K A and others (Particle Data Group) 2014 *Chin. Phys.* C **38** 090001
[41] Rosner J L *et al.* (CLEO) 2006 *Phys. Rev. Lett.* **96** 092003
[42] Blank M and Krassnigg A 2011 *Phys. Rev.* D **84** 096014
[43] Maris P 2007 *AIP Conf. Proc.* **892** 65–71
[44] Bedolla M A, Cobos-Martinez J and Bashir A 2015 *Phys. Rev.* D **92** 054031
[45] Krassnigg A and Blank M 2011 *Phys. Rev.* D **83** 096006
[46] Li Y, Maris P, Zhao X and Vary J P 2015 (*Preprint* 1509.07212)
[47] Segovia J, Ortega P G, Entem D R and Fernandez F 2016 (*Preprint* 1601.05093)
[48] Godfrey S and Rosner J L 2001 *Phys. Rev.* D **64** 097501
[49] Bhagwat M S, Krassnigg A, Maris P and Roberts C D 2007 *Eur. Phys. J.* A **31** 630–637

Journal of Physics: Conference Series **742** (2016) 012033 doi:10.1088/1742-6596/742/1/012033

Exploring neutron detection with HADES

L Chlad[1,2] for the HADES collaboration

[1] Nuclear Physics Institute of the CAS, Řež 130, 250 68 Řež, Czech Republic
[2] Charles University in Prague, Ke Karlovu 3, 121 16 Prague 2, Czech Republic

E-mail: chlad@ujf.cas.cz

Abstract. The HADES experiment at GSI Helmholtzzentrum für Schwerionenforschung in Darmstadt (Germany) is fixed target experiment using SIS-18 accelerator to study collisions of protons, heavy-ions or secondary pions with target nuclei. HADES was designed and provides very accurate measurement of di-electrons and charged hadrons. The pion induced reactions provide unique opportunity to study exclusive reactions with neutrons in the final state. Using the inclusive channel $\pi^- + \mathrm{p} \rightarrow \pi^- + \pi^+ + \mathrm{n}$ we can optimize the selection criteria for neutron hits in TOF/RPC. Dedicated simulations are compared with preliminary results of real data analysis for reaction channels with two neutral particles in the final state.

1. Introduction

A very successful model of atomic nuclei was developed already back in 1950's and it is called nuclear shell model that can predict the nucleus structure with good precision for a wide range of A and Z. According to this model neutrons and protons are affected only by mean field of all nucleons, otherwise they are independent. However, since 1990's the experiments with electron induced knock out of proton from nuclei have shown that only around 70 % of protons are really independent and the rest of them are correlated in pairs [1]. There are two types of correlations: long-range (distance between nucleons is few fm) and short-range (wave-functions of nucleons overlap, shortly SRC). Because the SRC are much stronger between protons and neutrons than the same kinds of nucleons, they are responsible for the high momentum tail in the nucleons momentum distribution above the Fermi momentum as it is explained in [2].

The aim to contribute to SRC studies motivates the test of HADES response on neutrons. It is also important that HADES recently collected data with pion beam (more in section 2) because it provides interactions with very few tracks and therefore the probability of accidental fake hits in TOF/RPC is rather low. The first step of our analysis aims on selection criteria for candidates of neutron hits. For this purpose we use the inclusive channel $\pi^- + \mathrm{p} \rightarrow \pi^- + \pi^+ + \mathrm{n}$ where we are able to determine missing neutron momentum from reconstructed tracks of charged pions. From this analysis we also find out what is the detection efficiency for neutrons and we compare these results with dedicated simulations. Then we proceed with testing of the selection criteria using η production channel.

2. Pion beam experiment with HADES spectrometer

The High Acceptance DiElectron Spectrometer (HADES) located at GSI Helmholtzzentrum für Schwerionenforschung (Darmstadt, Germany) is currently operating at SIS-18 accelerator which

Content from this work may be used under the terms of the Creative Commons Attribution 3.0 licence. Any further distribution of this work must maintain attribution to the author(s) and the title of the work, journal citation and DOI.
Published under licence by IOP Publishing Ltd

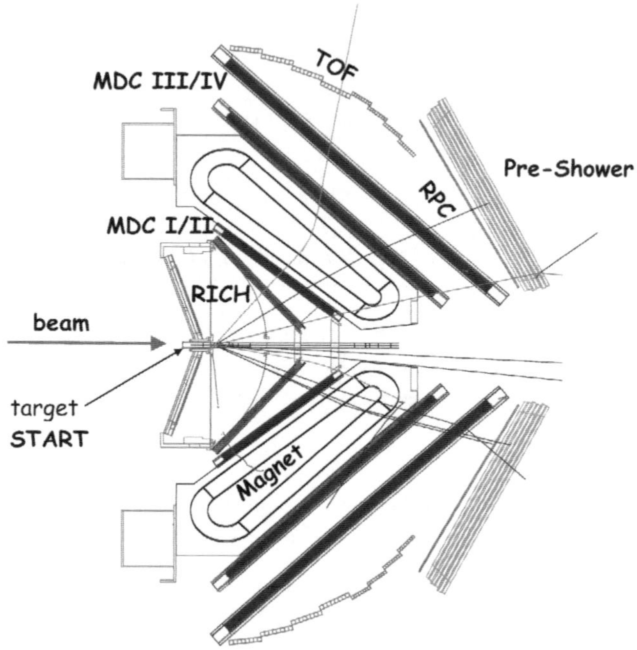

Figure 1. Schematic layout of HADES spectrometer with examples of particle tracks.

provides primary beams in the energy range $1 - 2\,\mathrm{GeV/u}$ for heavy-ions and up to $3.5\,\mathrm{GeV}$ for protons. Moreover the facility can also provide secondary pion beam.

The spectrometer is covering polar angle $18^\circ < \theta < 85^\circ$ and almost full azimuthal angle, see Figure 1. It is divided into 6 identical sectors (the geometry is dictated by the shape of toroidal magnetic field generated by superconducting coils) and it consists of several detectors: diamond-based START detector, Ring Imaging CHerenkov detector, Multi-wire Drift Chambers, time-of-flight walls and Pre-Shower detector. Detail description of technical parameters and actual performance can be found in [3].

The physics program is primarily focused on the dilepton spectra from decays of light vector mesons (ρ,ω,ϕ) [4; 5] but thanks to very precise measurement of charged particles, weak decays of hadrons containing strange quarks are also studied intensively [6].

The pion beamtime was divided into two parts with different physics scope. In the first part the beam momentum $p_{\mathrm{beam}} = 1.7\,\mathrm{GeV/c}$ was selected with respect to the ϕ meson production threshold. Typically 100 000 pions/s passed through the targets made of tungsten and carbon. During the second part, beam momentum scan (namely $p_{\mathrm{beam}} = \{656, 690, 748, 800\}$ MeV/c) was performed in order to cover the energy region which corresponds to excitation of baryonic resonances $N(1520)$ and $N(1535)$. Their coupling to the ρ meson is of special interest because it has a direct influence on in-medium modifies of ρ meson spectral function. The beam intensity for these energies was higher than in the first part (around 300 000 pions/s). Polyethylene and carbon targets were used to enable subtraction of carbon contribution inside polyethylene.

3. Selection criteria optimization

Given the composition of the HADES apparatus, one can search for neutron detection signal under certain assumptions. Because of the zero electric charge of neutrons their tracks are not bend in the magnetic field and also they do not leave any trace inside the drift chambers. Therefore the only way neutrons can be detected is the time of flight information provided by one of two detectors - for lower polar angles ($15^\circ < \theta < 45^\circ$) there are Resistive Plate Chambers and for higher polar angles ($40^\circ < \theta < 80^\circ$) it is the scintillator based Time Of Flight detector. RPC detects charged particles that come out from interaction of a neutron with metal cover of cells. In the plastic scintillators of the TOF detector neutron can scatter on nuclei and we detect the recoil nuclei or the light output from excited nuclei. It means that when we seek for the candidates of neutron detection in TOF/RPC we look for hits which cannot be matched with tracks in MDC.

To verify our "neutron candidate" hypothesis we take advantage of inclusive channel $\pi^- + \mathrm{p} \rightarrow \pi^- + \pi^+ + \mathrm{n}$ where we can calculate the missing neutron momentum if there are both charged pion tracks reconstructed in the event

$$\vec{p}_{\mathrm{expect}} = \vec{p}_{\mathrm{beam}} + \vec{p}_{\mathrm{target}} - \vec{p}_{\pi^-} - \vec{p}_{\pi^+}. \tag{1}$$

We decided to use 5° window around the predicted vector in both azimuthal and polar angle. The good candidates we declared as a "neutron hits" and then we focused on their time and position. The time of flight spectrum of neutron hits one can see in Figure 2. We also compare the magnitude of measured and expected neutron momentum for neutron hits based on the equation

$$\beta c = \frac{l_{\mathrm{path}}}{t_{\mathrm{tof}}} \Rightarrow p_{\mathrm{measure}} = \frac{\beta c\, m_{\mathrm{n}}}{\sqrt{1-\beta^2}}. \tag{2}$$

The agreement between real data analysis and simulation is shown in Figure 3. On both of these figures the results from real data analysis (green) are compared to simulations using GCalor (red) and Geisha (blue) hadron package for Geant3.

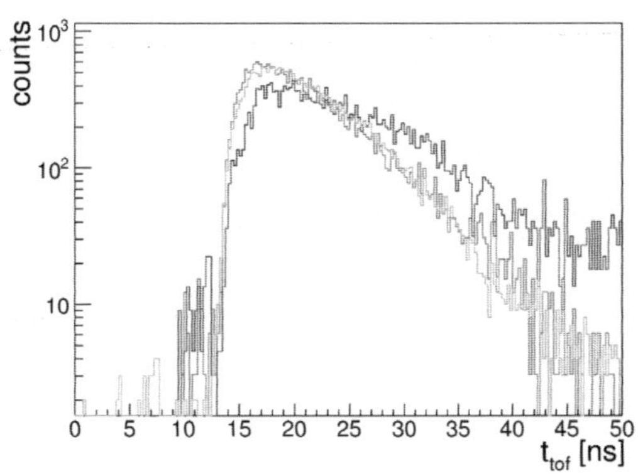

Figure 2. Time of flight spectrum for neutron hits after position matching.

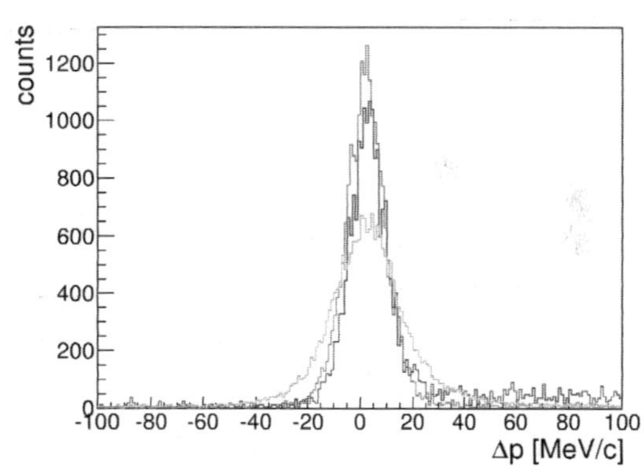

Figure 3. The difference between measured and expected neutron momentum.

Finally the neutron detection efficiency within HADES spectrometer is determined as a ratio of good candidates and number of events where both pion tracks are reconstructed and the missing mass of pions lies in the interval $900\,\mathrm{MeV/c^2} < m(\pi^-\pi^+)_{\mathrm{miss}} < 980\,\mathrm{MeV/c^2}$. The resulting efficiencies as a functions of polar angle and neutron momentum can be seen in Figure 4 and Figure 5.

4. Testing on η channel

After we established the selection criteria for neutron hits we can inspect their performance on channels with both neutral particles in the final state that would be otherwise inaccessible. We chose the η meson due to its production cross section in measured energy region and its decay mode $\eta \rightarrow \pi^+\pi^0\pi^-$ which we needed to trigger the data acquisition (two hits in TOF and/or RPC

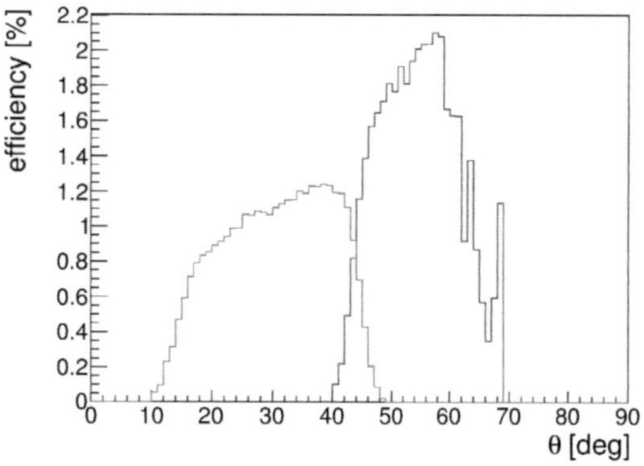

Figure 4. Neutron detection efficiency within TOF (blue) and RPC (red) detectors as a function of polar angle.

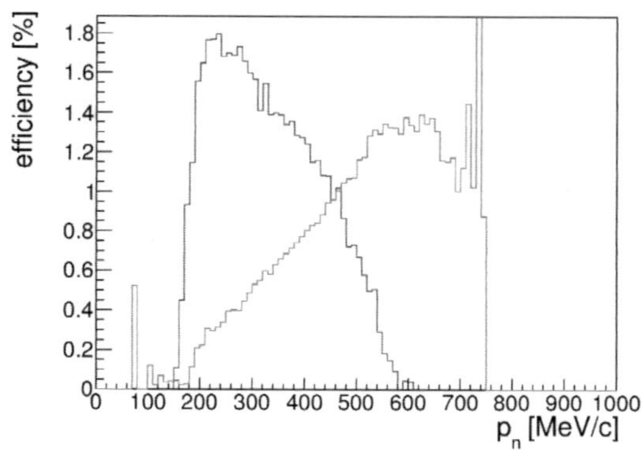

Figure 5. Neutron detection efficiency within TOF (blue) and RPC (red) detectors as a function of neutron momentum.

each in different sector of HADES). From the kinematics of the interaction $\pi^- + \mathrm{p} \to \eta + \mathrm{n}$ we can further restrict our search for neutron hit candidates to the RPC detector and also we can use cut on the time of flight $11\,\mathrm{ns} < t_{\mathrm{tof}} < 28\,\mathrm{ns}$ which improves the missing mass spectra displayed in Figure 6. The carbon target spectrum in that figure is scaled with factors obtained from analysis of the elastic scattering $\pi^-\mathrm{p} \to \pi^-\mathrm{p}$ where it is possible to scale $m\,(\pi^-\mathrm{p})_{\mathrm{miss}}$ missing mass carbon spectra to match the background in polyethylene. The subtracted spectrum is shown in Figure 7 and one can compare it with simulated spectrum in Figure 8 (using cocktail of interaction listed in Table 1).

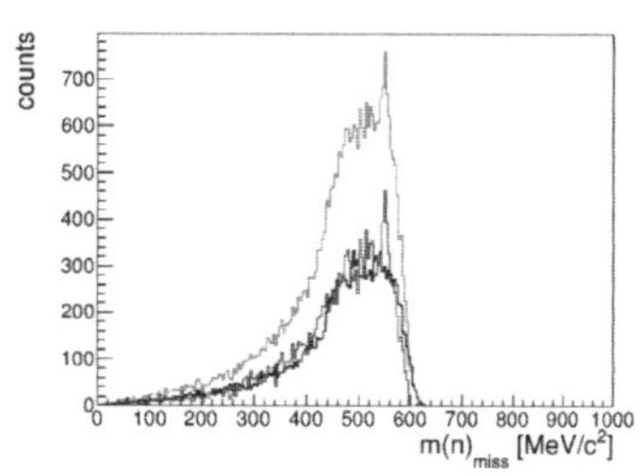

Figure 6. Neutron missing mass spectra for polyethylene (red) and carbon (black) target. The difference of these two spectra is equal to hydrogen contribution (blue) in polyethylene target.

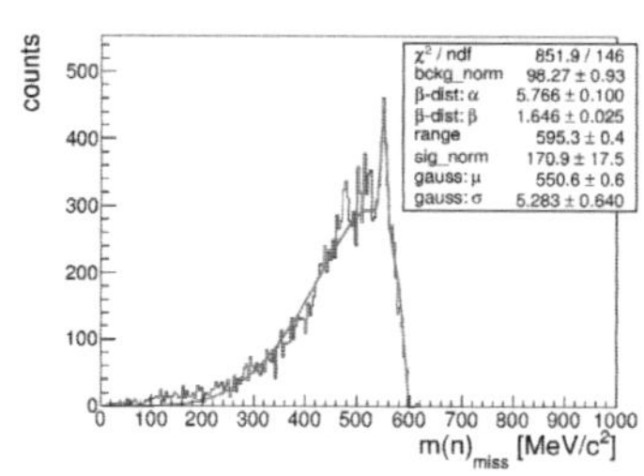

Figure 7. Hydrogen spectrum from figure 6 fitted with sum of beta and normal distributions.

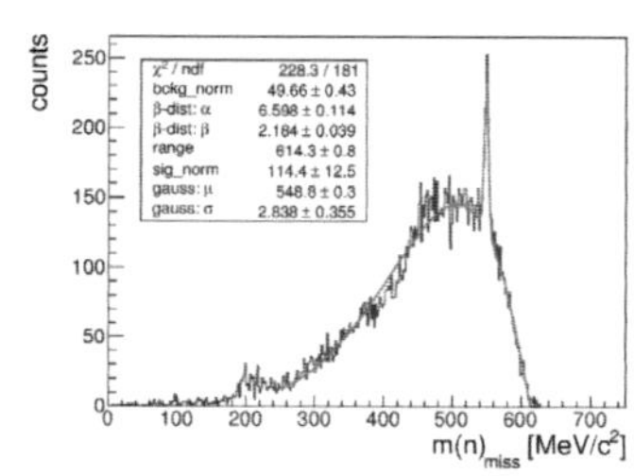

Figure 8. Simulated neutron missing mass spectrum fitted with sum of beta and normal distributions.

5. Conclusion

The study of short-range nucleon-nucleon correlations is important topic in order to understand the structure of nuclei. With respect to this a pioneer study on possible neutron detection with present layout of HADES spectrometer was carried on. We found out that the detection efficiency of neutrons is in magnitude of few percent and that these results is in reasonable

Table 1. Cross section data for signal and background channels in mb.

$\pi^- + \mathrm{p} \rightarrow$	p_{beam}[MeV/c] 666.8	699.7	748.4	799.1	source
$\mathrm{n} + \pi^0$	8.71	8.42	7.17	5.28	[7]
$\mathrm{n} + \pi^0 + \pi^0$	1.53	2.29	2.50	2.66	[8]
$\mathrm{n} + \pi^0 + \pi^0 + \pi^0$	0.003	0.009	0.014	0.020	[9]
$\mathrm{n} + \eta$	0.0	1.5	2.6	2.6	[10]
$\mathrm{n} + \pi^- + \pi^+$	5.49	5.96	6.19	6.60	[8]
$\mathrm{p} + \pi^- + \pi^0$	2.43	4.22	4.83	4.82	[8]
$\mathrm{n} + \pi^0 + \pi^+ + \pi^-$	0.8	0.8	0.8	0.8	estimation

agreement with dedicated simulations. We also observed that our method of identifying neutron hits in TOF/RPC works in case of two neutral particles in the final state.

Acknowledgments

This work has been supported by GACR 13-06759S of the Czech Science Foundation and grant LM201549 of the Ministry of Education of the Czech Republic.

References

[1] Subedi R *et al.* 2008 *Science* **320** 1476–1478 (*Preprint* arXiv:0908.1514 [nucl-ex])

[2] Hen O *et al.* 2014 *Science* **346** 614–617 (*Preprint* arXiv:1412.0138 [nucl-ex])

[3] Agakishiev G *et al.* (HADES) 2009 *Eur. Phys. J.* A **41** 243–277 (*Preprint* arXiv:0902.3478 [nucl-ex])

[4] Agakishiev G *et al.* (HADES) 2011 *Phys. Rev.* C **84** 014902 (*Preprint* arXiv:1103.0876 [nucl-ex])

[5] Agakishiev G *et al.* (HADES) 2012 *Phys. Lett.* B **715** 304–309 (*Preprint* arXiv:1205.1918 [nucl-ex])

[6] Agakishiev G *et al.* (HADES) 2014 *Phys. Rev.* C **90** 054906 (*Preprint* arXiv:1404.7011 [nucl-ex])

[7] Baldini A, Flaminio V, Moorhead W G and Morrison D R O 1988 *Total Cross-Sections for Reactions of High Energy Particles* 1st ed (*Landolt-Bornstein: New Series - Elementary Particles, Nuclei and Atoms* vol 12) (Heidelberg, Germany: Springer-Verlag)

[8] Manley D M, Arndt R A, Goradia Y and Teplitz V L 1984 *Phys. Rev.* D **30**(5) 904–936

[9] Starostin A *et al.* (Crystal Ball) 2003 *Phys. Rev.* C **67**(6) 068201

[10] Prakhov S *et al.* (Crystal Ball) 2005 *Phys. Rev.* C **72**(1) 015203

Journal of Physics: Conference Series **742** (2016) 012034 doi:10.1088/1742-6596/742/1/012034

Dilepton production with the SMASH model

Janus Weil[1], Jan Staudenmaier[1,2], Hannah Petersen[1,2,3]

[1] Frankfurt Institute for Advanced Studies, 60438 Frankfurt, Germany

[2] Institut für Theoretische Physik, Goethe-Universität, 60438 Frankfurt, Germany

[3] GSI Helmholtzzentrum für Schwerionenforschung GmbH, 64291 Darmstadt, Germany

E-mail: weil@fias.uni-frankfurt.de

Abstract. In this work the SMASH model is presented ("Simulating Many Accelerated Strongly-Interacting Hadrons"), a next-generation hadronic transport approach, which is designed to describe the non-equilibrium evolution of hadronic matter in heavy-ion collisions. We discuss first dilepton spectra obtained with SMASH in the few-GeV energy range of GSI/FAIR, where the dynamics of hadronic matter is dominated by the production and decay of various resonance states. In particular we show how electromagnetic transition form factors can emerge in a transport picture under the hypothesis of vector-meson dominance.

1. Introduction

Lepton pairs are a useful probe for the regions of high density and temperature that are produced in a heavy-ion collision. They have been measured experimentally by detectors like NA60 [1] and HADES [2]. In this work, the production of dileptons is investigated within a hadronic transport approach, SMASH, which essentially solves the Boltzmann equation for a hadron resonance gas. Here we focus on a few aspects of the model that are important for dilepton production. A full description of SMASH will be provided in an upcoming publication [3].

At low energies (of a few GeV) the particle interactions in SMASH rely on cross sections that are dominated by the excitation and decay of hadronic resonance states. A typical dilepton spectrum contains contributions from direct and Dalitz decays of resonances such as:

- $\rho, \omega, \phi \to e^+e^-$
- $\pi^0, \eta, \eta' \to e^+e^-\gamma$
- $\omega, \phi \to e^+e^-\pi$
- $N^*, \Delta \to e^+e^-N$

Here we mainly focus on the contributions from the ω meson. The major production channels of the ω in pp and AA collisions at low energies are assumed to be processes like $NN \to NN^* \to NN\omega$ or $NN \to \Delta N^* \to NN\pi\omega$, where the $N^* \to \omega N$ couplings for the N^* states are taken from the PDG database [4]. The dominant ω decay channel is $\omega \to 3\pi$, which in SMASH is treated via the two-step decay chain $\omega \to \pi\rho \to 3\pi$, the so-called GSW process (Gell-Mann, Sharp, Wagner) [5]. This is done in order to be able to fulfill detailed balance with $1 \leftrightarrow 2$ and $2 \leftrightarrow 2$ processes only (avoiding the technically more challenging $1 \leftrightarrow 3$ processes). This assumption for the ω meson is also used in other related works [6, 7]. Further it should be noted that SMASH uses the decay-width parametrizations of Manley et al. [8], which are also employed in the GiBUU transport model [9]. The above assumptions lead to the fact that in

Content from this work may be used under the terms of the Creative Commons Attribution 3.0 licence. Any further distribution of this work must maintain attribution to the author(s) and the title of the work, journal citation and DOI.
Published under licence by IOP Publishing Ltd

Journal of Physics: Conference Series **742** (2016) 012034 doi:10.1088/1742-6596/742/1/012034

SMASH the ω meson is exclusively produced via the decay of N^* resonances in pp collisions, while in AA collisions also the production via $\pi\rho$ and $\pi\pi$ collisions is possible.

In the following we first investigate the ω Dalitz decay into $e^+e^-\pi^0$ (which proceeds as a two-step chain via an intermediate ρ) and then the Dalitz decay $N^* \to e^+e^-N$, which can proceed via an intermediate ρ or ω meson and is also treated as a two-step process.

2. ω Dalitz decay

We investigate the electromagnetic Dalitz decay of the ω meson, $\omega \to \pi^0 e^+ e^-$, under the assumption that this decay proceeds in two steps as $\omega \to \pi^0\rho^0 \to \pi^0 e^+ e^-$. As mentioned above, this assumption is motivated mainly from the fact that the SMASH model currently emulates the $\omega \to 3\pi$ decay by $\omega \to \pi\rho$, in order to fulfill detailed balance without the need of introducing three-body collisions. Furthermore, this assumption is in fact equivalent to the two-step VMD treatment that was employed in earlier works for the electromagnetic decays of N^* and Δ^* baryons [10] and later extended also to the $\Delta(1232)$ [11].

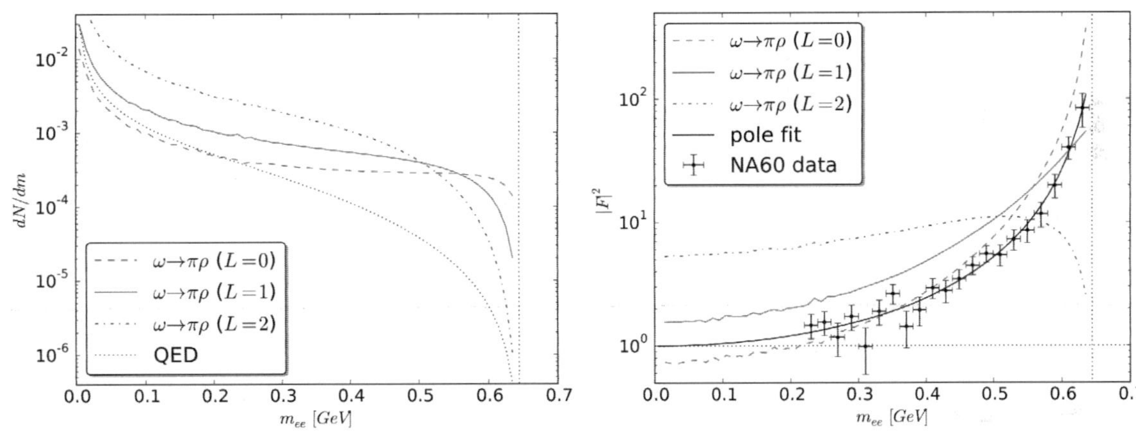

Figure 1. $\omega \to \pi^0 e^+ e^-$ decay spectrum (left) and form factor (right) for a branching ratio of 89% for $\omega \to \pi\rho$ and different angular momenta. The form factor is compared to data from the NA60 experiment [1].

In Fig. 1 (left) we show the dilepton mass spectrum originating from the process $\omega \to \pi\rho \to \pi e^+e^-$, using three different values for the angular momentum L in the $\omega \to \pi\rho$ decay and a branching ratio of $\mathrm{BR}(\omega \to \pi\rho) = 89\%$, such as to fully saturate the 3π decay. The values of $L = 0, 1, 2$ are the only ones allowed by angular momentum conservation, but in fact only $L = 1$ is allowed by parity conservation, which demands that L is odd in order to fulfill $P_\omega = P_\pi \cdot P_\rho \cdot (-1)^L$, where all three mesons have negative intrinsic parity. We actually show all three cases in order to illustrate the effect of the angular momentum. The plot also shows the QED case, which corresponds to a point-like vertex in the ω Dalitz decay [12]. All curves assume that the decaying ω meson is on the pole mass of $m_\omega = 782$ MeV. The vertical dotted line indicates the maximum dilepton mass of $m_\omega - m_\pi \approx 644$ MeV.

The right panel of Fig. 1 shows the form factor of the ω Dalitz decay (which is obtained simply by dividing the "two-step VMD" curves by the QED baseline) compared to the data taken by NA60 [1] and a fit of the data using the pole approximation

$$|F|^2 = (1 - m^2/\Lambda^2)^{-2}, \tag{1}$$

with $\Lambda = 668$ MeV. We note that the form factor is by definition normalized to one at the real-photon point ($m = 0$). However, this is not given a priori for our three cases. Instead their

Journal of Physics: Conference Series **742** (2016) 012034 doi:10.1088/1742-6596/742/1/012034

normalization is determined by the branching ratio, which we fix to 89% for now (i.e. the upper limit given by the 3π decay). While the $L = 2$ case is apparently in strong conflict with the data, $L = 0$ seems to be the best match for the data from a phenomenological point of view. However, it is physically forbidden, since parity conservation only allows $L = 1$. The curve for $L = 1$ touches the data only at the highest masses, while it overshoots the data over most of the spectrum and also overestimates the photon-point value. This is obviously due to the large branching ratio employed for $\omega \to \pi\rho$. From comparing the $L = 1$ form factor to the NA60 data, one has to conclude that the assumption that the whole $\omega \to 3\pi$ decay proceeds via an intermediate $\pi\rho$ state was too strong. Therefore, we reduce the $\omega \to \pi\rho$ branching, in order to avoid overshooting the form factor data. In particular we scale it down from 89% to 57% to obtain the right photon-point normalization. As shown in Fig. 2, this results in a reasonable description of the experimental data. A significant deviation is visible only for the two highest data points.

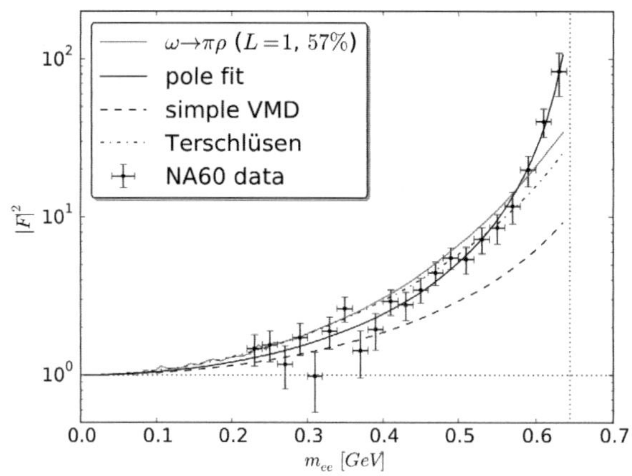

Figure 2. $\omega \to \pi^0 e^+ e^-$ transition form factor with reduced $\omega \to \pi\rho$ (p-wave) coupling.

For comparison we also show a 'simple' VMD curve, as presented in the NA60 paper [1], which only includes a plain ρ propagator in pole approximation. Our treatment of $\omega \to \pi\rho \to \pi e^+ e^-$ is of course also a VMD-like model, but in contrast to the 'simple' VMD it includes a hadronic form factor in the ω-π-ρ vertex.

In our model the electromagnetic transition form factor of the ω meson is essentially determined by the mass distribution of the ρ meson in the $\omega \to \pi\rho$ decay, which is given by the following formula (leaving out constant factors):

$$\frac{d\Gamma_{\omega\to\pi\rho}}{dm_\rho} \propto \mathcal{A}_\rho(m_\rho) \cdot p_F(m_\omega, m_\rho, m_\pi) \cdot B_L^2(p_F R) \quad (2)$$

Following from eq. (172)-(173) in [9], this equation illustrates that there are three factors that influence the mass distribution of the ρ meson in the decay of the ω:

(i) the spectral function $\mathcal{A}_\rho$ of the ρ meson, which we assume to be a Breit-Wigner function with mass-dependent width [13],

(ii) the phase-space factor p_F, i.e. the center-of-mass momentum in the π-ρ final state and

(iii) the Blatt-Weisskopf factor B_L, which depends on the angular momentum L and includes a cutoff parameter R. For $L = 1$, $B_1^2(x) = x^2/(1 + x^2)$ is obtained.

The Blatt-Weisskopf function contains both a phase-space factor as well as a hadronic form factor. The latter is characterized by the cutoff parameter R, which is taken to be $R = 1$ fm.

Our model suggests that the ω meson decays into $\pi\rho$ with a branching ratio of around 57%, which makes up more than half of the 3π final state and at the same time provides a rather good explanation of the electromagnetic transition form factor of the ω meson. We also show in Fig. 2 the result of the effective-Lagrangian approach of Terschlüsen et al. [14], which in fact is rather close to our result. Further we note that preliminary calculations seem to indicate that our approach can also be applied to the $\phi \to e^+e^-\pi^0$ decay and yields results compatible with recent data from the KLOE-2 collaboration [15].

3. Baryonic Dalitz decays

Next we discuss the baryonic Dalitz decays, using as an example the process $pp \to e^+e^-X$ at $E_{kin} = 3.5$ GeV, as measured by the HADES collaboration [2]. Our simulations do not yet include a detector-acceptance filtering, therefore we can not directly compare to data but only want to discuss qualitative effects here.

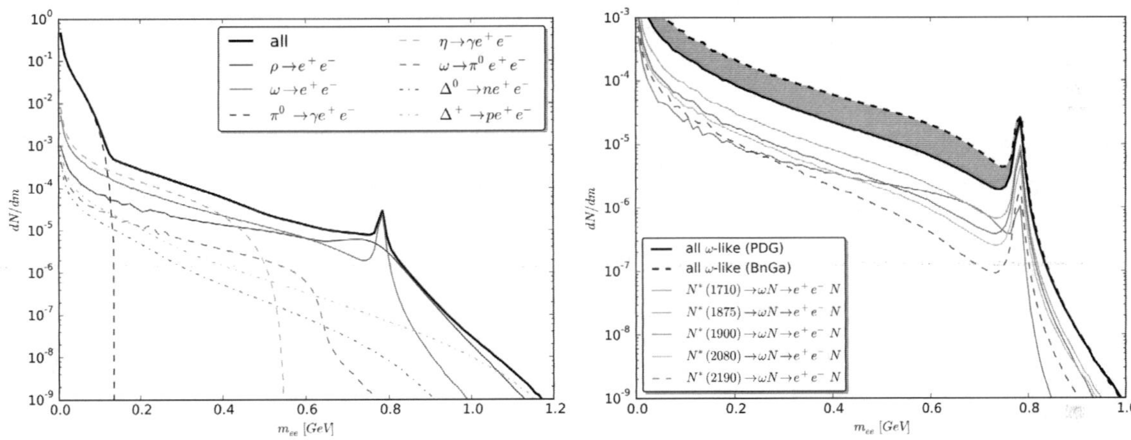

Figure 3. Dilepton mass spectra for p+p at 3.5 GeV. Left: Total inclusive spectrum. Right: $\omega \to e^+e^-$ only, with contributions from different baryon resonances.

Figure 3 (left) shows the inclusive dilepton mass spectrum simulated with the SMASH model. We note that the π^0, η and Δ Dalitz decays follow the standard treatment employed in many other transport approaches (see e.g. [13]), the first two with a parametrized form factor, the latter in QED approximation. The $\rho \to e^+e^-$ contribution consists mainly of $N^* \to \rho N \to e^+e^-N$ and $\Delta^* \to \rho N \to e^+e^-N$ processes, since almost all ρ mesons in pp collisions are produced via baryonic resonance decays in SMASH. This is why the ρ contribution has a Dalitz-like shape with contributions below the 2π threshold, which has already been seen in earlier GiBUU simulations [10]. In addition to the baryonic Dalitz decays, the ρ contains an additional feed-down from the $\omega \to \pi\rho$ decay, as discussed in the previous section. The most interesting feature of the SMASH dilepton spectrum is actually the $\omega \to e^+e^-$ contribution. In addition to the peak at the ω pole mass of 782 MeV, it also contains a strong Dalitz-like tail in the low-mass region. This is due to the fact that the ω in SMASH is produced mainly via the excitation and decay of baryonic resonances in pp collisions, just like the ρ meson. Due to isospin arguments, the production of the ω proceeds only via N^* resonances, while the ρ can also be produced from Δ^* decays.

The right panel of Fig. 3 shows the ω contribution to the dilepton spectrum, with all the contributions from different N^* states. The branching ratios for $N^* \to \omega N$ are taken from the current PDG database by default [4]. We also show how the spectrum changes when using the branching ratios from a recent analysis of the Bonn-Gatchina group [16]. This analysis yields strong ω couplings for sub-threshold resonances like the $N^*(1700)$ and $N^*(1720)$ that do not have an ωN mode in the current PDG, which further enhances the low-mass Dalitz tail (while the yield at the ω pole mass and above is basically unchanged). Due to the missing acceptance filtering, we can not yet comment whether these effects are compatible with experimental data. The general fact that effects from baryonic Dalitz decays exist not only for the ρ, but also for the ω meson is not extremely surprising. However, the sheer size of the effect is rather large in our preliminary simulations, and certainly requires further investigation.

4. Conclusions

We have shown that the first simulations of dilepton spectra with the SMASH model yield some interesting effects concerning the ω meson. Firstly, our model for the $\omega \to \pi\rho$ p-wave decay shows a rather good agreement with NA60 data for the electromagnetic ω transition form factor and suggests a branching ratio of roughly 57% for the GSW process $\omega \to \pi\rho$. Secondly we have investigated effects of the baryonic coupling of the ω meson, which can have an impact on dilepton spectra from pp as well as AA collisions. Our preliminary results indicate that such effects could be surprisingly large, but need further investigation before final conclusions can be drawn.

Acknowledgements

The authors thank S. Leupold and P. Salabura for useful comments and S. Damjanovic for providing the NA60 data points. J.W. and H.P. acknowledge funding of a Helmholtz Young Investigator Group VH-NG-822 from the Helmholtz Association and GSI. This work was supported by the Helmholtz International Center for the Facility for Antiproton and Ion Research (HIC for FAIR) within the framework of the Landes-Offensive zur Entwicklung Wissenschaftlich-Oekonomischer Exzellenz (LOEWE) program launched by the State of Hesse. Computational resources have been provided by the Center for Scientific Computing (CSC) at the Goethe-University of Frankfurt.

References

[1] Arnaldi R *et al.* (NA60) 2009 *Phys. Lett. B* **677** 260–266 (*Preprint* 0902.2547)
[2] Agakishiev G *et al.* (HADES) 2012 *Eur. Phys. J. A* **48** 64 (*Preprint* 1112.3607)
[3] Weil J *et al. (in preparation)*
[4] Olive K A *et al.* (Particle Data Group) 2014 *Chin. Phys. C* **38** 090001
[5] Gell-Mann M, Sharp D and Wagner W G 1962 *Phys. Rev. Lett.* **8** 261
[6] Muhlich P and Mosel U 2006 *Nucl. Phys. A* **773** 156–172 (*Preprint* nucl-th/0602054)
[7] Santini E, Cozma M D, Faessler A, Fuchs C, Krivoruchenko M I and Martemyanov B 2008 *Phys. Rev. C* **78** 034910 (*Preprint* 0804.3702)
[8] Manley D M and Saleski E M 1992 *Phys. Rev. D* **45** 4002–4033
[9] Buss O, Gaitanos T, Gallmeister K, van Hees H, Kaskulov M, Lalakulich O, Larionov A B, Leitner T, Weil J and Mosel U 2012 *Phys. Rept.* **512** 1–124 (*Preprint* 1106.1344)
[10] Weil J, van Hees H and Mosel U 2012 *Eur. Phys. J. A* **48** 111 (*Preprint* 1203.3557)
[11] Weil J, Endres S, van Hees H, Bleicher M and Mosel U 2014 *Proceedings, 20th International Conference on Particles and Nuclei (PANIC 14)* pp 290–293 (*Preprint* 1410.4206)
[12] Landsberg L G 1985 *Phys. Rept.* **128** 301–376
[13] Weil J 2013 *Vector Mesons in Medium in a Transport Approach* Ph.D. thesis Giessen U. URL http://geb.uni-giessen.de/geb/volltexte/2013/10253/
[14] Terschlüsen C, Strandberg B, Leupold S and Eichstädt F 2013 *Eur. Phys. J. A* **49** 116 (*Preprint* 1305.1181)
[15] Anastasi A *et al.* (KLOE-2) 2016 *Phys. Lett. B* **757** 362–367 (*Preprint* 1601.06565)
[16] Denisenko I, Anisovich A V, Crede V, Eberhardt H, Klempt E, Nikonov V A, Sarantsev A V, Schmieden H, Thoma U and Wilson A 2016 *Phys. Lett. B* **755** 97–101 (*Preprint* 1601.06092)

Journal of Physics: Conference Series **742** (2016) 012035 doi:10.1088/1742-6596/742/1/012035

Disentangling $\bar{P}$ANDA's time-based data stream

M. Tiemens on behalf of the PANDA Collaboration
KVI - Center For Advanced Radiation Technology, University of Groningen, Zernikelaan 25, 9747 AA Groningen, The Netherlands
E-mail: m.tiemens@rug.nl

Abstract. To allow the online selection of events, the readout of the $\bar{P}$ANDA detector will reconstruct particles online. Several algorithms to perform this task in one of the subsystems, the electromagnetic calorimeter, are being developed. The algorithms are discussed, and a simple test case shows that they appear to behave similarly in terms of the ability to reconstruct events, and the time it takes to do this. However, some peculiarities, like the fact that the algorithms show the best performance in what is expected to be the worst-case scenario, still require additional investigation.

1. Introduction

This work discusses some software developments for the future $\bar{P}$ANDA experiment. For an introduction, the reader is invited to read other publications on the experiment in this proceedings series. The design of the detector is shown in Figure 1. This paper will mainly focus on one subsystem, the electromagnetic calorimeter (EMC), shown in purple in Figure 1. The device consists of four parts: a Barrel, containing 11,360 scintillation crystals and a Backward and Forward Endcap, containing 524 and 3856 crystals, respectively. The faces of the crystals are approximately 2×2 cm. This is complemented by a sampling (shashlik) calorimeter, placed in the forward spectrometer a few meter downstream from the interaction point, to cover very small polar angles.

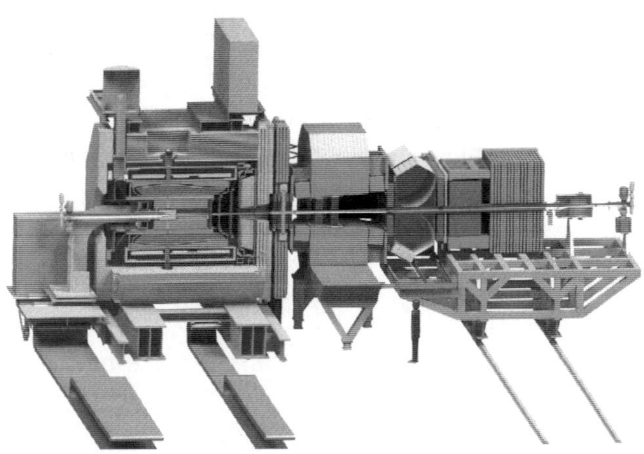

Figure 1: Technical drawing of the $\bar{P}$ANDA detector. *Courtesy:* A. Cebulla, FZ Jülich, 2014.

Content from this work may be used under the terms of the Creative Commons Attribution 3.0 licence. Any further distribution of this work must maintain attribution to the author(s) and the title of the work, journal citation and DOI.
Published under licence by IOP Publishing Ltd

Journal of Physics: Conference Series **742** (2016) 012035 doi:10.1088/1742-6596/742/1/012035

An important feature of the detector is the ability to reconstruct detected decays online, and perform the event selection based on those reconstructions. To be able to do this, complete information on the final-state particles is required. Some common particles, like photons and electrons (and positrons), are reconstructed using input from the EMC. Three algorithms to perform these reconstructions will be discussed below. Two of these are eligible for implementation on the online platform, providing vital information for the event selection.

2. Cluster Forming

An interaction between a proton and an antiproton will be referred to as an *event*. In the $\bar{P}$*ANDA* experiment, events are generated at a rate of $2 \times 10^7/s$ at the design luminosity. The interaction may lead to the production of secondary particles. The majority of final-state particles at the end of the decay chain of these reactions will create a particle shower inside the EMC crystals. Due to conservation of momentum, the shower spreads out laterally, which may cause the shower to create hits in neighbouring crystals as well. In this context, a *hit* is a crystal with an energy deposition above the detection threshold. The formation of particle showers creates groups of hits in the calorimeter, which are called *clusters*.

3. Cluster Finding

To find the four-momentum of the impinging particle, which is needed to perform the online event reconstruction, it suffices to add the individual hits in a cluster. However, the high interaction rate, combined with the large variety of intermediate states that can be directly formed, having different decay times, may lead to pile-up and event mixing. These features complicate the task of assigning hits to the correct clusters considerably. As these features occur in the time domain, the timestamp of each hit will play a key role in disentangling the hits. For this reason, the simulation that has been designed to reproduce this structure is called *time-based* simulation. The distribution of events over time follows a Poisson distribution, with the mean time between two events determined by the interaction rate. This creates a bunched structure in the final data stream. The size of the bunches, called *timebunches*, can be controlled by a time threshold. Tuning this parameter can help put hits from a single (or few) event(s) into a single timebunch, improving the assignment of hits to clusters. If multiple events are present in a timebunch, their corresponding hits are likely located in different parts of the detector, because apart from being forward boosted, the decays are isotropic. There are multiple methods under development to search for clusters in these timebunches, but the main aim is to develop a method that can be easily implemented to process the data online. As the data in the EMC is produced at a rate of 80 GB/s, this algorithm needs to consume as little resources as possible. The next sections describe the established ("standard") method, which cannot be implemented online, and the proposed online versions. The final part will compare the methods.

3.1. Method 1: Standard Cluster Finding

The currently implemented version in the offline software package PandaRoot[1] takes the stream of hits, and treats each new hit as a seperate cluster, unless it neighbours[1] to an existing one. In that case, the hit is absorbed in that cluster.

Advantages:

\+ This method is widely used, and well tested.

Disadvantages:

[1] As the time domain is important, in all references, "neighbour" means close in space *and* time.

- Either each cluster size has to be recalculated when a new hit is processed, or a loop over each existing hit is needed to check for neighbours, making the method relatively slow.
- When two hits from the same cluster are processed as first and second hit in the stream, they will be put into separate clusters when they are not direct neighbours. However, in general, the central hit in a cluster will have the lowest timestamp, and will therefore be processed first, so the probability for this failure to occur is minimal.

3.2. Method 2: Online Cluster Finding

The first proposed version to be eligible for online usage loops over all pairs of hits in the input stream to establish neighbour relations between them. It then uses this information to merge them into clusters.

Advantages:

+ Low processing time, as it is only needed to loop over hits once.

Disadvantages:

- There are still many iterations involved, making the solution possibly less suitable for implementation online.

3.3. Method 3: Distributed Cluster Finding

Before proceeding to the method, a brief introduction to the readout chain is required.

3.3.1. The Readout Chain of the P̄ANDA Detector The light created in each crystal is collected at the backface by photosensors, which are read out by digitisation modules. Each module can read out 16 crystals, and sends the extracted information to a data concentration stage, which performs further signal processing. Each Data Concentrator (DC) reads out 8 digitisation modules, and hence has access to 128 crystals –a sizeable portion of the calorimeter. The DCs send their information to the Event Building stage, which performs the online event reconstruction.

3.3.2. Distributed Cluster Finding As the DCs have access to a section of the EMC, it may be possible to look for clusters at this level. This will reduce the load at the later stages. In distributed cluster finding clusters are searched for in such a section using the Online method. However, as can be seen in Figure 2, it is possible that a cluster will be split at the edge of two or more sections, in which case they will need to be merged later on (see, for example, the white-and-gray cluster in Figure 2). Therefore, for the clusters found at this stage, called *preclusters*, only their (x, y, t) location[2] and radius will be calculated (i.e., no energies or advanced cluster properties). The radius is the circumcircle of the precluster (coloured circles in Figure 2). In the next stage, preclusters are merged if the distance between them is smaller than the sum of their radii.

Advantages:

+ Each DC processes a small sample of the dataset, reducing the processing time.
+ The next stage can use preclusters instead of hits, which also reduces the processing time.

Disadvantages:

- Some additional operations are needed, such as the calculation of the precluster location and radius.

[2] The idea is to use a 2D map of the EMC, where each crystal is a single point, so no z information is needed.

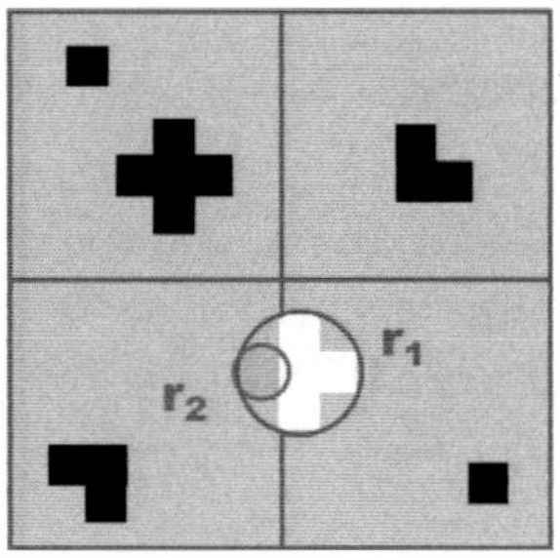

Figure 2: Illustration of precluster forming in Data Concentrators (yellow squares). Hits are indicated by black, gray, and white squares. More details can be found in the text.

- If preclusters have very elliptical shapes, the radius will be greatly overestimated. However, as the incoming particles hit the crystals orthogonal[3], the shower shape is expected to be (approximately) circular.

4. Comparing Methods

The next task is to compare these methods. Two tests on an artificial simulated data set, 5000 $p\bar{p} \to \gamma\gamma$ events at a centre-of-mass energy of 2.081 GeV/c^2, are carried out. It must be noted that the results of these tests are very preliminar, as both the algorithms and the simulation framework are still under development. For each method, three interaction-rate scenarios are considered: High Rate (HR, $2 \times 10^7/s$), Low Rate (LR, $2 \times 10^5/s$), and Event-Based (EB, completely disjoint events).

4.1. Test 1: Ability to reconstruct events

The first test checks how many from the original 5000 events can be recovered. Because of the forward boosted particle production, the raw cluster energies do not give much information. The invariant mass of the parent $p\bar{p}$ system is taken for comparison. The spectrum (see Figure 3) shows a peak around that value, which is fitted and integrated to obtain the number of reconstructed events. The results are shown in Figure 4a. The high rate scenario consistently shows higher yields, implying that it is more efficient than the ideal (event-based) case. As the high rate case suffers from pile-up and event mixing effects, this result is unexpected. A check with the information from the event generator will need to be performed to see if the reconstructed events correspond to the simulated events. However, due to technical difficulties, that information is not yet available.

Quantitatively, all methods obtain similar yields. However, the Online Cluster Finding method appears to be slightly less effective, as can also be seen in Figure 3.

4.2. Test 2: Time needed to reconstruct events

As the data is produced at a high rate (see section 3), the runtime of the algorithms is another key parameter to be checked. The results are shown in Figure 4b for the HR case, relative to method 1 (Standard). As expected, the Online method is the fastest, and the two steps of Distributed Cluster Finding (precluster finding and merging) are separately the fastest. However, no solid conclusions can be drawn at this point, because this test was performed using a desktop CPU, and the final algorithm will run on an FPGA[4]. The performance likely differs on such a device. Hardware prototypes containing FPGAs are available, but the cluster finding algorithm has not

[3] The crystals in the EMC are oriented to face inwards, pointing just behind the interaction point (to avoid particles to get undetected in the gaps between the crystals).

[4] A Field-Programmable Gate Array (FPGA) is an integrated circuit containing an array of programmable logic blocks, connected by a collection of reconfigurable interconnects.

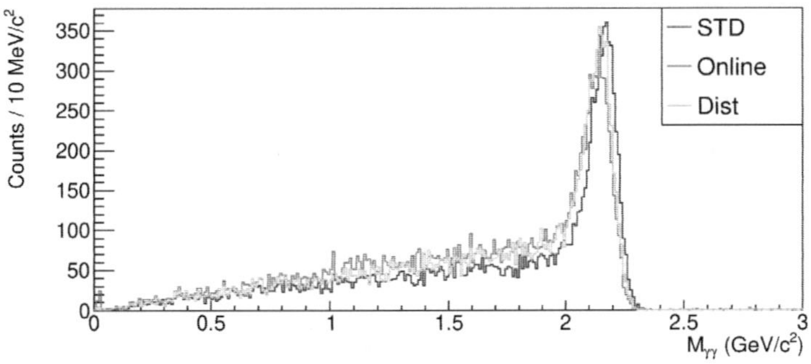

Figure 3: Comparison between the three cluster finding methods for the high rate scenario, by taking the invariant mass of all two-photon combinations. Legend: STD = Standard Cluster Finding, Online = Online Cluster Finding, Dist = Distributed Cluster Finding.

yet been implemented on them. That will be done when the algorithm has been debugged and streamlined. Hence, at this moment, the comparison on CPUs is the best that can be done.

5. Summary

This paper discussed three methods to search for clusters in the EMC of the future $\bar{P}$ANDA experiment. All methods performed similarly in both event reconstruction and runtime, with the Online algorithm being the fastest overall. The two steps in Distributed Cluster Finding, precluster finding and merging, are separately the fastest, as expected. The yield was higher in the high rate scenario, which is not what would be expected. Further investigations are needed for this purpose.

After debugging, the algorithms will be tested using more complex decays, such as $h_c \to \gamma\eta_c \to 7\gamma$, and by including background signals from e.g. a DPM background generator. The effect of the so-called bump splitting, where two overlapping clusters, that are identified as one, are separated, also has to be investigated. Ideally, this should be included, but it will consume additional resources which may not be available. It should therefore be investigated to what extend including this procedure gives an improvement, and then design a lighter version.

References

[1] Spataro S *et al* ($\bar{P}$ANDA Collaboration) 2010 *IOP Publishing - Journal of Physics: Conference Series* **331** 3

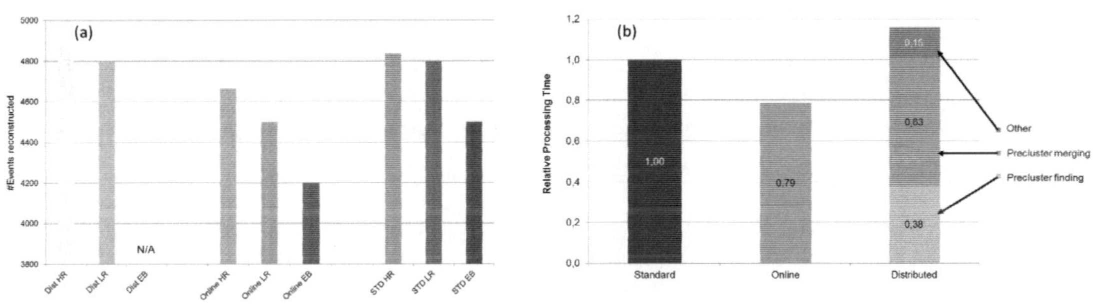

Figure 4: Comparison between the three cluster finding methods: (a) Between the number of reconstructed events in three scenarios, (b) Between the runtimes for event reconstruction, relative to the Standard method, for the high rate scenario.

The CGEM-IT of the BESIII experiment: project update and test results in magnetic field

G Mezzadri
on behalf of the BESIIICGEM consortium
University of Ferrara & INFN Ferrara, Ferrara, IT
E-mail: gmezzadr@fe.infn.it

Abstract. The BESIII experiment is a multi-purpose detector operating on the electron-positron collider BEPCII in Beijing. Since 2008, the world's largest sample of J/ψ, ψ' were collected. Due to increasing luminosity, the inner drift chamber is showing signs of aging. In 2014, an upgrade was proposed by the Italian collaboration based on the Cylindrical Gas Electron Multipliers (CGEM) technology, developed within the KLOE-II experiment, but with several new features and innovations. In this contribution, an overview of the project will be presented. Preliminary results of a beam test will be shown, with particular focus on the detector performance in magnetic field, with different configurations of electric field. A new readout mode, the μTPC readout, will also be described. The project has been recognized as a Significant Research Project within the Executive Programme for Scientific and Technological Cooperation between Italy and P.R.C for the years 2013-2015, and more recently has been selected as one of the project funded by the European Commission within the call H2020-MSCA-RISE-2014.

1. Introduction

Since the start of its operations in 2008, the Beijing Electron Spectrometer III (BESIII) has collected the largest sample of data in the energy region between 2 and 4.6 GeV, thanks of the high luminosity of the Beijing Electron-Positron Collider (BEPC – II). The complete physics program of BESIII [1] covers charmonium physics, open charm decay studies, search of exotic XYZ states and light hadron spectroscopy. Starting from the beryllium beam pipe to the outside, a Multi-layer Drift Chamber (MDC) is operating as tracking system, Time-Of-Flight (TOF) detectors are used to extract time information of the passage of particle, that combined with the energy information collected in the CsI(Tl) eletromagnetic calorimeter (ECAL), allows the particle identification (PID). A solenoidal superconducting magnet provides a uniform 1 Tesla magnetic field, while in its return yoke, Resistive Plate Chambers (RPC) operate as muon detectors.[2]

The innermost layers of the MDC exhibit aging effects. To compensate the increment of spark rate caused by these effects, the first layers must operate at lower relative gains. Since this deterioration seems to be constant in time, and is measured to be net 4% per year, the replacement is mandatory by 2018. In 2014, Italian collaborators proposed a possible upgrade based on Cylindrical Gas Electron Multipliers (CGEM), a Micro Pattern Gas Detector (MPGD) technology already developed within the KLOE-II collaboration [3], but with several innovation

Content from this work may be used under the terms of the Creative Commons Attribution 3.0 licence. Any further distribution of this work must maintain attribution to the author(s) and the title of the work, journal citation and DOI.
Published under licence by IOP Publishing Ltd

and new features with respect to the former case. The Conceptual Design Report [4] was approved by the BESIII board.

Several requirements, based on both spatial and momentum resolution of the new chamber (therein called CGEM-IT), were proposed by the BESIII Executive and Technical Board and are listed in Table 1 for further reference.

Requirements
Inner radius = 78 mm (min)
Outer radius = 179 mm (max)
93% of 4π solid angle
$\sigma_{xy} \sim 130\,\mu$m (per layer)
$\sigma_z < 1$ mm (per layer)
$X_0 < 1.5\%$
Trigger rate $\sim 10^4$Hz/cm^2

Table 1. List of all the requirements provided by the BESIII collaboration for the new inner tracker

In the following sections, a brief report on the CGEM-IT project will be described. Also, data from a recent beam test will be discussed in order to understand if we are able to push the technology to match the requirements.

2. The CGEM-IT project

Based on the experience gained in building the KLOE-II CGEM, we started developing a three layer triple-GEM detector with analog readout. This is the most crucial difference between the KLOE-II case and the BESIII CGEM-IT. This type of readout is the best compromise between the spatial resolution and the number of instrumentable channels. The charge centroid method allows to improve the spatial resolution over the one achievable with digital information, which has a limit of $pitch/\sqrt{12}$. With our project we will match the momentum resolution requirements, while we will improve by at least by a factor 2 the resolution along the beam axis. With this improved resolution, the performance of BESIII in recostructing primary and secondary vertices will largely improve, leading to better background rejection. Moreover, decays with a more complex topology will be finally accessible, boosting the open charm decay and exotic states physics program that BESIII will run likely until 2022.

From the technological point of view, we will deploy two innovations with respect to the state-of-art in GEM detectors. Given the strict requirement on the material budget, we will use a PMI-based foam called Rohacell [5] to give mechanical rigidity to the anode and the cathode. The jagged anode (in Figure 2) will allow us to reduce up to 30% the interstrip capacitance. This is essential because our strip pitch is quite large ($pitch = 650\,\mu m$) and long ($length = 82\,cm$): these leads to parasitic capacitance effect by design. If it is not kept under control, the noise will raise until the signal will be undetectable.

The first cylindrical prototype was assembled in INFN Laboratori Nazionali of Frascati at the end of February and the beginnning of March 2016. Currently, the gas is fluxing inside the detector that will be ready to be tested before of the end of the year with a custom readout electronics.

3. Beam test results in magnetic field

To prove the features of the final detector design, in-beam test of two prototypes was performed in June 2015 at H4 line in Prevessin, North Area, CERN. The only differences between the two

Journal of Physics: Conference Series **742** (2016) 012036 doi:10.1088/1742-6596/742/1/012036

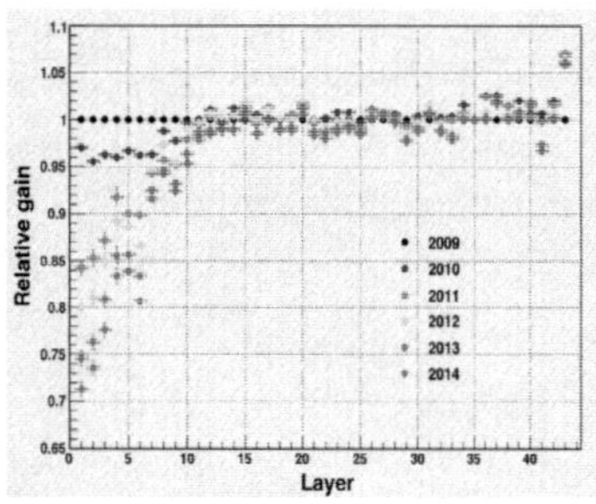

Figure 1. Relative gain as a function of the layer ID. The different colored lines show different years. To compensate the aging effect, first layers operate with reduced relative gain year by year.

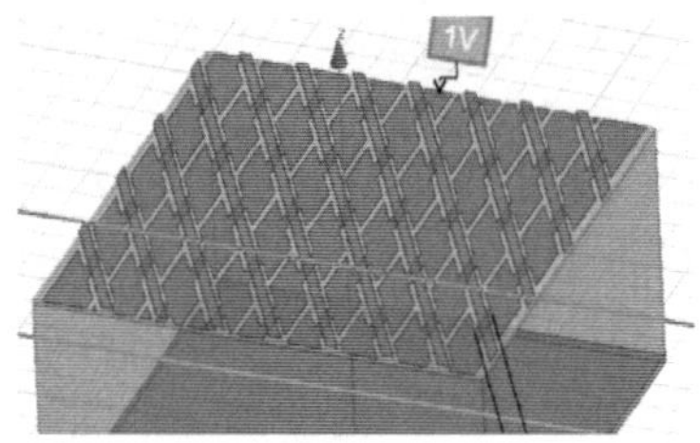

Figure 2. Pictorical example of a jagged anode that it is intended to be deployed in CGEM-IT project.

chambers were the anode and the drift gaps. In the so called prototype 1, the anode was linear and the gap was $5\,mm$, while in the prototype 2 we installed a jagged anode with a $3\,mm$ gap. A dipole magnet GOLIATH can provide magnetic field up to 1.5 T with both field polarisations. During three weeks of data taking within RD51 Collaboration [6] several configurations were tested. In the following, two of the most interesting results in magnetic field will be discussed. All the results discussed are obtained by means of the charge centroid method.

The confidential resolution is extracted as the width of the residual distribution. In Figure 3 the confidential resolution as a function of the magnetic field strength and polarisation is presented. As expected, the extracted width increases with the field strength. This effect is mainly caused by the presence of the distorsion of the diffusion electron cloud by the Lorentz angle, that is due to the presence of a magnetic field. This Lorentz angle represents the difference of the mean position of the cloud direction with and without magnetic field, as shown in Figure 4 in a Garfield [7] simulation. This distortion enlarges the average cluster and reduces the mean charge per strip. The charge centroid method is not anymore optimal. A possible solution to improve the resolution will be discussed in the next section.

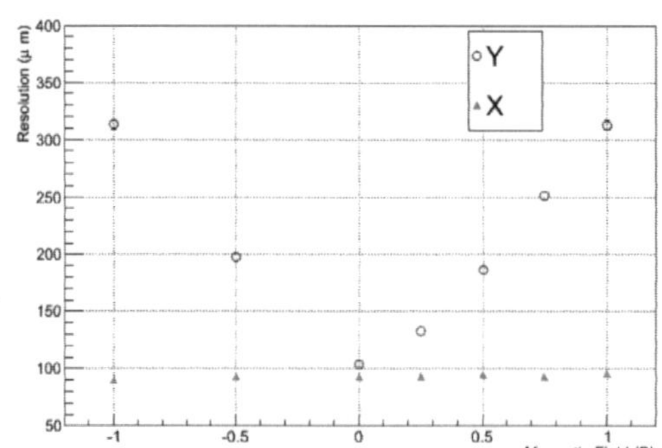

Figure 3. Residual distribution of the bending (white circle, Y) and non-bending (red triangle, X) view of one prototype with respect of the magnetic field in $ArCO_2$ (70/30) gas mixture. (color online)

Journal of Physics: Conference Series **742** (2016) 012036 doi:10.1088/1742-6596/742/1/012036

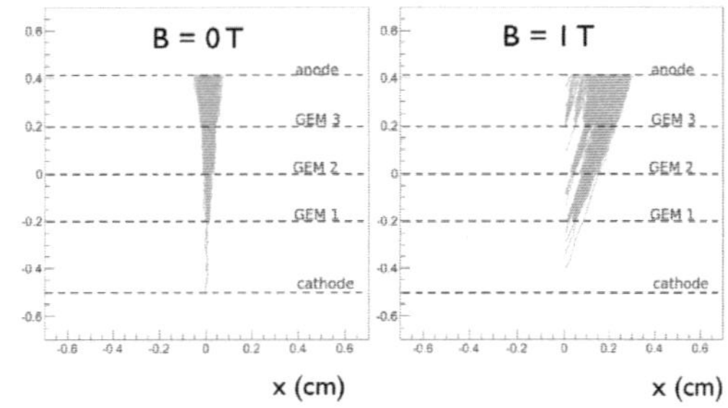

Figure 4. Garfield simulation of the effect of the magnetic field on to the diffusion cloud. The angle between the mean position without (left) and with (right) magnetic field is called Lorentz angle.

With the optimization of gas mixture and drift field it is possible to achieve a resolution below $\sigma = 200\,\mu m$, as shown in Figure 5. In Ar/Iso (90/10), the Lorentz angle decreases as the drift field become stronger. The distorsion due to the magnetic field are thereby reduced, and so the resolution improves. Despite being one of the best results obtained by triple-GEM detectors in magnetic field, it is still not enough to match the requirements. To overcome this limit, a new readout mode, the μTPC, has been proposed.

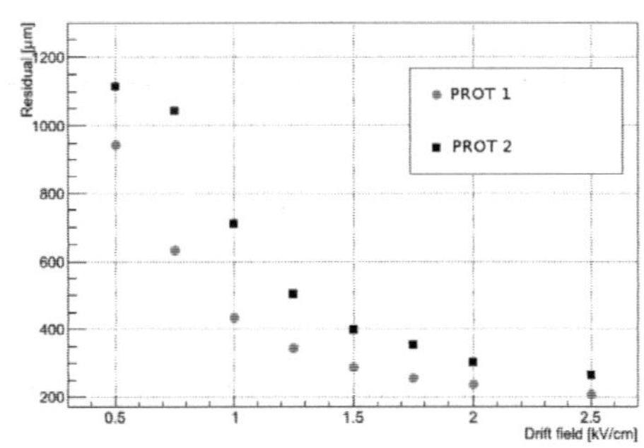

17

Figure 5. Residual distribution of the bending view with respect to the drift field in argon/isobutane (90/10) gas mixture. The only differences between the two prototypes are the different anodes and the gaps.

4. A new readout mode for GEM detector: the μTPC mode

In the case of a very broad distribution, *i.e.* with a very large incident angle or with magnetic field, the charge centroid method does not provide sufficient position resolution. The ATLAS collaboration has recently developed for their Small Wheel upgrade [8] a new type of readout mode, based on the time information that in this case is more precise than the charge one. Since the approach is similar to what is done in the large Time Projection Chambers (TPCs), the readout mode is called μTPC, as the drift gap is only few millimeters wide.

Based on ATLAS approach on MicroMegas [9], a similar procedure is developed for GEM detectors. The idea is to extract the differential time of arrival of the maximum value of the inducted charge in each strip. In this way it is possible to extract, being the drift velocity well-known, the position of a particular primary ionization cluster inside the drift gap.

From the ATLAS results with different angles [9], as depicted in Figure 6, we are very confident that, once the algorithm will be ready, it will allow to match the requirements. Preliminary studies indicate the improvement od the resolution and first will be available soon.

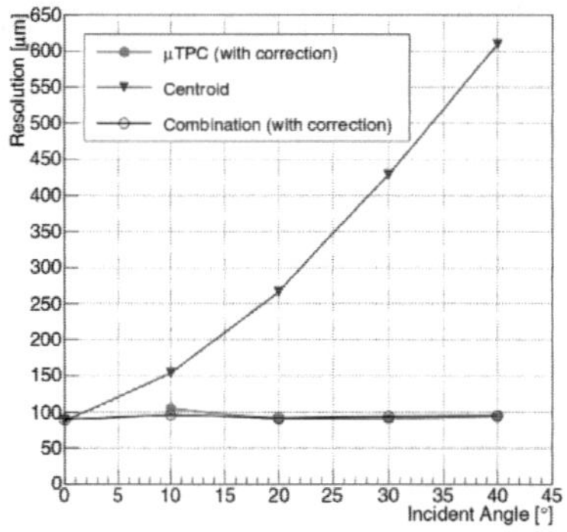

Figure 6. Comparison between the μTPC readout (red) and the charge centroid method (blue) with respect to the incident angle [9] in ATLAS MicroMegas studies. The black line represents the results obtained combining the two readouts. (color online)

5. Conclusion and outlook

To overcome the aging effect of the inner part of the BESIII MDC, we proposed an upgrade based on CGEM technology. The future chamber will deploy innovative features with respect to the present GEM detector, allowing for reach $\sigma_{xy} \sim 130\,\mu m$ position resolution in transverse plane and better than $\sigma_z < 200\,\mu m$ along the beam direction. The first cylindrical prototype has been built and it will be tested in-beam test before the end of the year.

Results in magnetic field show that is possible to optimize the combination of gas mixture and drift field to reach a position resolution around $200\,\mu m$ in magnetic field. This is one the world's best results for GEM detectors in magnetic field with charge centroid method. However, it is still not sufficient to match the required performance. This is due to the spread of the drift cloud. To pass the limits of the charge centroid method, a new readout, named μTPC, is being tested. Based on ATLAS results with MicroMegas at large angles, itr is expected to reach the required resolution in 1 T magnetic field. Preliminary studies on this new method show that it is feasible to be apply it in BESIII CGEM-IT and it will improve the spatial resolution. Soon the very first result will be available.

References

[1] D. M. Asner *et al.*, Physics at BES-III, arXiv:0809.1869v1 [hep-ex] (2008)
[2] M. Ablikim *et al.* (BESIII Collaboration), Design and Construction of the BESIII Detector, arXiv:0911.4906v1 (2009)
[3] KLOE-2 Collaboration, Technical Design Report of the Inner Tracker for the KLOE-2 experiment, arXiv:1992,2572v1 [physics.ins-det] (2010)
[4] M. Ablikim *et al.* (BESIII Collaboration), Conceptual Design Report of BESIII Cylindrical GEM Inner Tracker, BESIII internal note (2014)
[5] http://www.rohacell.com/product/rohacell/en/Pages/default.aspx
[6] http://rd51-public.web.cern.ch/RD51-Public/
[7] https://garfieldpp.web.cern.ch/garfieldpp/
[8] T. Kawamoto, B. Di Girolamo *et al.*, New Small Wheel Technical Design Report, CERN-LHCC-2013-006; ATLAS-TDR-020
[9] T. Alexopoulos, Performance studies of resistive Micromegas detector for the upgrade of the ATLAS Muon Spectrometer, 4th LNF Workshop on Cylindrical GEM Detectors (16-18 November 2015)

Journal of Physics: Conference Series **742** (2016) 012037 doi:10.1088/1742-6596/742/1/012037

A Custom Probe Station for Microstrip Detector Quality Assurance of the CBM Experiment

I. Panasenko[1,2], E. Lavrik[1], A. Lymanets[3] and H. R. Schmidt[1]

[1] Physics Institute - Eberhard Karls University Tuebingen, Auf der Morgenstelle 14, D-72076 Tuebingen, Germany

[2] Kiev Institute for Nuclear Research, Nauky Ave 47, Kiev, Ukraine

[3] GSI Helmholtz Centre for Heavy Ion Research, Planckstrasse 1, 64291 Darmstadt, Germany

E-mail: `iaroslav.panasenko@uni-tuebingen.de`

Abstract. The double-sided silicon microstrip sensors with $58\mu m$ pitch are the main building blocks of the Silicon Tracking System (STS) — the central detector of the Compressed Baryonic Matter (CBM) Experiment. The STS will employ about 1200 such sensors arranged on eight traking stations. Electrical characterization of the sensors is necessary to ensure their compliance with the specifications. For this purpuse a custom probe station is being developed at Tuebingen University. One of the main requirements is a high accuracy and a repeatability better than 1 μm to allow an automatic, succesive positioning on all 1024 pads of a sensor, as well as a positioning range in accordance with the size of STS sensors. The probe station is controlled via dedicated software developed at Tuebingen University. It allows to inspect the required $\sim$ 10% of the sensors on the series production stage with characterization time 4-5 hours per one double-sided sensor. The construction of the probe station and first measurements are discussed in this paper.

1. Introduction

The CBM experiment at FAIR will investigate the properties of nuclear matter under extreme temperatures and densities created in ultrarelativistic heavy-ion collisions. Figure 1 shows the CBM detector configuration for hadron and electron measurements. Its core detector — the Silicon Tracking System [1] inside the dipole magnet — will allow to reconstruct charged particle tracks with high precision and determine their momentum. The detection of rare probes requires the STS to be capable to measure at interaction rates of up to 10 MHz for Au-Au collisions. The track multiplicity will reach up to 700 within the detector aperture covering the polar angle 2.5^0 and 25^0 which corresponds to a rapidity range from mid rapidity to close to beam rapidity. The high track density as well as stringent requirements for the momentum resolution (around 1% at p $\geq$ 1 GeV/c) require a setup with high granularity and low material budget. Therefore front-end read-out electronics, cooling and mechanical infrastructure are located outside of the detector acceptance.

2. Microstrip Sensors for the CBM Silicon Tracking System

The STS will be constructed of 8 detection layers entirely covered by silicon microstrip sensors [1]. Figure 2 shows the concept design of the STS. About 1200 sensors will be mounted on light-weight carbon structures and the signal will be routed to the FEE by ultra-thin cables. Sensors of

Content from this work may be used under the terms of the Creative Commons Attribution 3.0 licence. Any further distribution of this work must maintain attribution to the author(s) and the title of the work, journal citation and DOI.
Published under licence by IOP Publishing Ltd

Figure 1. CBM SIS100 experimental setup for hadron and electron measurements. The STS is placed inside the dipole magnet. Downstream are the RICH, TRD and TOF detectors, and calorimeters. The MUCH detector can be moved into the acceptance after the RICH is removed.

4 different sizes have been chosen for different areas of the detection layers to keep the maximum strip occupancy at the level of few percent. Thus, short strips are chosen close to the beam pipe where the particles densities are highest. All sensors have the same width and number of strips but differ in the lengths of strips. The main sensor parameters are:

- n-type silicon
- thickness: 285/320 $\pm$ 15 μm (vendor dependent)
- double-sided segmentation
- 1024 strips per side of 58 μm pitch, strip width $\sim$ 18 μm
- strip lengths: 2/4/6/12 cm
- strip angle w.r.t. vertical edge: 7.5 degree (p-side), 0 degree (n-side)
- AC coupled read-out
- 2nd metal read-out routing lines interconnecting the AC layers of the corner strips
- outer dimensions: 6.2 cm width; 2.2, 4.2, 6.2 and 12.4 cm height.

The prototype sensors have been produced by two companies: CiS, Erfurt [2], Germany and Hamamatsu Photonics, Japan [3]. Figure 3 shows the microscope view of a recently produced prototype sensor and figure 4 represents a simplified model of the silicon microstrip sensor as an extended network of capacitors and resistors (omitted for simplicity) which is used for determination of strip parameters.

3. Electrical Quality Assurance of Silicon Microstrip Sensors

The complicated fabrication technology of double-sided microstrip sensors and the large-scale sensor production demands a reliable and efficient quality assurance (QA) program for the CBM-STS sensors to guarantee the expected operation of the whole detector system. The electrical QA program for sensors consists of:

- **Basic tests**: IV, CV curves for every received sensor.

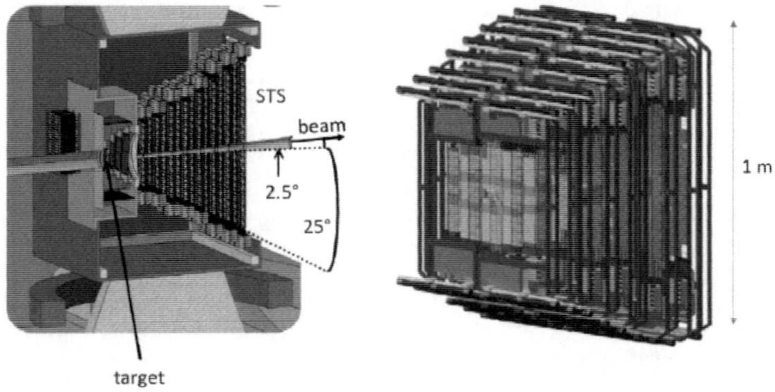

Figure 2. Left: Design of the Silicon Tracking System, showing a cut through the detector in the dipole magnet. Right: complete STS, with the eight tracking stations.

- **Subset tests**: pinholes in coupling capacitors, strip metal and implant shorts and opens, strip leakage current — on ~ 10% of the sensors which pass the basic tests.
- **Specific tests**: coupling and interstrip capacitances, polysilicon resistors — on a small fraction of sensors to measure in much more details electrical parameters.

To perform all those tests two quality test centers are being equipped in GSI Detector Laboratory and at University of Tuebingen, respectively. Both centers have the same set of the measurement instrumentation: SourceMeter Keithley 2410, Picoammeter Keithley 6487, LCR-Meter QuadTech 7600, Switching matrix Keithley 708B.

Current-voltage (IV) and capacitance-voltage (CV) measurements are the first and simple estimation of the sensor quality. This measurements describe the sensor as a single entity and allow to extract full depletion voltage (V_{FD}), bulk capacitance (C_{bulk}) and leakage current ($I_{leakage}$) at full depletion voltage. Figure 6 shows the CV characteristic plotted in a double logarithmic scale for the sensor labeled as CBM06C6-w18. C_{bulk} measured between backplane and bias ring by LCR meter with $Cs - Rs$ function at 1 kHz frequency. Full depletion for most of the tested prototype sensors is reached at $\approx 70\ V$ and the capacitance saturates at $\approx 1.2\ nF$.

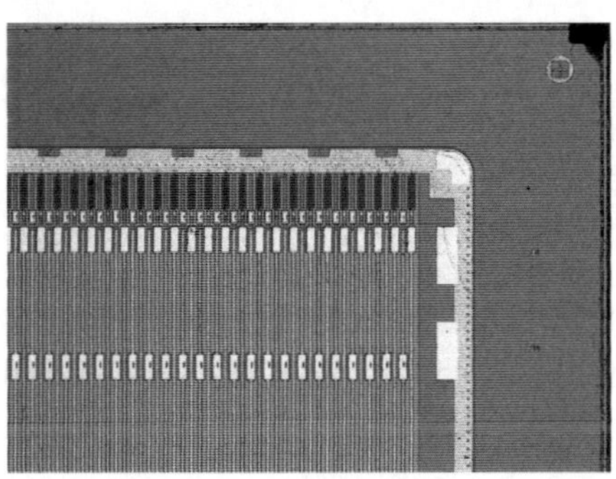

Figure 3. Microscope view of the corner of the n-side of 6.2 x 6.2 cm prototype microstrip sensor from CiS.

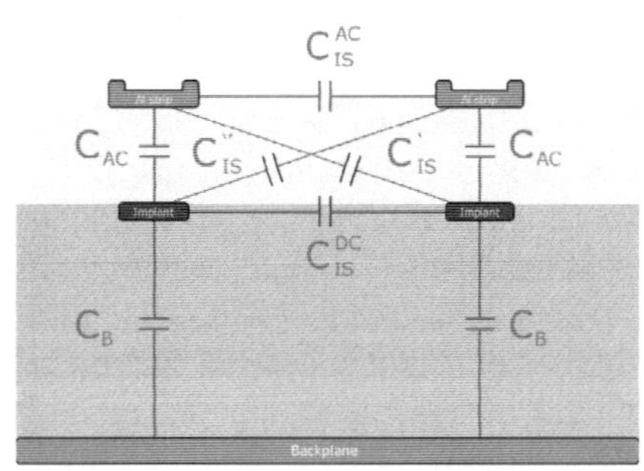

Figure 4. Sensor model for capacitance measurements. *Blue* - n-bulk silicon, *yellow* - dielectric layer between aluminum read-out strips and implant strips.

For strip-by-strip characterization a custom made, automated probe station has been developed and set up in the clean room at University of Tuebingen [4], while commercial wafer prober (Suss PA300PS) is employed at GSI [5]. In addition to high accuracy of positioning and repeatability ($< 1\mu m$) the custom made probe station provides large travel range of both chuck positioning and optical systems that allows to contact needles to pads at opposite corners of the silicon sensor. This feature makes it possible to perform some specific tests like probing of the 2nd metal lines on the sensors' p-side. Figure 5 shows the concept design of the custom probe station. A switching scheme has been implemented to allow acquirement of several electrical parameters at each step. Instruments (HV-source, Amp-meter, LCR-meter) are connected via a switching matrix to the needles which contact the sensor to perform different measurements. For each test the switching matrix has to be reconfigured.

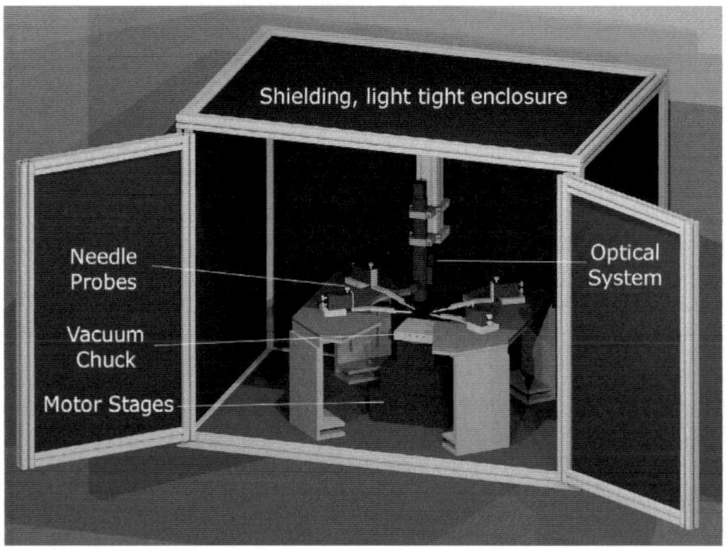

Figure 5. Concept design of the custom probe station. The microstrip sensor is held by the custom vacuum chuck. The box is closed during the measurements.

LabVIEW based [6] software provides automation for all kind of measurements. Its stepping procedure for subset tests probes every strip of the sensor in order to check the capacitor dielectric for pinholes, to look for strip metal and implant shorts and opens and to determine the coupling capacitance of the readout strip. Figure 7 shows the results of one of the subset test — pinhole check performed on the CBM06C6-w18 sensor. The current through the capacitor for most of the strips is much less than 1 nA, i.e. no pinholes are present. The group of the strips reveals a twice higher value of current than normal which is due to metal shorts between pairs of strips. This measurement has been done by contacting the strip's DC and AC pads with two needles and measuring the current after the test voltage of 10 V is applied. A source of the revealed defective strips is a scratch on the sensor surface. Such kind of defect degrades the spatial resolution of the system and has to be determined. Further specific tests (e.g. interstrip capacitance, posysilicon resistance, coupling capacitor breakdown) have also been implemented.

A time estimate for the complete testing protocol (consisting of 4 QA tests) is 4-5 hours per sensor. The specification requires < 1 % bad strips on every sensor to be accepted for module assembling. We define bad strip as a strip with at least one of the electrical parameters outside the specified cuts. Current prototype sensors fulfill this requirement [7].

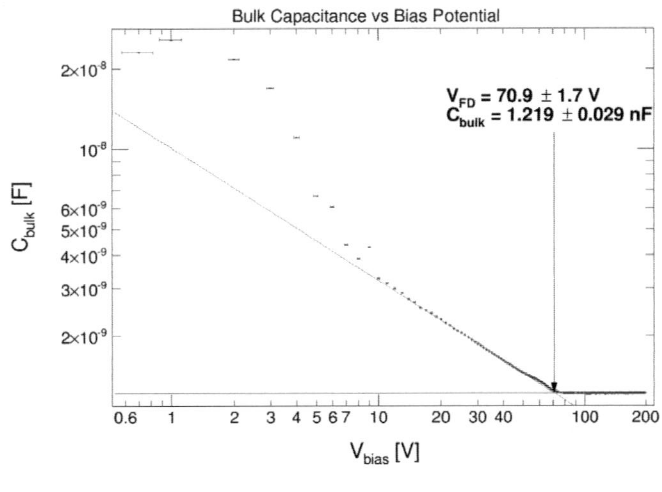

Figure 6. Capacitance-voltage measurements.

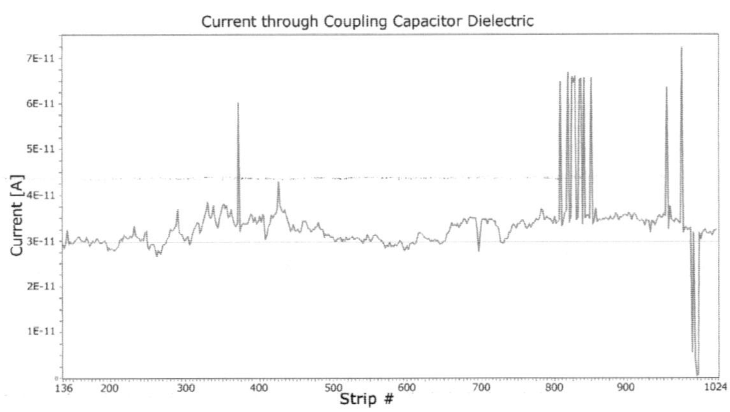

Figure 7. Pinholes test results for p-side strips of the prototype sensor.

Conclusions

A quality assurance program with detailed characterization procedures has been developed for CBM-STS microstrip sensors. For this purpose the custom made high precision probe station has been set-up at the University of Tuebingen. Prototype sensors for CBM experiment were successfully tested using this probe station — the results are in compliance with CBM detectors specifications. The probe station allows to inspect required $\sim$ 10% of the sensors on series production stage. Characterization of one double-sided sensor with 1024 strips on every side takes $\approx$ 5 hours.

Acknowledgments

This work is supported by BMBF under grant 05P12VTFCE.

References

[1] J. Heuser et al., *Technical Design Report for the CBM Silicon Tracking System*, GSI Report 2013-14, http://repository.gsi.de/record/54798
[2] http://www.cismst.org
[3] http://www.hamamatsu.com
[4] I. Panasenko, A. Lymanets and H. R. Schmidt, *A custom made wafer prober for quality assurance of CBM-STS sensors*, CBM Progress Report 2014, p. 31.
[5] P. Larionov and U. Frankenfeld, *Automatized quality assurance of sensors for the CBM Silicon Tracking System*, CBM Progress Report 2013, p. 27.
[6] http://www.ni.com/labview/
[7] I. Panasenko, *Quality assurance of silicon microstrip sensors for the CBM experiment*, DPG Meeting 2016, Darmstadt.